Introduction to

Robotics

Mechanics & Control

John J. Craig

Stanford University

ADDISON-WESLEY PUBLISHING COMPANY

Reading, Massachusetts • Menlo Park, California
Don Mills, Ontario • Wokingham, England • Amsterdam
Sydney • Singapore • Tokyo • Mexico City • Bogotá
Santiago • San Juan

Library of Congress Cataloging-in-Publication Data
Craig, John J., 1955–
Introduction to robotics.

Bibliography: p.
Includes index.
1. Robotics. I. Title.
TJ211.C67 1986 629.8′92 85–13527
ISBN 0–201–10326–5

BCDEFGHIJ-MA-89876

PREFACE

Scientists often have the feeling that through their work they are learning about some aspect of themselves. Physicists see this connection in their work, as do the psychologists, or chemists. In the study of robotics, the connection between the field of study and ourselves is unusually obvious. And, unlike a science that seeks only to analyze, robotics as presently pursued takes the engineering bent toward synthesis. Perhaps it is for these reasons that the field fascinates so many of us.

The study of robotics concerns itself with the desire to synthesize some aspects of human function by the use of mechanisms, sensors, actuators, and computers. Obviously, this is a huge undertaking which seems certain to require a multitude of ideas from various "classical" fields.

Presently different aspects of robotics research are carried out by experts in various fields. It is usually not the case that any single individual has the entire area of robotics in his or her grasp. A partitioning of the field is natural to expect. At a relatively high level of abstraction, splitting robotics into four major areas seems reasonable: mechanical manipulation, locomotion, computer vision, and artificial intelligence.

This book introduces the science and engineering of mechanical manipulation. This subdiscipline of robotics has its foundations in several classical fields. The major relevant fields are mechanics, control theory, and computer science. In this book, Chapters 1 through 7 cover topics from mechanical engineering and mathematics, Chapters 8 and 9 cover control theoretical material, and Chapter 10 might be classed as computer science material. Additionally, the book emphasizes computational aspects of the problems throughout; for example, each chapter which is predominantly concerned with mechanics has a brief section devoted to computational considerations.

This book has evolved from class notes used to teach "Introduction to Robotics" at Stanford University during the autumns of 1983 and 1984. At Stanford, this course is the first in a three quarter sequence where the second quarter covers computer vision and the third covers artificial intelligence, locomotion, and advanced topics.

This book is appropriate for a junior/senior undergraduate or first year graduate level course. It is helpful if the student has had one basic course in statics and dynamics, a course in linear algebra, and can program in a high level language. Additionally, for the material of Chapters 8 and 9 it is helpful, though not absolutely necessary, that the student have an introductory course in control theory. One aim of the book is to present material in a simple, intuitive way. Specifically, the audience need not be strictly mechanical engineers, though much of the material is taken from that field. At Stanford, many electrical engineers, computer scientists, and mathematicians found the notes quite readable.

While this book is directly of use to those engineers developing robotic systems, the material should be viewed as important background material for anyone who will be involved with robotics. In much the same way that software people have usually studied at least some hardware, people not directly involved with the mechanics and control of robots should have some background such as that offered by this text.

The book is organized as 10 chapters. At Stanford one chapter was covered per week to provide the material for the 10 week quarter. At this pace, all of the topics cannot be covered in great depth. In some ways, the book is organized with this in mind; for example, most chapters present only one approach to solving the problem at hand. One of the challenges of writing this book has been in trying to do justice to the topics covered within the time constraints of usual teaching situations. One method employed to this end was to consider only the material which directly impacts on the study of mechanical manipulation. At the end of Chapter 1 several references are listed, including a listing of research oriented journals which publish in the robotics area.

At the end of each chapter is a set of exercises. Each exercise has been assigned a difficulty factor, indicated in square brackets following the exercise's number. Difficulties vary between [00] and [50], where [00] is trivial and [50] is an unsolved research problem.* Of course, what one

---

* I have adopted the same scale as in *The Art of Computer Progamming*, by D. Knuth, Addison-Wesley, publisher.

person finds difficult, another may find easy, so some readers may find them misleading in some cases. Nevertheless, an effort has been made to appraise the difficulty of the exercises.

Additionally, at the end of each chapter there is a programming assignment in which the student applies the subject matter of the corresponding chapter to a simple three-jointed planar manipulator. This simple manipulator is complex enough to demonstrate nearly all the principles of general manipulators, while not bogging down the student with too much complexity. Each programming assignment builds upon the previous ones, until, at the end of the course, the student has an entire library of manipulator software.

Chapter 1 is an introduction to the field of robotics. It introduces some background material, the adopted notation of the book, a few fundamental ideas, and previews the material in following chapters.

Chapter 2 covers the mathematics used to describe positions and orientations in 3-space, This is extremely important material since, by definition, mechanical manipulation concerns itself with moving objects (parts, tools, the robot itself) around in space. We need ways to describe these actions in a way which is easily understood and as intuitive as possible.

Chapters 3 and 4 deal with the geometry of mechanical manipulators. They introduce the branch of mechanical engineering known as kinematics, the study of motion without regard to the forces which cause it. In these chapters we deal with the kinematics of manipulators, but restrict ourselves to only static positioning problems.

Chapter 5 expands our investigation of kinematics to velocities and static forces.

In Chapter 6 we deal for the first time with the forces and moments required to cause motion of a manipulator. This is the problem of manipulator dynamics.

Chapter 7 is concerned with describing motions of the manipulator in terms of trajectories through space.

In Chapter 8 we study methods of controlling a manipulator (usually with a digital computer) so that it will faithfully track a desired position trajectory through space.

Chapter 9 covers the relatively new field of active force control with a manipulator. That is, we discuss how to control the application of forces by the manipulator. This mode of control is important when the manipulator comes into contact with the environment around it, such as when washing a window with a sponge.

Finally, Chapter 10 overviews methods of programming robots, specifically the elements needed in a robot programming system, and the particular problems associated with programming industrial robots.

I would like to thank the many people who have contributed their time to helping me with this book. First, my thanks to the students of ME219 in the autumn of 1983 and 1984 who suffered through the first drafts and found many errors, and provided many suggestions. Professor Bernard Roth has contributed in many ways, both through constructive criticism of the manuscript and by providing me with an environment in which to complete it. I owe a debt to my previous mentors in robotics: Marc Raibert, Carl Ruoff, and Tom Binford. Many others around Stanford have helped in various ways—many thanks to John Mark Agosta, Mike Ali, Chuck Buckley, Joel Burdick, Jim Callan, Monique Craig, Subas Desa, Ashitava Ghosal, Chris Goad, Ron Goldman, Johann Jäger, Jeff Kerr, Oussama Khatib, Jim Kramer, Dave Lowe, Dave Marimont, Madhusudan Raghavan, Ken Salisbury, Donalda Speight, and Sandy Wells. I am grateful for the support of the Systems Development Foundation through much of the book's writing. Finally I wish to thank Tom Robbins and his staff at Addison-Wesley, Gary Grosso at Textset, Inc., and several anonymous reviewers.

Stanford, California J.C.

CONTENTS

6

7

1

INTRODUCTION

1.1 Introduction

The history of industrial automation is characterized by periods of rapid change in popular methods. Either as a cause or, perhaps, an effect, such periods of change in automation techniques seem closely tied to world economics. Use of the **industrial robot**, which became identifiable as a unique device in the 1960's, along with computer aided design (CAD) systems, and computer aided manufacturing (CAM) systems, characterize the latest trends in the automation of the manufacturing process [1]. These technologies are leading industrial automation through another transition, the scope of which is still unknown.

This book focuses on the mechanics and control of the most important form of the industrial robot, the **mechanical manipulator**. Exactly what constitutes an industrial robot is sometimes debated. Devices such as that shown in Fig. 1.1 are always included, while numerically controlled (NC) milling machines are usually not. The distinction lies somewhere in the sophistication of the programmability of the device—if a mechanical device can be programmed to perform a wide variety of applications, it is probably an industrial robot. Machines which are for the most part relagated to one class of task are considered **fixed automation**. For the purposes of this text, the distinctions need not be debated as most material is of a basic nature that applies to a wide variety of programmable machines.

By and large, the study of the mechanics and control of manipulators is not a new science, but merely a collection of topics taken from "classical" fields. Mechanical engineering contributes methodologies for the study of machines in static and dynamic situations. Mathematics supplies tools for describing spatial motions and other attributes of manipulators. Control theory provides tools for designing and evaluating algorithms to realize desired motions or force application. Electrical engineering techniques are brought to bear in the design of sensors and interfaces for industrial robots, and computer science contributes a basis for programming these devices to perform a desired task.

FIGURE 1.1 The Cincinnati Milacron 776 manipulator has six rotational joints and is popular in spot welding applications. Courtesy of Cincinnati Milacron.

1.2 The mechanics and control of mechanical manipulators

Introduction

The following sections introduce some terminology and briefly preview each of the topics which will be covered in the text.

Description of position and orientation

In the study of robotics we are constantly concerned with the location of objects in three dimensional space. These objects are the links of the manipulator, the parts and tools with which it deals, and other objects in the manipulator's environment. At a crude but important level, these objects are described by just two attributes: their position and their orientation. Naturally, one topic of immediate interest is the manner in which we represent these quantities and manipulate them mathematically.

In order to describe the position and orientation of a body in space we will always attach a coordinate system, or **frame**, rigidly to the object. We then proceed to describe the position and orientation of this frame with respect to some reference coordinate system (see Fig. 1.2).

Since any frame can serve as a reference system within which to express the position and orientation of a body, we often think of *transforming* or *changing the description of* these attributes of a body from one frame to another. In Chapter 2 we discuss conventions and methodologies for dealing with the description of position and orientation, and the mathematics of manipulating these quantities with respect to various coordinate systems.

Forward kinematics of manipulators

Kinematics is the science of motion which treats motion without regard to the forces which cause it. Within the science of kinematics one studies the position, velocity, acceleration, and all higher order derivatives of the position variables (with respect to time or any other variable(s)). Hence, the study of the kinematics of manipulators refers to all the geometrical and time based properties of the motion.

Manipulators consist of nearly rigid **links** which are connected with **joints** which allow relative motion of neighboring links. These joints are usually instrumented with position sensors which allow the relative

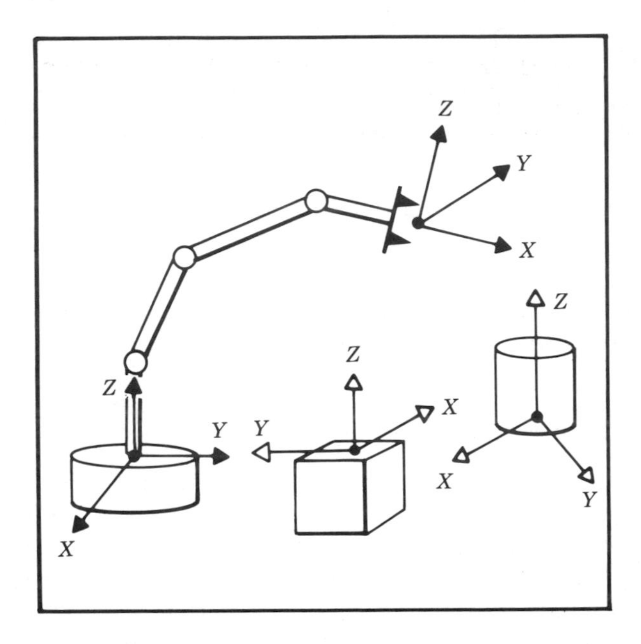

FIGURE 1.2 Coordinate systems or "frames" are attached to the manipulator and objects in the environment.

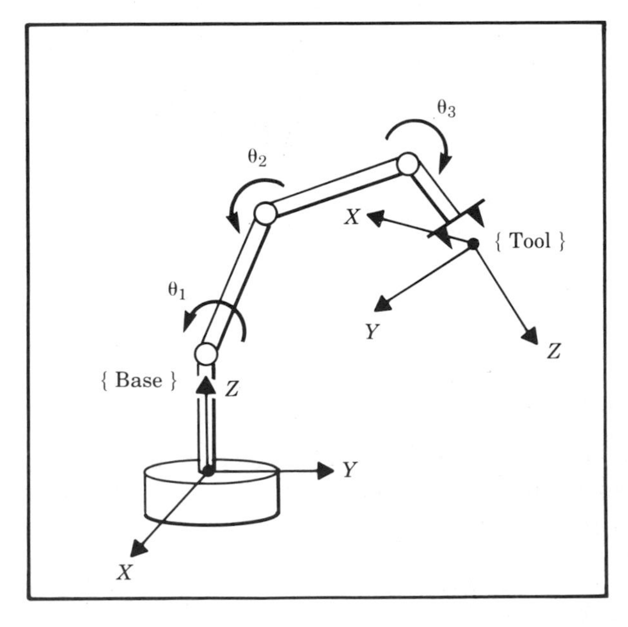

FIGURE 1.3 Kinematic equations describe the tool frame relative to the base frame as a function of the joint variables.

position of neighboring links to be measured. In the case of rotary or **revolute** joints, these displacements are called **joint angles**. Some manipulators contain sliding, or **prismatic** joints in which the relative displacement between links is a translation, sometimes called the **joint offset**.

The number of **degrees of freedom** that a manipulator possesses is the number of independent position variables which would have to be specified in order to locate all parts of the mechanism. This is a general term used for any mechanism. For example, a four bar linkage has only one degree of freedom (even though there are three moving members). In the case of typical industrial robots, because a manipulator is usually an open kinematic chain, and because each joint position is usually defined with a single variable, the number of joints equals the number of degrees of freedom.

At the free end of the chain of links which make up the manipulator is the **end-effector**. Depending on the intended application of the robot, the end-effector may be a gripper, welding torch, electro-magnet, or other device. We generally describe the position of the manipulator by giving a description of the **tool frame**, which is attached to the end-effector, relative to the **base frame** which is attached to the nonmoving base of the manipulator (see Fig. 1.3).

A very basic problem in the study of mechanical manipulation is that of **forward kinematics**. This is the static geometrical problem of computing the position and orientation of the end-effector of the manipulator. Specifically, given a set of joint angles, the forward kinematic problem is to compute the position and orientation of the tool frame relative to the base frame. Sometimes we think of this as changing the representation of manipulator position from a **joint space** description into a **Cartesian space** description.* This problem will be explored in Chapter 3.

Inverse kinematics of manipulators

In Chapter 4 we will consider the problem of **inverse kinematics**. This problem is posed as follows: Given the position and orientation of the end-effector of the manipulator, calculate all possible sets of joint angles which could be used to attain this given position and orientation (see Fig. 1.4). This is a fundamental problem in the practical use of manipulators.

The inverse kinematic problem in not as simple as the forward kinematics. Because the kinematic equations are nonlinear, their solution

* By *Cartesian space* we mean the space in which the position of a point is given with 3 numbers, and in which the orientation of a body is given with 3 numbers. It is sometimes called *task space* or *operational space*.

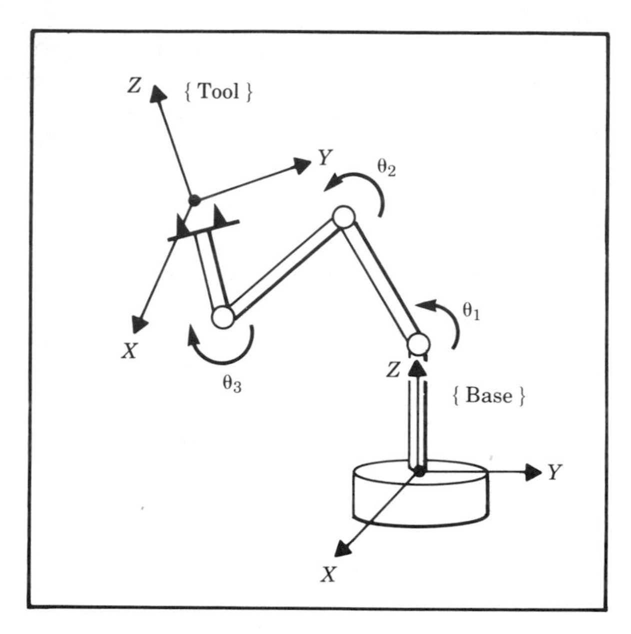

FIGURE 1.4 For a given position and orientation of the tool frame, values for the joint variables can be calculated using the inverse kinematics.

is not always easy or even possible in a closed form. Also, the questions of existence of a solution, and of multiple solutions, arise.

The existence or nonexistence of a kinematic solution defines the **workspace** of a given manipulator. The lack of a solution means that the manipulator cannot attain the desired position and orientation because it lies outside of the manipulator's workspace.

Velocities, static forces, singularities

In addition to dealing with static positioning problems, we may wish to analyze manipulators in motion. Often in performing velocity analysis of a mechanism it is convenient to define a matrix quantity called the **Jacobian** of the manipulator. The Jacobian specifies a **mapping** from velocities in joint space to velocities in Cartesian space (see Fig. 1.5). The nature of this mapping changes as the configuration of the manipulator varies. At certain points, called **singularities**, this mapping is not invertible. An understanding of the phenomenon is important to designers and users of manipulators.

Manipulators do not always move through space; sometimes they are also required to contact a workpiece or work surface and apply a static force. In this case the problem arises: Given a desired contact force and moment, what set of **joint torques** are required to generate them? Once

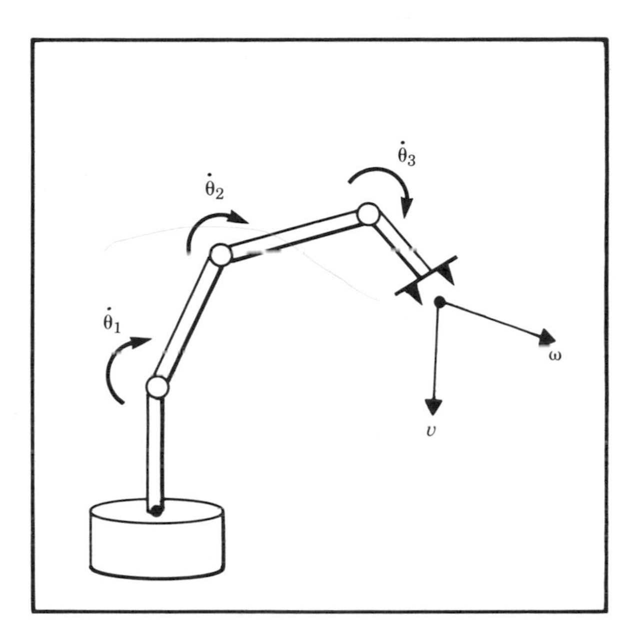

FIGURE 1.5 The geometrical relationship between joint rates and velocity of the end-effector can be described in a matrix called the Jacobian.

again, the Jacobian matrix of the manipulator arises quite naturally in the solution of this problem.

Dynamics

Dynamics is a huge field of study devoted to studying the forces required to cause motion. In order to accelerate a manipulator from rest, glide at a constant end-effector velocity, and finally decelerate to a stop, a complex set of torque functions must be applied by the joint actuators.* The exact form of the required functions of actuator torque depend on the spatial and temporal attributes of the path taken by the end-effector as well as the mass properties of the links and payload, friction in the joints, etc. One method of controlling a manipulator to follow a desired path involves calculating these actuator torque functions using the dynamic equations of motion of the manipulator.

A second use of the dynamic equations of motion is in **simulation**. By reformulating the dynamic equations so that acceleration is computed as a function of actuator torque, it is possible to simulate how a manipulator would move under application of a set of actuator torques (see Fig. 1.6).

* We use *joint actuators* as the generic term for devices which power a manipulator, for example: electric motors, hydraulic and pneumatic actuators, muscles, etc.

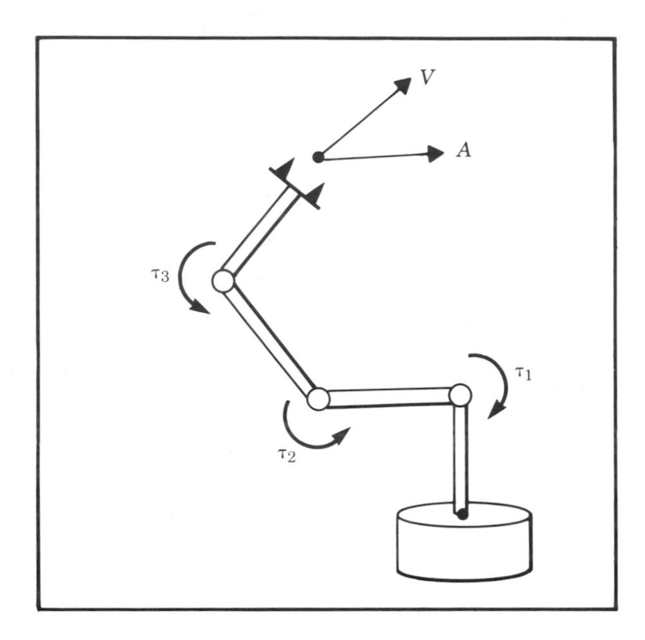

FIGURE 1.6 The relationship between torques applied by the actuators and the resulting motion of the manipulator is embodied in the dynamic equations of motion.

In Chapter 6 we develop dynamic equations of motion which may be used to control or simulate the motion of manipulators.

Trajectory generation

A common way of causing a manipulator to move from here to there in a smooth, controlled fashion is to cause each joint to move as specified by a smooth function of time. Commonly, each joint starts and ends its motion at the same time, so that the manipulator motion appears coordinated. Exactly how to compute these motion functions is the problem of **trajectory generation** (see Fig. 1.7).

Often a path is described not only by a desired destination but also by some intermediate locations, or **via points**, through which the manipulator must pass en route to the destination. In such instances the term **spline** is sometimes used to refer to a smooth function which passes through a set of via points.

In order to force the end-effector to follow a straight line (or other geometric shape) through space the desired motion must be converted to an equivalent set of joint motions. This **Cartesian trajectory generation** will also be considered in Chapter 7.

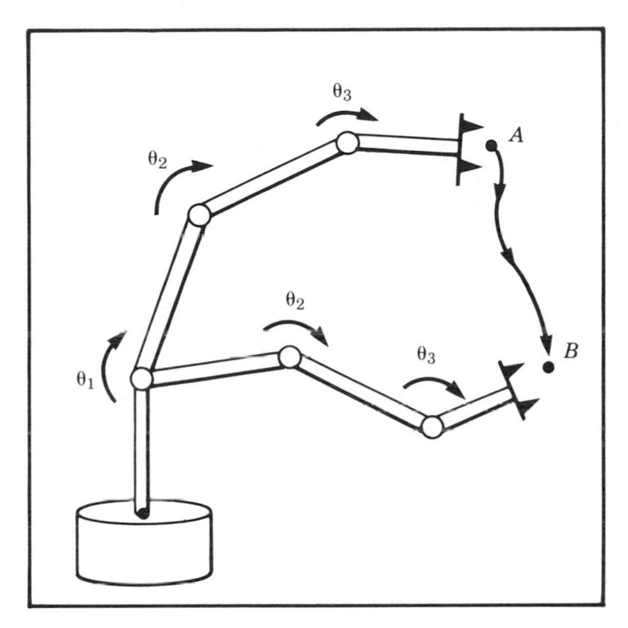

FIGURE 1.7 In order to move the end effector through space from point A to point B we must compute a trajectory for each joint to follow.

Position control

Some manipulators are equipped with stepper motors or other actuators which can directly execute a desired trajectory. However, the vast majority of manipulators are driven by actuators which supply a force or a torque to cause motion of the links. In this case, an algorithm is needed which computes torques which will cause the desired motion. The problem of dynamics is central to the design of such algorithms, but does not in itself constitute a solution. A primary concern of a **position control system** is to automatically compensate for errors in knowledge of the parameters of a system, and to suppress disturbances which try to perturb the system from the desired trajectory. To accomplish this, position and velocity **sensors** are monitored by the **control algorithm** which computes torque commands for the actuators (see Fig. 1.8). In Chapter 8 we apply our knowledge of manipulator dynamics with some control theoretic ideas to investigate the design of manipulator position control algorithms.

Force control

The ability for a manipulator to control forces of contact when it touches parts, tools, or worksurfaces seems to be of great importance

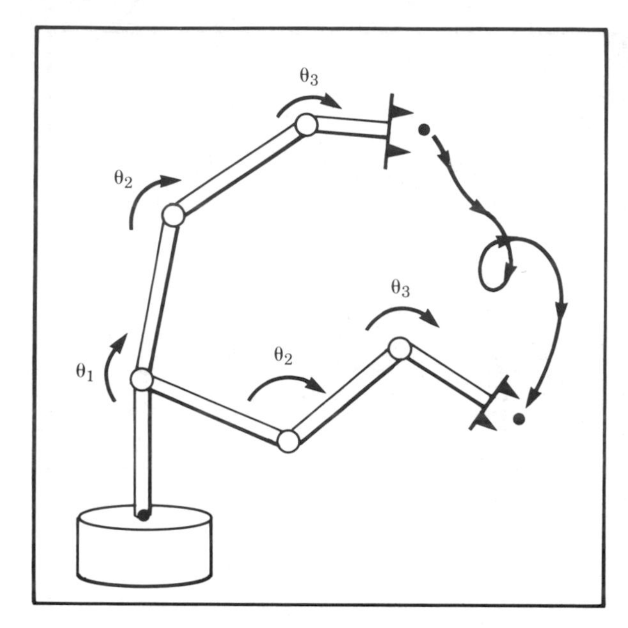

FIGURE 1.8 In order to cause the manipulator to follow the desired trajectory, a position control system must be implemented. Such a system uses feedback from joint sensors to keep the manipulator on course.

in applying manipulators to many real-world tasks. **Force control** is complementary to position control in that we usually think of one or the other to be applicable in a certain situation. When a manipulator is moving in free space, only position control makes sense, since there is no surface to react against. When a manipulator is touching a rigid surface however, position control schemes can cause excessive forces to build up at the contact or may cause contact to be lost with the surface when it was desired for some application. Since manipulators are rarely constrained by reaction surfaces in all directions simultaneously, using a mixed or **hybrid** control is required, with some directions controlled by a **position control law** and remaining directions controlled by a **force control law** (see Fig. 1.9). Chapter 9 introduces a methodology for implementing such a force control scheme.

Programming robots

A **robot programming language** serves as the interface between the human user and the industrial robot. Central questions arise such as: How are motions through space described easily by the programmer? How are multiple manipulators programmed so that they can work in parallel? How are sensor-based actions described in a language?

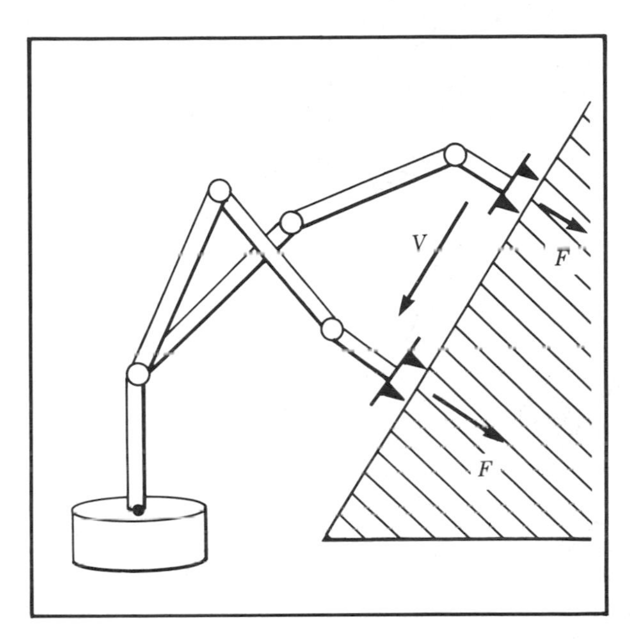

FIGURE 1.9 In order for a manipulator to slide across a surface while applying a constant force, a hybrid position-force control system must be used.

Robot manipulators differentiate themselves from **fixed automation** by being "flexible" which means programmable. Not only are the movements of manipulators programmable, but through the use of sensors and communications with other factory automation, manipulators can *adapt* to variations as the task proceeds (see Fig. 1.10).

The sophistication of the user interface is becoming extremely important as manipulators and other programmable automation are applied to more and more demanding industrial applications. The problem of programming manipulators encompasses all the issues of "traditional" computer programming, and so is an extensive subject in itself. Additionally, some particular attributes of the manipulator programming problem cause additional issues to arise. Some of these topics will be discussed in Chapter 10.

1.3 Notation

Notation is always an issue in science and engineering. In this book, we use the following conventions:

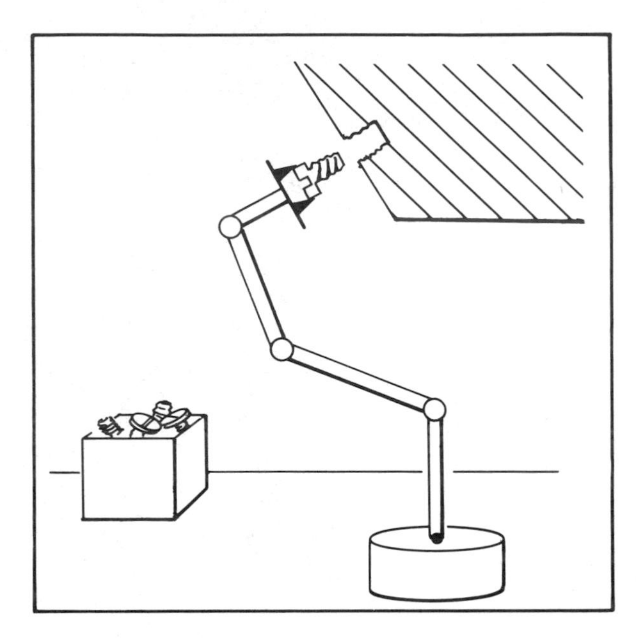

FIGURE 1.10 Desired motions of the manipulator and end-effector, desired contact forces, and complex manipulation strategies can be described in a *robot programming language*.

1. Usually variables written in uppercase represent vectors or matrices. Lowercase variables are scalars.
2. Leading subscripts and superscripts give information regarding which coordinate system a quantity is written in. For example, ${}^{A}P$ represents a position vector written in coordinate system $\{A\}$, and ${}^{A}_{B}R$ is a rotation matrix* which specifies the relationship between coordinate systems $\{A\}$ and $\{B\}$.
3. Trailing superscripts are used (as widely accepted) for indicating the inverse or transpose of a matrix, e.g. R^{-1}, R^{T}.
4. Trailing subscripts are not subject to any strict convention, but may indicate a vector component (e.g. x, y, or z) or may be used as a description as in P_{bolt}, the position of a bolt.
5. We will use many trigonometric functions. Our notation for the cosine of an angle θ_1 may take any of the forms: $\cos\theta_1 = c\theta_1 = c_1$.

Vectors are taken as column vectors; hence row vectors will have the transpose indicated explicitly.

A note on vector notation in general: Many mechanics texts treat vector quantities at a very abstract level and routinely use vectors

* This term will be introduced in Chapter 2.

defined relative to different coordinate systems in expressions. The clearest example is that of addition of vectors which are given or known relative to differing reference systems. This is often very convenient and leads to compact and somewhat elegant formulas. For example, consider the angular velocity, ${}^{0}\omega_4$, of the last body in a series connection of four rigid bodies (as in the links of a manipulator) relative to the fixed base of the chain. Since angular velocities sum vectorially, we may write a very simple vector equation for the angular velocity of the final link:

$$ {}^{0}\omega_4 = {}^{0}\omega_1 + {}^{1}\omega_2 + {}^{2}\omega_3 + {}^{3}\omega_4. \tag{1.1} $$

However, unless these quantities are expressed with respect to a common coordinate system, they cannot be summed, and so while elegant, equation (1.1) has hidden much of the "work" of the computation. For the particular case of the study of mechanical manipulators, statements like that of (1.1) hide the chore of bookkeeping of coordinate systems, which is often the very idea which we need to deal with in practice.

Therefore, in this book, we carry frame of reference information in the notation for vectors, and we do not sum vectors unless they are in the same coordinate system. In this way, we derive expressions which solve the "bookkeeping" problem, and may be applied directly to actual numerical computation.

References

[1] B. Roth, "Principles of Automation," Future Directions in Manufacturing Technology, Based on the Unilever Research and Engineering Division Symposium held at Port Sunlight, April 1983, Published by Unilever Research, UK.

General reference books

[2] R. Paul, "Robot Manipulators," MIT Press, 1982.
[3] M. Brady et al, "Robot Motion," MIT Press, 1983.
[4] G. Beni, and S. Hackwood, editors, "Recent Advances in Robotics," Wiley, 1985.
[5] R. Dorf, "Robotics and Automated Manufacturing," Reston, 1983.
[6] A. Critchlow, "Introduction to Robotics," Macmillan, 1985.
[7] W. Synder, "Industrial Robots: Computer Interfacing and Control," Prentice Hall, 1985.
[8] Y. Koren, "Robotics for Engineers," McGraw Hill, 1985.
[9] V. Hunt, "Industrial Robotics Handbook," Industrial Press Inc., 1983.
[10] J. Engelberger, "Robots in Practice," AMACOM, 1980.

General reference journals and magazines

[11] *Robotics Today.*
[12] *Robotics World.*
[13] *The Industrial Robot.*
[14] *IEEE Transactions on Robotics and Automation.*
[15] *IEEE Transactions on System, Man, and Cybernetics.*
[16] *IEEE Transactions on Automatic Control.*
[17] *International Journal of Robotics Research.*
[18] *ASME Journal of Dynamic Systems, Measurement, and Control.*

Exercises

1.1 [20] Make a chronology of major events in the development of industrial robots over the past 30 years. See references section.

1.2 [20] Make a chart showing the major applications of industrial robots (e.g. spot welding, assembly, etc.) and the percentage of installed robots in use presently in each application area. See references section.

1.3 [20] Make a chart of the major industrial robot vendors and their market share, either in the U.S. or worldwide. See references section.

1.4 [10] In a sentence or two, define: kinematics, workspace, trajectory.

1.5 [10] In a sentence or two, define: frame, degree of freedom, position control.

1.6 [10] In a sentence or two, define: force control, robot programming language.

1.7 [20] Make a chart indicating how labor costs have risen over the past 20 years.

1.8 [20] Make a chart indicating how the computer performance/price ratio has increased over the past 20 years.

Programming Exercise (Part 1)

1. Familiarize yourself with the computer you will use to do the programming exercises at the end of each chapter. Make sure you can create and edit files, and compile and execute programs.

2

SPATIAL DESCRIPTIONS AND TRANSFORMATIONS

2.1 Introduction

Robotic manipulation, by definition, implies that parts and tools will be moved around in space by some sort of mechanism. This naturally leads to the need of representing positions and orientations of the parts, tools, and of the mechanism itself. To define and manipulate mathematical quantities which represent position and orientation we must define coordinate systems and develop conventions for representation. Many of the ideas developed here in the context of position and orientation will form a basis for our later consideration of linear and rotational velocities as well as forces and torques.

We adopt the philosophy that somewhere there is a **universe coordinate system** to which everything we discuss can be referenced. We will describe all positions and orientations with respect to the universe coordinate system or with respect to other Cartesian coordinate systems which are (or could be) defined relative to the universe system.

2.2 Descriptions: positions, orientations, and frames

A **description** is used to specify attributes of various objects with which a manipulation system deals. These objects are parts, tools, or perhaps the manipulator itself. In this section we discuss the description of positions, orientations, and of an entity which contains both of these descriptions, frames.

Description of a position

Once a coordinate system is established we can locate any point in the universe with a 3×1 **position vector**. Because we will often define many coordinate systems in addition to the universe coordinate system, vectors must be tagged with information identifying which coordinate system they are defined within. In this book vectors are written with a leading superscript indicating the coordinate system to which they are referenced (unless it is clear from context), for example, ${}^{A}P$. This means that the components of ${}^{A}P$ have numerical values which indicate distances along the axes of $\{A\}$. Each of these distances along an axis can be thought of as the result of projecting the vector onto the corresponding axis.

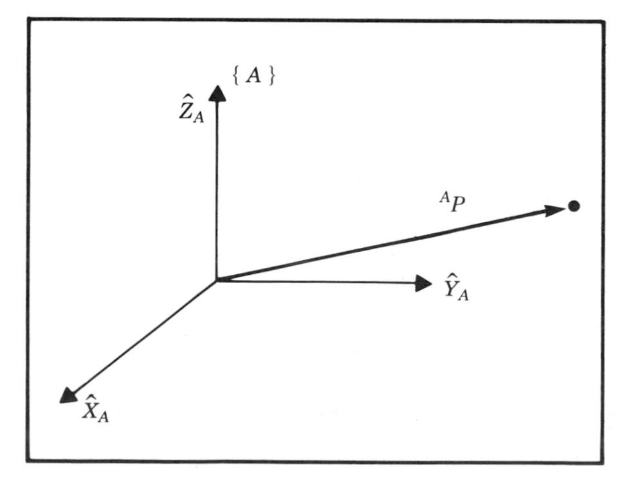

FIGURE 2.1 Vector relative to frame example.

Figure 2.1 pictorially represents a coordinate system, $\{A\}$, with 3 mutually orthogonal unit vectors with solid heads. A point ${}^{A}P$ is represented with a vector and can equivalently be thought of as a position in space, or simply as an ordered set of 3 numbers. Individual elements of a vector are given subscripts x, y, and z:

$$ {}^{A}P = \begin{bmatrix} p_x \\ p_y \\ p_z \end{bmatrix}. \qquad \textbf{(2.1)} $$

In summary, we will describe the position of a point in space with a position vector. Other 3-tuple descriptions of the position of points, such as spherical or cylindrical coordinate representations are discussed in the exercises at the end of the chapter.

Description of an orientation

Often we will find it necessary not only to represent a point in space, but also to describe the **orientation** of a body in space. For example, if vector ${}^{A}P$ in Fig. 2.2 locates the point directly between the finger tips of a manipulator hand, the complete location of the hand is still not specified until its orientation is also given. Assuming that the manipulator has a sufficient number of joints* the hand could be *oriented* arbitrarily while keeping the finger tips at the same position in space. In order to describe the orientation of a body we will *attach a coordinate system to the body and then give a description of this coordinate system relative to the reference system.* In Fig. 2.2, coordinate system $\{B\}$ has been attached to the body in a known way. A description of $\{B\}$ relative to $\{A\}$ now suffices to give the orientation of the body.

Thus, positions of points are described with vectors and orientations of bodies are described with an attached coordinate system. One way to describe the body-attached coordinate system, $\{B\}$, is to write the unit vectors of its three principal axes† in terms of the coordinate system $\{A\}$.

We denote the unit vectors giving the principle directions of coordinate system $\{B\}$ as $\hat{X}_B$, $\hat{Y}_B$, and $\hat{Z}_B$. When written in terms of coordinate system $\{A\}$ they are called ${}^{A}\hat{X}_B$, ${}^{A}\hat{Y}_B$, and ${}^{A}\hat{Z}_B$. It will be convenient if we stack these three unit vectors together as the columns of a 3×3 matrix, in the order ${}^{A}\hat{X}_B$, ${}^{A}\hat{Y}_B$, ${}^{A}\hat{Z}_B$. We will call this matrix a **rotation matrix**, and because this particular rotation matrix describes $\{B\}$ relative to $\{A\}$, we name it with the notation ${}^{A}_{B}R$. The choice of leading sub- and superscripts in the definition of rotation matrices will

* How many are "sufficient" will be discussed in Chapters 3 and 4.

† It is often convenient to use three, although any two would suffice since the third can always be recovered by taking the cross product of the two given.

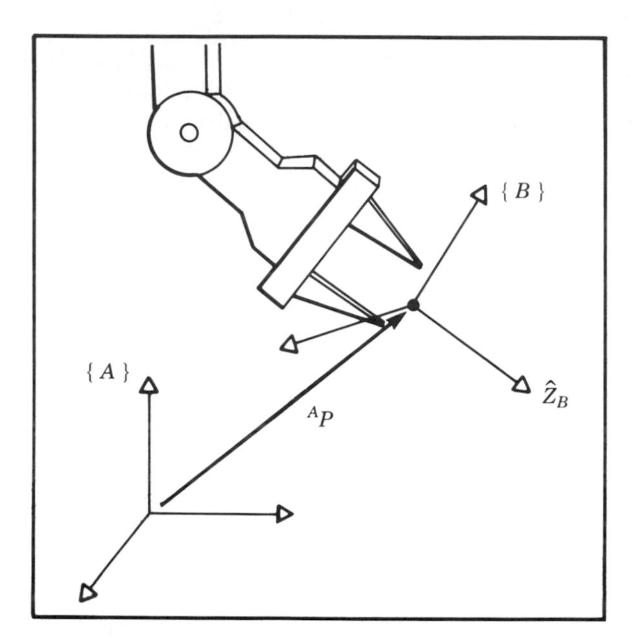

FIGURE 2.2 Locating an object in position and orientation.

become clear in following sections.

$$
{}^A_B R = \begin{bmatrix} {}^A\hat{X}_B & {}^A\hat{Y}_B & {}^A\hat{Z}_B \end{bmatrix} = \begin{bmatrix} r_{11} & r_{12} & r_{13} \\ r_{21} & r_{22} & r_{23} \\ r_{31} & r_{32} & r_{33} \end{bmatrix}. \quad \textbf{(2.2)}
$$

In summary, a set of three vectors may be used to specify an orientation. For convenience we will construct a 3×3 matrix which has these three vectors as its columns. Hence, whereas the position of a point is represented with a vector, the orientation of a body is represented with a matrix. In Section 2.8 we will consider some other descriptions of orientation which require only three parameters.

Description of a frame

The information needed to completely specify the whereabouts of the manipulator hand in Fig. 2.2 is a position and an orientation. The point on the body whose position we describe could be chosen arbitrarily, however: *For convenience, the point whose position we will describe is chosen as the origin of the body-attached frame.* The situation of a position and an orientation pair arises so often in robotics that we define an entity called a **frame** which is a set of four vectors giving position and orientation information. For example, in Fig. 2.2 one vector locates the finger tip position and three more describe its orientation. Equivalently, the description of a frame can be thought of as a position vector and a rotation matrix. Note that a frame is a coordinate system, where in addition to the orientation we give a position vector which locates its origin relative to some other imbedding frame. For example, frame $\{B\}$ is described by ${}^{A}_{B}R$ and ${}^{A}P_{BORG}$, where ${}^{A}P_{BORG}$ is the vector which locates the origin of the frame $\{B\}$:

$$\{B\} = \left\{ {}^{A}_{B}R,\ {}^{A}P_{BORG} \right\}. \tag{2.3}$$

In Fig. 2.3 there are three frames that are shown along with the universe coordinate system. Frames $\{A\}$ and $\{B\}$ are known relative to the universe coordinate system and frame $\{C\}$ is known relative to frame $\{A\}$.

In Fig. 2.3 we introduce a *graphical representation* of frames which is convenient in visualising frames. A frame is depicted by three arrows representing unit vectors defining the principal axes of the frame. An arrow representing a vector is drawn from one origin to another. This vector represents the position of the origin at the head of the arrow in terms of the frame at the tail of the arrow. The direction of this locating arrow tells us, for example, in Fig. 2.3, that $\{C\}$ is known relative to $\{A\}$ and not vice versa.

In summary, a frame can be used as a description of one coordinate system relative to another. A frame encompasses the ideas of representing both position and orientation, and so may be thought of as a generalization of those two ideas. Positions could be represented by a frame whose rotation matrix part is the identity matrix and whose position vector part locates the point being described. Likewise, an

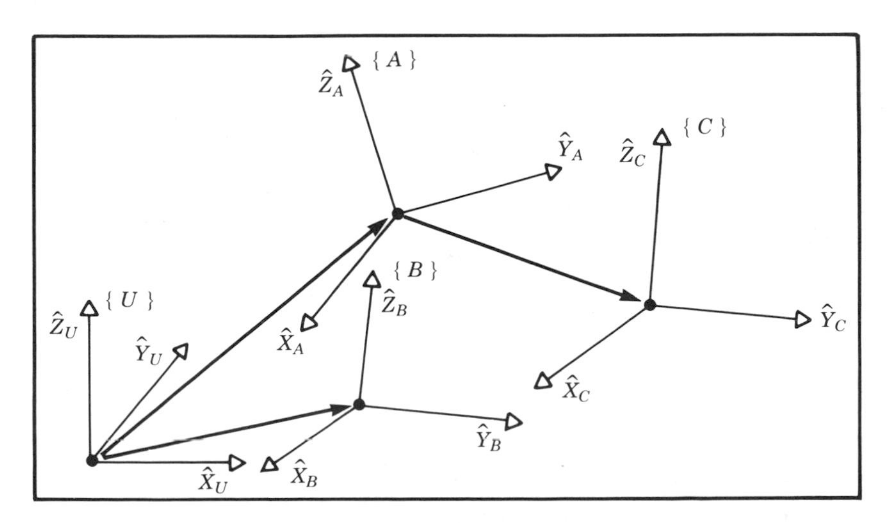

FIGURE 2.3 Example of several frames.

orientation could be represented with a frame whose position vector part was the zero vector.

2.3 Mappings: changing descriptions from frame to frame

In a great deal of the problems in robotics, we are concerned with expressing the same quantity in terms of various reference coordinate systems. Having introduced descriptions of positions, orientations, and frames in the previous section, we now consider the mathematics of **mapping** in order to change descriptions from frame to frame.

Mappings involving translated frames

In Fig. 2.4 we have a position defined by the vector ${}^{B}P$. We wish to express this point in space in terms of frame $\{A\}$, when $\{A\}$ has the same orientation as $\{B\}$. In this case, $\{B\}$ differs from $\{A\}$ only by a *translation* which is given by ${}^{A}P_{BORG}$, a vector which locates the origin of $\{B\}$ relative to $\{A\}$.

Because both vectors are defined relative to frames of the same orientation, we calculate the description of point P relative to $\{A\}$,

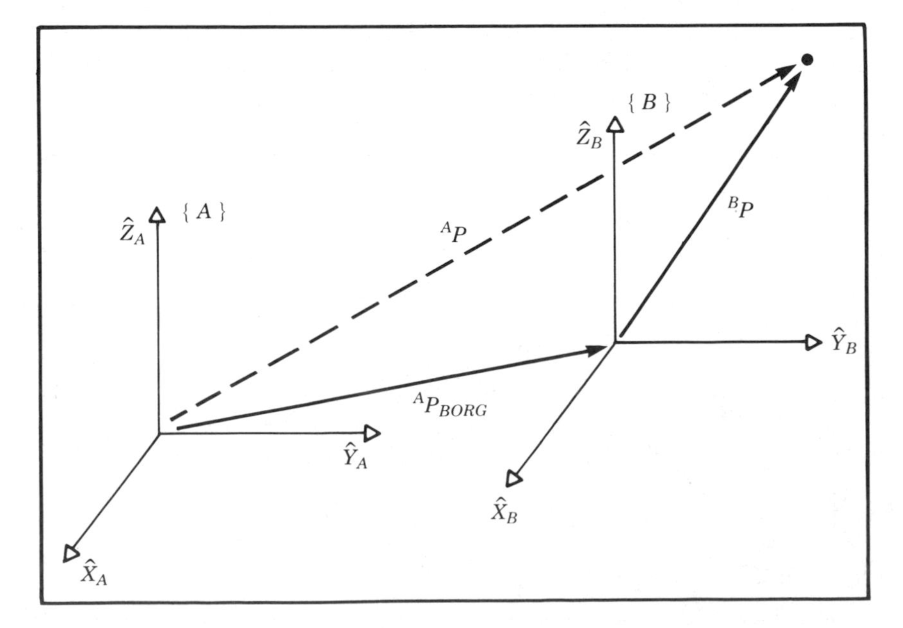

FIGURE 2.4 Translational mapping.

${}^{A}P$, by vector addition:

$$ {}^{A}P = {}^{B}P + {}^{A}P_{BORG}. \tag{2.4} $$

Note that only in the special case of equivalent orientations may we add vectors which are defined in terms of different frames.

In this simple example we have illustrated mapping a vector from one frame to another. This idea of mapping, or changing the description from one frame to another, is an extremely important concept. The quantity itself (here, a point in space) is not changed, only its description is changed. This is illustrated in Fig. 2.4, where the point described by ${}^{B}P$ is not translated, but remains the same, and instead we have computed a new description of the same point, but now with respect to system $\{A\}$.

We say that the vector ${}^{A}P_{BORG}$ defines this mapping, since all the information needed to perform the change in description is contained in ${}^{A}P_{BORG}$ (along with the knowledge that the frames had equivalent orientation).

Mappings involving rotated frames

In Section 2.2 we introduced the notion of describing an orientation by three unit vectors denoting the principal axes of a body-attached coordinate system. For convenience we stack these three unit vectors together as the columns of a 3×3 matrix. We will call this matrix a rotation matrix, and if this particular rotation matrix describes $\{B\}$ relative to $\{A\}$, we name it with the notation ${}^{A}_{B}R$.

Note that by our definition, the columns of a rotation matrix all have unit magnitude, and further, these unit vectors are orthogonal. From linear algebra [1] we know that the inverse of a matrix with orthonormal columns is equal to its transpose:

$$ {}^{A}_{B}R = {}^{B}_{A}R^{-1} = {}^{B}_{A}R^{T}. \tag{2.5} $$

Therefore, since the columns of ${}^{A}_{B}R$ are the unit vectors of $\{B\}$ written in $\{A\}$, then the *rows* of ${}^{A}_{B}R$ are the unit vectors of $\{A\}$ written in $\{B\}$.

So a rotation matrix can be interpreted as a set of three column vectors or as a set of three row vectors as follows:

$$ {}^{A}_{B}R = \begin{bmatrix} {}^{A}\hat{X}_{B} & {}^{A}\hat{Y}_{B} & {}^{A}\hat{Z}_{B} \end{bmatrix} = \begin{bmatrix} {}^{B}\hat{X}_{A}^{T} \\ {}^{B}\hat{Y}_{A}^{T} \\ {}^{B}\hat{Z}_{A}^{T} \end{bmatrix}. \tag{2.6} $$

As in Fig. 2.5, the situation will arise often where we know the definition of a vector with respect to some frame, $\{B\}$, and we would like to know its definition with respect to another frame, $\{A\}$, where the origins of

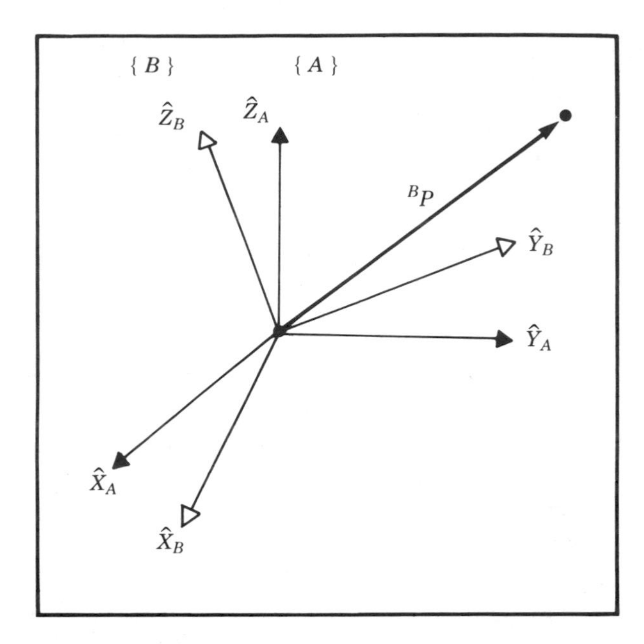

FIGURE 2.5 Rotating the description of a vector.

the two frames are coincident. This computation is possible when a description of the orientation of $\{B\}$ is known relative to $\{A\}$. This orientation is given by the rotation matrix ${}^A_BR$, whose columns are the unit vectors of $\{B\}$ written in $\{A\}$.

In order to calculate ${}^AP$, we note that the components of any vector are simply the projections of that vector onto the unit directions of its frame. The projection is calculated with the vector dot product. Thus we see that the components of ${}^AP$ may be calculated as

$$\begin{aligned} {}^Ap_x &= {}^B\hat{X}_A \cdot {}^BP, \\ {}^Ap_y &= {}^B\hat{Y}_A \cdot {}^BP, \\ {}^Ap_z &= {}^B\hat{Z}_A \cdot {}^BP. \end{aligned} \tag{2.7}$$

In order to express (2.7) in terms of a rotation matrix multiplication, we note from (2.6) that the *rows* of ${}^A_BR$ are ${}^B\hat{X}_A$, ${}^B\hat{Y}_A$, and ${}^B\hat{Z}_A$. So (2.7) may be written compactly using a rotation matrix as

$$ {}^AP = {}^A_BR\, {}^BP. \tag{2.8}$$

Equation (2.8) implements a mapping, that is, it changes the description of a vector from ${}^BP$, which describes a point in space relative to $\{B\}$, into ${}^AP$, which is a description of the same point, but expressed relative to $\{A\}$.

We now see that our notation is of great help in keeping track of mappings and frames of reference. A helpful way of viewing the notation we have introduced is to imagine that leading subscripts cancel the leading superscripts of the following entity, for example the Bs in (2.8).

EXAMPLE 2.1

Figure 2.6 shows a frame $\{B\}$ which is rotated relative to frame $\{A\}$ about $\hat{Z}$ by 30 degrees. Here, $\hat{Z}$ is pointing out of the page.

Writing the unit vectors of $\{B\}$ in terms of $\{A\}$ and stacking them as the columns of the rotation matrix we obtain

$$ {}^{A}_{B}R = \begin{bmatrix} 0.866 & -0.500 & 0.000 \\ 0.500 & 0.866 & 0.000 \\ 0.000 & 0.000 & 1.000 \end{bmatrix}. \quad \textbf{(2.9)} $$

Given

$$ {}^{B}P = \begin{bmatrix} 0.0 \\ 2.0 \\ 0.0 \end{bmatrix}. \quad \textbf{(2.10)} $$

We calculate ${}^{A}P$ as

$$ {}^{A}P = {}^{A}_{B}R \; {}^{B}P = \begin{bmatrix} -1.000 \\ 1.732 \\ 0.000 \end{bmatrix}. \quad \textbf{(2.11)} $$

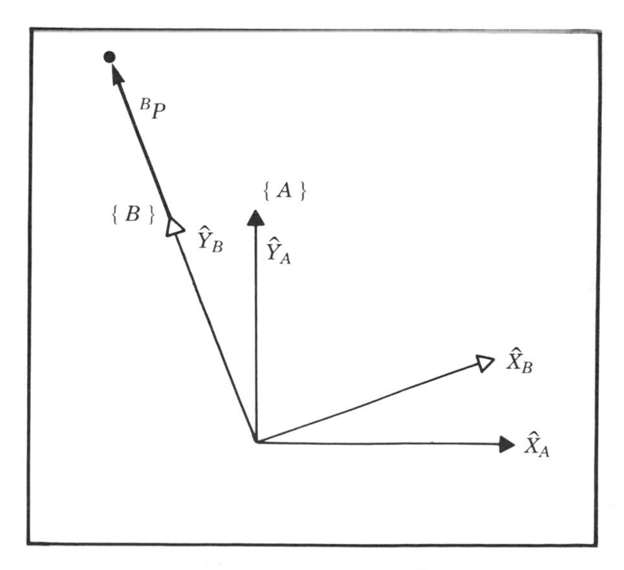

FIGURE 2.6 $\{B\}$ rotated 30 degrees about $\hat{Z}$.

Here ${}^{A}_{B}R$ acts as a mapping which is used to describe ${}^{B}P$ relative to frame $\{A\}$, ${}^{A}P$. As introduced in the case of translations, it is important to remember that, viewed as a mapping, the original vector P is not changed in space. Rather, we compute a new description of the vector relative to another frame. ■

Mappings involving general frames

Very often we know the description of a vector with respect to some frame, $\{B\}$, and we would like to know its description with respect to another frame, $\{A\}$. We now consider the general case of mapping. Here the origin of frame $\{B\}$ is not coincident with that of frame $\{A\}$ but has a general vector offset. The vector that locates $\{B\}$'s origin is called ${}^{A}P_{BORG}$. Also $\{B\}$ is rotated with respect to $\{A\}$ as described by ${}^{A}_{B}R$. Given ${}^{B}P$, we wish to compute ${}^{A}P$, as in Fig. 2.7.

We can first change ${}^{B}P$ to its description relative to an intermediate frame which has the same orientation of $\{A\}$, but whose origin is coincident with the origin of $\{B\}$. This is done by premultiplying by ${}^{A}_{B}R$ as we saw in Section 2.3. We then account for the translation between origins by simple vector addition as we saw in Section 2.3, yielding

$$ {}^{A}P = {}^{A}_{B}R \, {}^{B}P + {}^{A}P_{BORG}. \tag{2.12} $$

Equation (2.12) describes a general transformation mapping of a vector from its description in one frame to a description in a second frame. Note the following interpretation of our notation as exemplified in (2.12): the

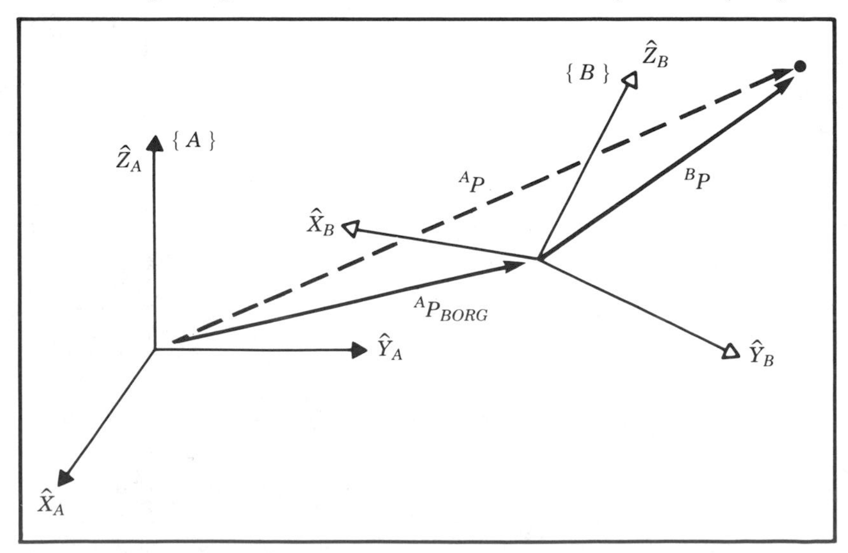

FIGURE 2.7 General transform of a vector.

Bs cancel leaving all quantities as vectors written in terms of A, which may then be added.

The form of (2.12) is not as appealing as the conceptual form,

$$^{A}P = {}^{A}_{B}T \, {}^{B}P. \tag{2.13}$$

That is, we would like to think of a mapping from one frame to another as an operator in matrix form. This aids in writing compact equations as well as being conceptually clearer than (2.12). In order that we can write the mathematics given in (2.12) in the matrix operator form suggested by (2.13), we define a 4×4 matrix operator, and use 4×1 position vectors, so that (2.13) has the structure

$$\begin{bmatrix} {}^{A}P \\ \hline 1 \end{bmatrix} = \left[\begin{array}{ccc|c} & {}^{A}_{B}R & & {}^{A}P_{BORG} \\ \hline 0 & 0 & 0 & 1 \end{array}\right] \begin{bmatrix} {}^{B}P \\ \hline 1 \end{bmatrix}. \tag{2.14}$$

That is,

1. A "1" is added as the last element of the 4×1 vectors.
2. A row "[0 0 0 1]" is added as the last row of the 4×4 matrix.

We adopt the convention that a position vector is 3×1 or 4×1 depending on whether it appears multiplied by a 3×3 matrix or by a 4×4 matrix. It is readily seen that (2.14) implements:

$$\begin{aligned} {}^{A}P &= {}^{A}_{B}R \, {}^{B}P + {}^{A}P_{BORG} \\ 1 &= 1. \end{aligned} \tag{2.15}$$

The 4×4 matrix in (2.14) is called a **Homogeneous transform**. For our purposes it can be regarded purely as a construction used to cast the rotation and translation of the general transform into a single matrix form. In other fields of study it can be used to compute perspective and scaling operations (when the last row is other than "[0 0 0 1]"). The interested reader should see [2].

Often we will write equations like (2.13) without any notation indicating that this is a homogeneous representation, because it is obvious from context. Note that while homogeneous transforms are useful in writing compact equations, a computer program to transform vectors would generally not use them because of time wasted multiplying ones and zeros. Thus, this representation is mainly for our convenience when thinking and writing equations down on paper.

Just as we used rotation matrices to specify an orientation, we will use transforms (usually in homogeneous representation) to specify a frame. Note that while we have introduced homogeneous transforms in the context of mappings, they also serve as descriptions of frames. The description of frame $\{B\}$ relative to $\{A\}$ is ${}^{A}_{B}T$.

EXAMPLE 2.2

Figure 2.8 shows a frame $\{B\}$ which is rotated relative to frame $\{A\}$ about $\hat{Z}$ by 30 degrees, and translated 10 units in $\hat{X}_A$, and 5 units in $\hat{Y}_A$. Find ${}^AP$ where ${}^BP = [3.0\ 7.0\ 0.0]^T$.

The definition of frame $\{B\}$ is:

$$ {}^A_BT = \begin{bmatrix} 0.866 & -0.500 & 0.000 & 10.0 \\ 0.500 & 0.866 & 0.000 & 5.0 \\ 0.000 & 0.000 & 1.000 & 0.0 \\ 0 & 0 & 0 & 1 \end{bmatrix}. \tag{2.16} $$

Given:

$$ {}^BP = \begin{bmatrix} 3.0 \\ 7.0 \\ 0.0 \end{bmatrix}. \tag{2.17} $$

We use the definition of $\{B\}$ given above as a transformation,

$$ {}^AP = {}^A_BT\ {}^BP = \begin{bmatrix} 9.098 \\ 12.562 \\ 0.000 \end{bmatrix}. \quad \blacksquare \tag{2.18} $$

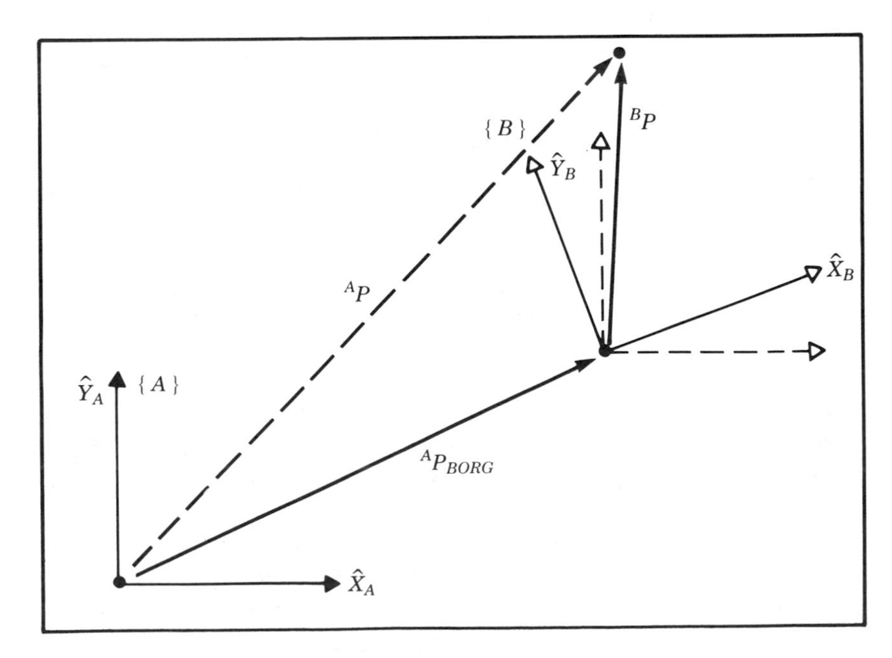

FIGURE 2.8 Frame $\{B\}$ rotated and translated.

2.4 Operators: translations, rotations, transformations

The same mathematical forms which we have used to map points between frames can also be interpreted as operators which translate points, rotate vectors, or both. In this section we illustrate this interpretation of the mathematics we have already developed.

Translational operators

A translation moves a point in space a finite distance along a given vector direction. Using this interpretation of actually translating the point in space, only one coordinate system need be involved. It turns out that translating the point in space is accomplished with the same mathematics as mapping the point to a second frame. Almost always, it is very important to understand which interpretation of the mathematics is being used. The distinction is as simple as this: When a vector is moved "forward" relative to a frame, we may consider either that the vector moved "forward" or that the frame moved "backward." The mathematics involved in the two cases is identical, only our view of the situation is different. Figure 2.9 indicates pictorially how a vector ${}^{A}P_1$ is translated by a vector ${}^{A}Q$. Here the vector ${}^{A}Q$ gives the information needed to perform the translation.

The result of the operation is a new vector ${}^{A}P_2$, calculated as

$$ {}^{A}P_2 = {}^{A}P_1 + {}^{A}Q. \tag{2.19}$$

To write this translation operation as a matrix operator, we use the notation

$$ {}^{A}P_2 = TRANS({}^{A}\hat{Q}, |Q|)\ {}^{A}P_1, \tag{2.20}$$

where we have defined the notation of the "$TRANS$" operator. Its first argument is a unit direction, and its second argument is a magnitude. The $TRANS$ operator may be thought of as a homogeneous transform of the special simple form:

$$ TRANS(\hat{Q}, |Q|) = \begin{bmatrix} 1 & 0 & 0 & q_x \\ 0 & 1 & 0 & q_y \\ 0 & 0 & 1 & q_z \\ 0 & 0 & 0 & 1 \end{bmatrix}. \tag{2.21}$$

Equations (2.4) and (2.19) implement the same mathematics. Note that if we had defined ${}^{B}P_{AORG}$ (instead of ${}^{A}P_{BORG}$) in Fig. 2.4 and had used it in (2.4) then we would have seen a sign change between (2.4) and

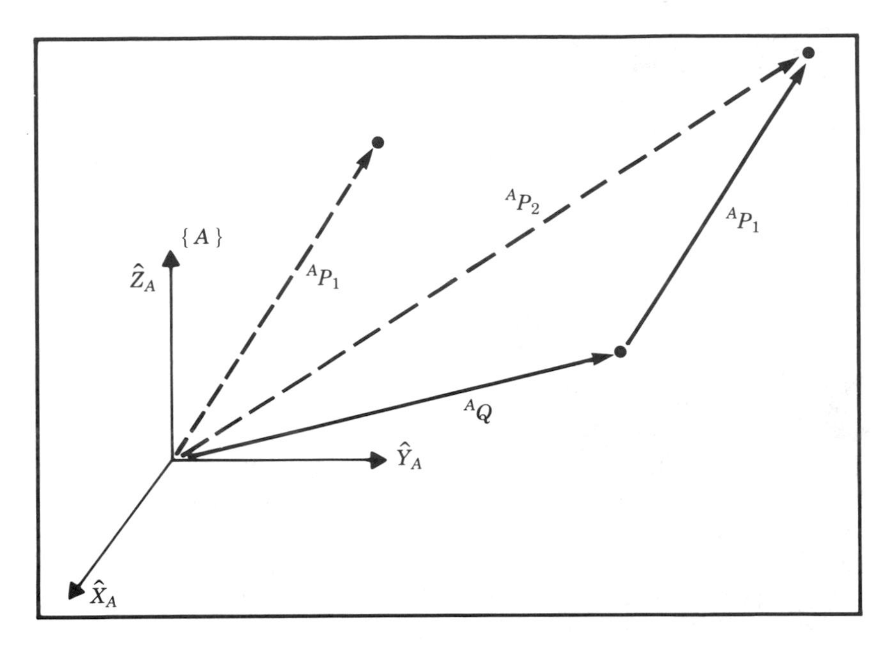

FIGURE 2.9 Translation operator.

(2.19). This sign change would indicate the difference between moving the vector "forward" and moving the coordinate system "backward." By defining the location of $\{B\}$ relative to $\{A\}$ (with $^AP_{BORG}$) we cause the mathematics of the two interpretations to be the same. Now that the "$TRANS$" notation has been introduced, we may also use it to describe frames, and also as a mapping.

Rotational operators

Another interpretation of a rotation matrix is as a *rotational operator* which operates on a vector $^AP_1$ and changes that vector to a new vector, $^AP_2$, by means of a rotation, R. Usually, when a rotation matrix is shown as an operator no sub- or superscripts appear since it is not viewed as relating two frames. That is, we may write

$$^AP_2 = R\ ^AP_1. \tag{2.22}$$

Again, as in the case of translations, the mathematics described in (2.8) and in (2.22) are the same; only our interpretation is different. This fact also allows us to see *how to obtain* rotational matrices which are to be used as operators:

The rotation matrix which rotates vectors through some rotation, R, is the same as the rotation matrix which describes a frame rotated by R relative to the reference frame.

Although a rotation matrix is easily viewed as an operator, we will also define another notation for a rotational operator which clearly indicates which axis is being rotated about:

$$ {}^{A}P_2 = ROT(\hat{K},\ \theta)\ {}^{A}P_1. \tag{2.23}$$

In this notation "ROT" is a rotational operator which performs a rotation about the axis direction $\hat{K}$ by an amount θ degrees. This operator may be written as a homogeneous transform whose position vector part is zero. For example, the operator which rotates about the $\hat{Z}$ axis by θ is written

$$ROT(\hat{Z},\ \theta) = \begin{bmatrix} \cos\theta & -\sin\theta & 0 & 0 \\ \sin\theta & \cos\theta & 0 & 0 \\ 0 & 0 & 1 & 0 \\ 0 & 0 & 0 & 1 \end{bmatrix}. \tag{2.24}$$

Of course, to rotate a position vector we could just as well use the 3×3 rotation matrix part of the homogeneous transform. The "ROT" notation, therefore, may be considered to represent a 3×3 or a 4×4 matrix. Later in this chapter we will see how to write the rotation matrix for a rotation about a general axis, $\hat{K}$.

EXAMPLE 2.3

Figure 2.10 shows a vector ${}^{A}P_1$. We wish to compute the vector obtained by rotating this vector about $\hat{Z}$ by 30 degrees. Call the new vector ${}^{A}P_2$.

The rotation matrix which rotates vectors by 30 degrees about $\hat{Z}$ is the same as the rotation matrix which describes a frame rotated 30 degrees about $\hat{Z}$ relative to the reference frame. Thus the correct rotational operator is

$$ROT(\hat{Z}, 30.0) = \begin{bmatrix} 0.866 & -0.500 & 0.000 \\ 0.500 & 0.866 & 0.000 \\ 0.000 & 0.000 & 1.000 \end{bmatrix}. \tag{2.25}$$

Given:

$${}^{A}P_1 = \begin{bmatrix} 0.0 \\ 2.0 \\ 0.0 \end{bmatrix}. \tag{2.26}$$

We calculate ${}^{A}P_2$ as

$${}^{A}P_2 = ROT(\hat{Z}, 30.0)\ {}^{A}P_1 = \begin{bmatrix} -1.000 \\ 1.732 \\ 0.000 \end{bmatrix}. \quad \blacksquare \tag{2.27}$$

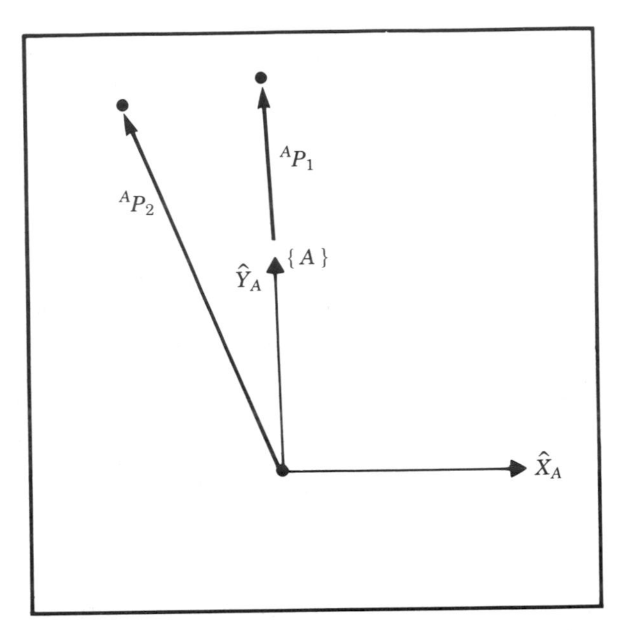

FIGURE 2.10 The vector $^AP_1$ rotated 30 degrees about $\hat{Z}$.

Equations (2.8) and (2.22) implement the same mathematics. Note that if we had defined ${}^B_AR$ (instead of ${}^A_BR$) in (2.8) then the inverse of R would appear in (2.22). This change would indicate the difference between rotating the vector "forward" versus rotating the coordinate system "backward." By defining the location of $\{B\}$ relative to $\{A\}$ (with ${}^A_BR$) we cause the mathematics of the two interpretations to be the same.

Transformation operators

As with vectors and rotation matrices, a frame has another interpretation as a *transformation operator*. In this interpretation, only one coordinate system is involved, and so the symbol T is used without sub- or superscripts. The operator T rotates and translates a vector $^AP_1$ to compute a new vector, $^AP_2$. Thus

$$^AP_2 = T\,^AP_1. \qquad \textbf{(2.28)}$$

Again, as in the case of rotations, the mathematics described in (2.13) and in (2.28) are the same, only our interpretation is different. This fact also allows us to see how to obtain homogeneous transforms which are to be used as operators:

The transform which rotates by R and translates by Q is the same as the transform which describes a frame rotated by R and translated by Q relative to the reference frame.

A transform is usually thought of as being in the form of a homogeneous transform with general rotation matrix and position vector parts.

EXAMPLE 2.4

Fig. 2.11 shows a vector ${}^A P_1$. We wish to rotate it about $\hat{Z}$ by 30 degrees, and translate it 10 units in $\hat{X}_A$, and 5 units in $\hat{Y}_A$. Find ${}^A P_2$ where ${}^A P_1 = [3.0\ 7.0\ 0.0]^T$.

The operator T, which performs the translation and rotation, is

$$T = \begin{bmatrix} 0.866 & -0.500 & 0.000 & 10.0 \\ 0.500 & 0.866 & 0.000 & 5.0 \\ 0.000 & 0.000 & 1.000 & 0.0 \\ 0 & 0 & 0 & 1 \end{bmatrix}. \tag{2.29}$$

Given:

$${}^A P_1 = \begin{bmatrix} 3.0 \\ 7.0 \\ 0.0 \end{bmatrix}. \tag{2.30}$$

We use T as an operator:

$${}^A P_2 = T\ {}^A P_1 = \begin{bmatrix} 9.098 \\ 12.562 \\ 0.000 \end{bmatrix}. \tag{2.31}$$

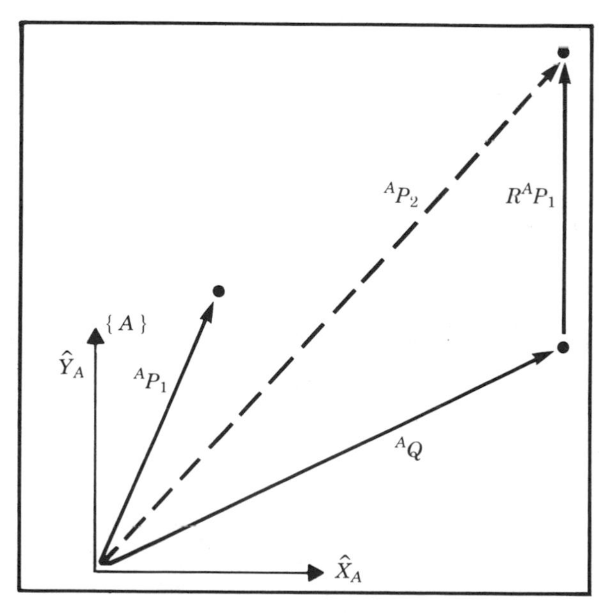

FIGURE 2.11 The vector ${}^A P_1$ rotated and translated to form ${}^A P_2$.

Note that this example is numerically exactly the same as Example 2.2, but the interpretation is quite different. ∎

2.5 Summary of interpretations

We have introduced concepts first for the case of translation only, then for the case of rotation only, and finally for the general case of rotation about a point and translation of that point. Having understood the general case of rotation and translation, we will not need to explicitly consider the two simpler cases since they are contained within the general framework.

As a general tool to represent frames we have introduced the *homogeneous transform*, a 4×4 matrix containing orientation and position information.

We have introduced three interpretations of this homogeneous transform:

1. It is a *description of a frame.* ${}^{A}_{B}T$ describes the frame $\{B\}$ relative to the frame $\{A\}$. Specifically, the columns of ${}^{A}_{B}R$ are unit vectors defining the directions of the principle axes of $\{B\}$, and ${}^{A}P_{BORG}$ locates the position of the origin of $\{B\}$.
2. It is a *transform mapping.* ${}^{A}_{B}T$ maps ${}^{B}P$ ${}^{A}P$.
3. It is a *transform operator.* T operates on ${}^{A}P_1$ to create ${}^{A}P_2$.

From this point on the terms *frame* and *transform* will both be used to refer to a position vector plus an orientation. *Frame* is the term favored when speaking of a description, and *transform* is used most frequently when use as a mapping or operator is implied. Note that transformations are generalizations of translations and rotations, so we will often use the term *transform* when speaking of a pure rotation (or translation).

2.6 Transformation arithmetic

In this section we look at the multiplication of transforms, and the inversion of transforms. These two elementary operations form a functionally complete set of transform operators.

Compound transformations

In Fig. 2.12, we have ${}^{C}P$ and wish to find ${}^{A}P$.

Frame $\{C\}$ is known relative to frame $\{B\}$, and frame $\{B\}$ is known relative to frame $\{A\}$. We can transform ${}^{C}P$ into ${}^{B}P$ as

$$ {}^{B}P = {}^{B}_{C}T \, {}^{C}P, \tag{2.32} $$

And then transform ${}^{B}P$ into ${}^{A}P$ as

$$ {}^{A}P = {}^{A}_{B}T \, {}^{B}P. \tag{2.33} $$

Combining (2.32) and (2.33) we get the following, not unexpected result:

$$ {}^{A}P = {}^{A}_{B}T \, {}^{B}_{C}T \, {}^{C}P, \tag{2.34} $$

from which we could define:

$$ {}^{A}_{C}T = {}^{A}_{B}T \, {}^{B}_{C}T. \tag{2.35} $$

Again, note that familiarity with the sub- and superscript notation makes these manipulations simple. In terms of the known descriptions of $\{B\}$ and $\{C\}$, we can give the expression for ${}^{A}_{C}T$ as

$$ {}^{A}_{C}T = \left[\begin{array}{ccc|c} & {}^{A}_{B}R \, {}^{B}_{C}R & & {}^{A}_{B}R \, {}^{B}P_{CORG} + {}^{A}P_{BORG} \\ \hline 0 & 0 & 0 & 1 \end{array} \right]. \tag{2.36} $$

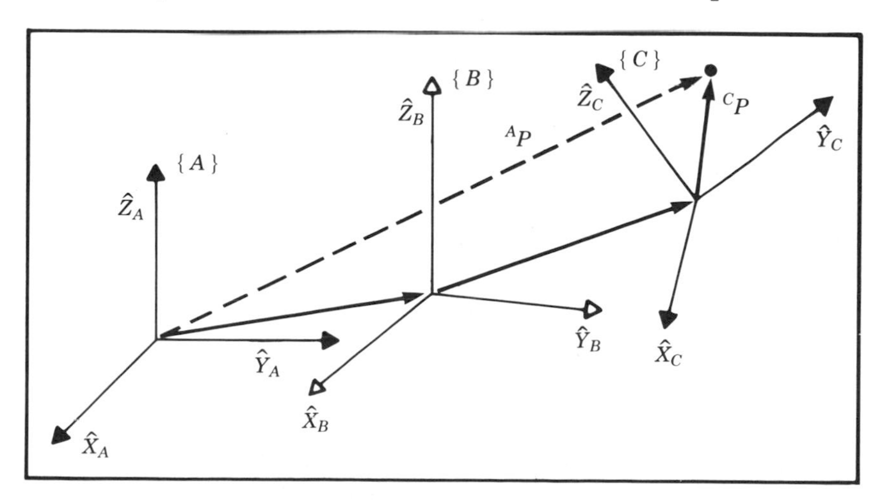

FIGURE 2.12 Compound frames: each is known relative to previous.

Inverting a transform

Consider a frame $\{B\}$ which is known with respect to a frame $\{A\}$, that is, we know the value of ${}^A_BT$. Sometimes we will wish to invert this transform, in order to get a description of $\{A\}$ relative to $\{B\}$, i.e. ${}^B_AT$. A straightforward way of calculating the inverse is to compute the inverse of the 4×4 homogeneous transform. However, if we do so, we are not taking full advantage of the structure inherent in the transform. It is easy to find a computationally simpler method of computing the inverse which does take advantage of this structure.

To find ${}^B_AT$ we must compute ${}^B_AR$ and ${}^BP_{AORG}$ from ${}^A_BR$ and ${}^AP_{BORG}$. First, recall from our discussion of rotation matrices that

$$ {}^B_AR = {}^A_BR^T. \tag{2.37} $$

Next, we change the description of ${}^AP_{BORG}$ into $\{B\}$ using Eq. 2.12:

$$ {}^B\left({}^AP_{BORG}\right) = {}^B_AR \; {}^AP_{BORG} + {}^BP_{AORG}. \tag{2.38} $$

Since the lefthand side of Eq. 2.38 must be zero, we have

$$ {}^BP_{AORG} = -{}^B_AR \; {}^AP_{BORG} = -{}^A_BR^T \; {}^AP_{BORG}. \tag{2.39} $$

Using (2.37) and (2.39) we can write the form of ${}^B_AT$ as

$$ {}^B_AT = \left[\begin{array}{ccc|c} & {}^A_BR^T & & -{}^A_BR^T \; {}^AP_{BORG} \\ \hline 0 & 0 & 0 & 1 \end{array}\right]. \tag{2.40} $$

Note that with our notation,

$$ {}^B_AT = {}^A_BT^{-1}. $$

Equation (2.40) is a general and extremely useful way of computing the inverse of a homogeneous transform.

EXAMPLE 2.5

Figure 2.13 shows a frame $\{B\}$ which is rotated relative to frame $\{A\}$ about $\hat{Z}$ by 30 degrees, and translated 4 units in $\hat{X}_A$, and 3 units in $\hat{Y}_A$. Thus, we have a description of ${}^A_BT$. Find ${}^B_AT$.

The frame defining $\{B\}$ is

$$ {}^A_BT = \begin{bmatrix} 0.866 & -0.500 & 0.000 & 4.0 \\ 0.500 & 0.866 & 0.000 & 3.0 \\ 0.000 & 0.000 & 1.000 & 0.0 \\ 0 & 0 & 0 & 1 \end{bmatrix}. \tag{2.41} $$

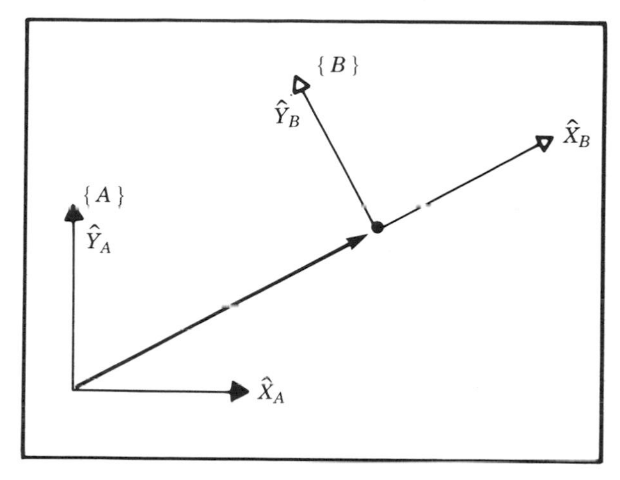

FIGURE 2.13 {B} relative to {A}.

Using (2.40) we compute

$$
{}^B_AT = \begin{bmatrix} 0.866 & 0.500 & 0.000 & -4.964 \\ -0.500 & 0.866 & 0.000 & -0.598 \\ 0.000 & 0.000 & 1.000 & 0.0 \\ 0 & 0 & 0 & 1 \end{bmatrix}. \quad \blacksquare \tag{2.42}
$$

2.7 Transform equations

Figure 2.14 indicates a situation in which a frame $\{D\}$ can be expressed as products of transformations in two different ways. First,

$$
{}^U_DT = {}^U_AT \; {}^A_DT, \tag{2.43}
$$

but also as:

$$
{}^U_DT = {}^U_BT \; {}^B_CT \; {}^C_DT. \tag{2.44}
$$

We may set these two descriptions of ${}^U_DT$ equal to form a **transform equation**:

$$
{}^U_AT \; {}^A_DT = {}^U_BT \; {}^B_CT \; {}^C_DT. \tag{2.45}
$$

Transform equations may be used to solve for transforms in the case of n unknown transforms and n transform equations. Consider (2.45) in the case that all transforms are known except ${}^B_CT$. Here we have one transform equation and one unknown transform; hence, we easily find its solution as

$$
{}^B_CT = {}^U_BT^{-1} \; {}^U_AT \; {}^A_DT \; {}^C_DT^{-1}. \tag{2.46}
$$

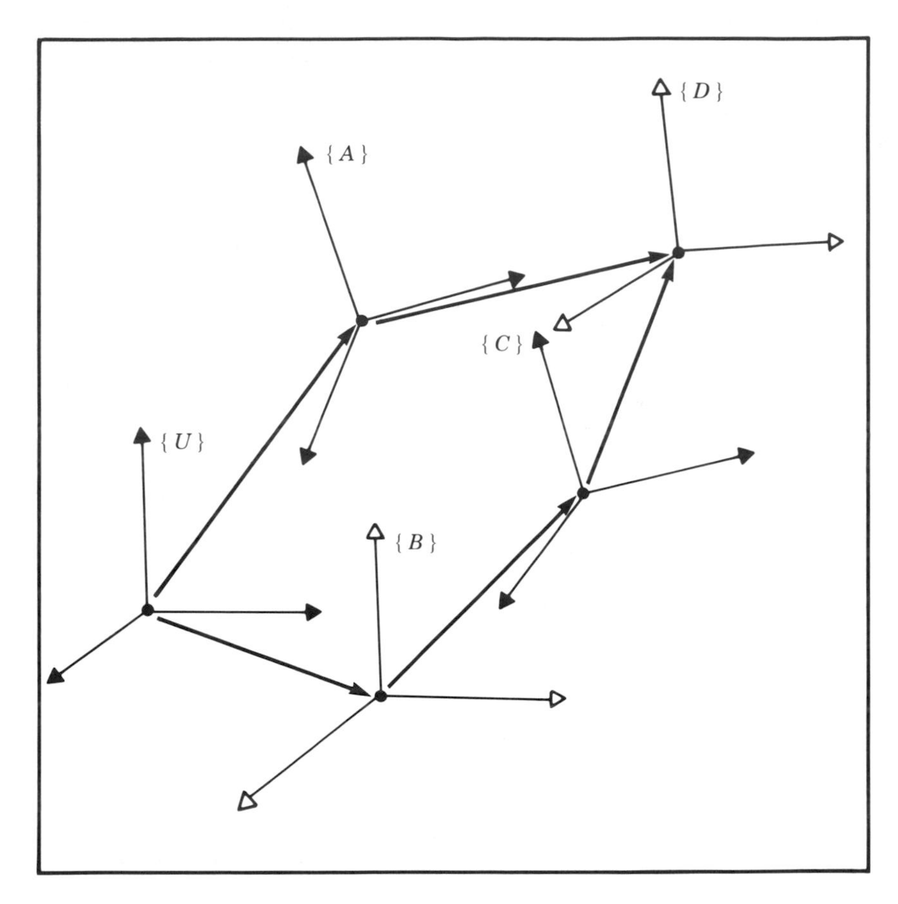

FIGURE 2.14 Set of transforms forming a loop.

Figure 2.15 indicates another similar situation.

Note that in all figures we have introduced a *graphical* representation of frames as an arrow pointing from one origin to another origin. The arrow's direction indicates which way the frames are defined: in Fig. 2.14, frame $\{D\}$ is defined relative to $\{A\}$, but in Fig. 2.15 frame $\{A\}$ is defined relative to $\{D\}$. In order to compound frames when the arrows line up, we simply compute the product of the transforms. If an arrow points the opposite way in a chain of transforms, we simply compute its inverse first. In Fig. 2.15 two possible descriptions of $\{C\}$ are

$$ {}^{U}_{C}T = {}^{U}_{A}T \, {}^{D}_{A}T^{-1} \, {}^{D}_{C}T \tag{2.47} $$

and:

$$ {}^{U}_{C}T = {}^{U}_{B}T \, {}^{B}_{C}T. \tag{2.48} $$

Again, we might equate (2.47) and (2.48) to solve for, say, ${}^{U}_{A}T$:

$$ {}^{U}_{A}T = {}^{U}_{B}T \, {}^{B}_{C}T \, {}^{D}_{C}T^{-1} \, {}^{D}_{A}T. \tag{2.49} $$

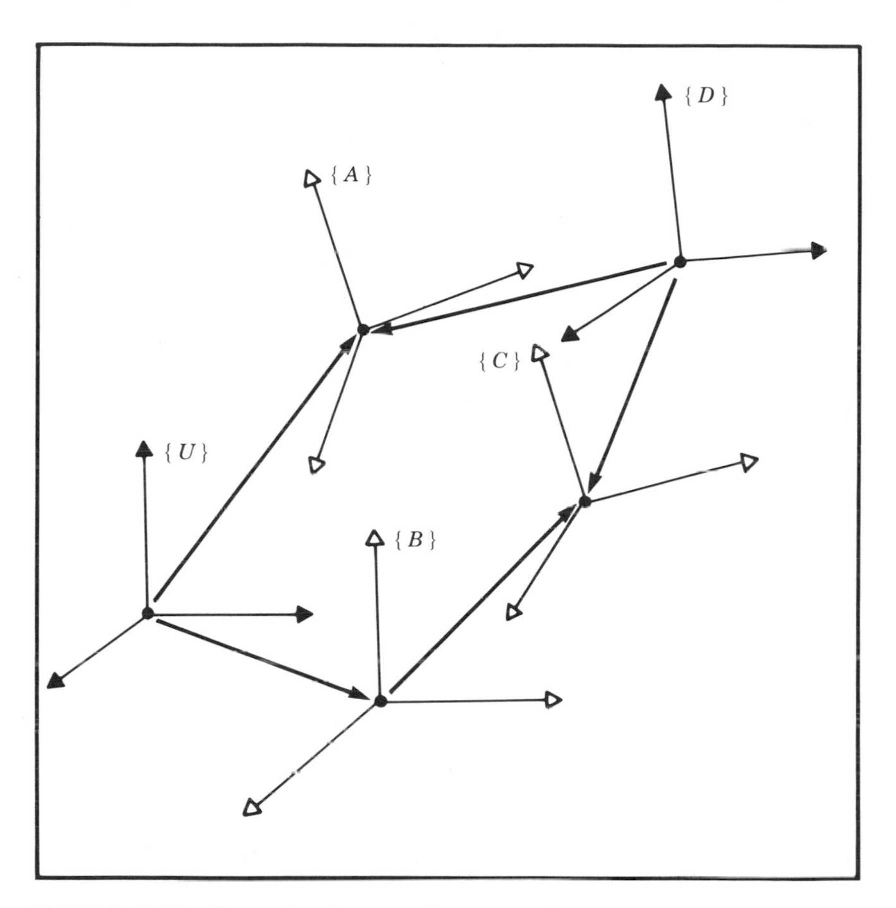

FIGURE 2.15 Example of a transform equation.

EXAMPLE 2.6

Assume we know the transform ${}^B_T T$ in Fig. 2.16, which describes the frame at the manipulator's fingertips $\{T\}$ relative to the base of the manipulator, $\{B\}$. Also, we know where the table top is located in space relative to the manipulator's base because we have a description of the frame $\{S\}$ which is attached to the table as shown, ${}^B_S T$. Finally, we know the location of the frame attached to the bolt lying on the table relative to the table frame, that is ${}^S_G T$. Calculate the position and orientation of the bolt relative to the manipulator's hand, ${}^T_G T$.

Guided by our notation (but, it is hoped, also by our understanding) we compute the bolt frame relative to the hand frame as

$$ {}^T_G T = {}^B_T T^{-1} \, {}^B_S T \, {}^S_G T. \quad \blacksquare \tag{2.50} $$

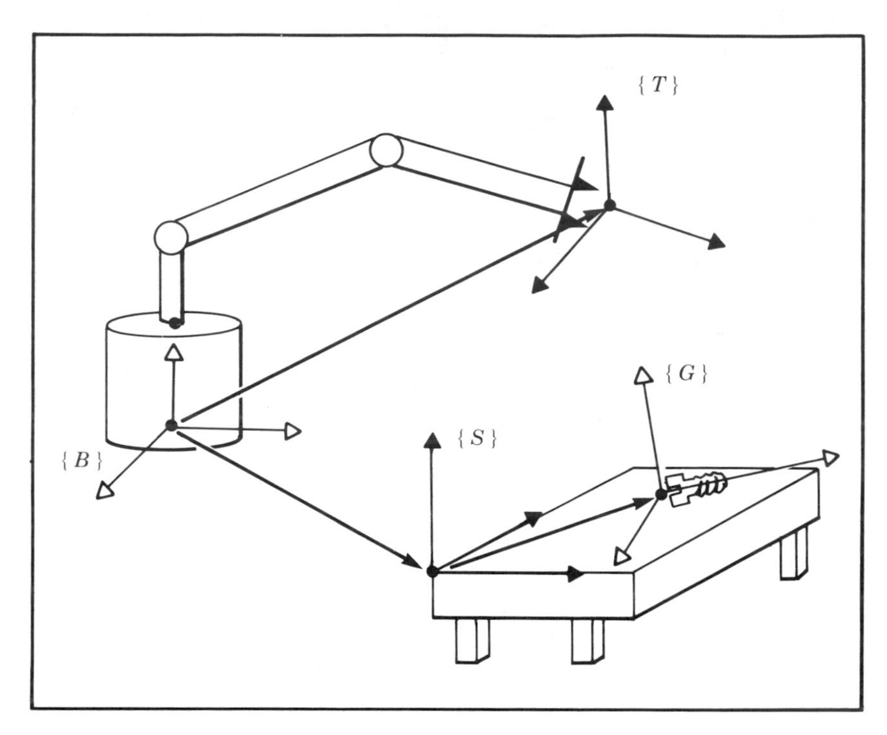

FIGURE 2.16 Manipulator reaching for a bolt.

2.8 More on representation of orientation

So far, our only means of representing an orientation is by giving a 3×3 rotation matrix. This representation contains nine parameters even though a body's orientation in space can be completely defined with just three parameters. Clearly, the nine elements of a rotation matrix are not all independent. In fact, given a rotation matrix, R, it is easy to write down the six dependencies between the elements. Imagine R as three columns, as we originally introduced it:

$$R = \begin{bmatrix} \hat{X} & \hat{Y} & \hat{Z} \end{bmatrix}. \tag{2.51}$$

As we know from Section 2.2, these three vectors are the unit axes of some frame written in terms of the reference frame. Since each is a unit vector, and since all three must be mutually perpendicular, we see that there are six constraints on the nine matrix elements:

$$
\begin{aligned}
|\hat{X}| &= 1, \\
|\hat{Y}| &= 1, \\
|\hat{Z}| &= 1, \\
\hat{X} \cdot \hat{Y} &= 0, \\
\hat{X} \cdot \hat{Z} &= 0, \\
\hat{Y} \cdot \hat{Z} &= 0.
\end{aligned} \tag{2.52}
$$

It is natural then to ask whether representations of orientation can be devised such that the representation is conveniently specified with three parameters. In this section we will present several such representations.

Whereas translations along three mutually perpendicular axes are quite easy to visualize, rotations seem less intuitive. Unfortunately, people have a hard time describing and specifying orientations in three dimensional space. One problem is that rotations don't generally commute. That is, ${}^{A}_{B}R\ {}^{B}_{C}R$ is not the same as ${}^{B}_{C}R\ {}^{A}_{B}R$.

EXAMPLE 2.7

Consider two rotations, one about $\hat{Z}$ by 30 degrees and one about $\hat{X}$ by 30 degrees.

$$
R_z = \begin{bmatrix} 0.866 & 0.500 & 0.000 \\ 0.500 & 0.866 & 0.000 \\ 0.000 & 0.000 & 1.000 \end{bmatrix} \tag{2.53}
$$

$$
R_x = \begin{bmatrix} 1.000 & 0.000 & 0.000 \\ 0.000 & 0.866 & -0.500 \\ 0.000 & 0.500 & 0.866 \end{bmatrix} \tag{2.54}
$$

$$
\begin{aligned}
R_z\, R_x &= \begin{bmatrix} 0.87 & -0.43 & 0.25 \\ 0.50 & 0.75 & -0.43 \\ 0.00 & 0.50 & 0.87 \end{bmatrix} \neq R_x \\
R_z &= \begin{bmatrix} 0.87 & -0.50 & 0.00 \\ 0.43 & 0.75 & -0.50 \\ 0.25 & 0.43 & 0.87 \end{bmatrix}
\end{aligned} \tag{2.55}
$$

This is not surprising since we use matrices to represent rotations and multiplication of matrices is not commutative in general. ■

Because rotations can be thought of either as operators or as descriptions of orientation, it is not surprising that different representations are favored for each of these uses. Rotation matrices are useful as operators. Their matrix form is such that when multiplied by a vector they perform the rotation operation. However, rotation matrices are somewhat unwieldy when used to specify an orientation. A human operator at a computer terminal who wishes to type in the specification of the desired orientation of a robot's hand would have a hard time to input a nine element matrix with orthonormal columns. A representation which requires only three numbers would be simpler. In the following sections we introduce several such representations.

Roll, pitch, and yaw angles about fixed axes

One method of describing the orientation of a frame $\{B\}$ is as follows:

> Start with the frame coincident with a known reference frame $\{A\}$. First rotate $\{B\}$ about $\hat{X}_A$ by an angle γ, then rotate about $\hat{Y}_A$ by an angle β, and then rotate about $\hat{Z}_A$ by an angle α.

Each of the three rotations takes place about an axis in the fixed reference frame, $\{A\}$. We will call this convention for specifying an orientation **roll, pitch, yaw** angles. The rotation about $\hat{X}$ by γ is sometimes called *roll*, rotation about $\hat{Y}$ by β called *pitch*, and about $\hat{Z}$ by α called *yaw*.

The derivation of the equivalent rotation matrix, ${}^A_B R_{rpy}(\gamma, \beta, \alpha)$, is straightforward because all rotations occur about axes of the reference

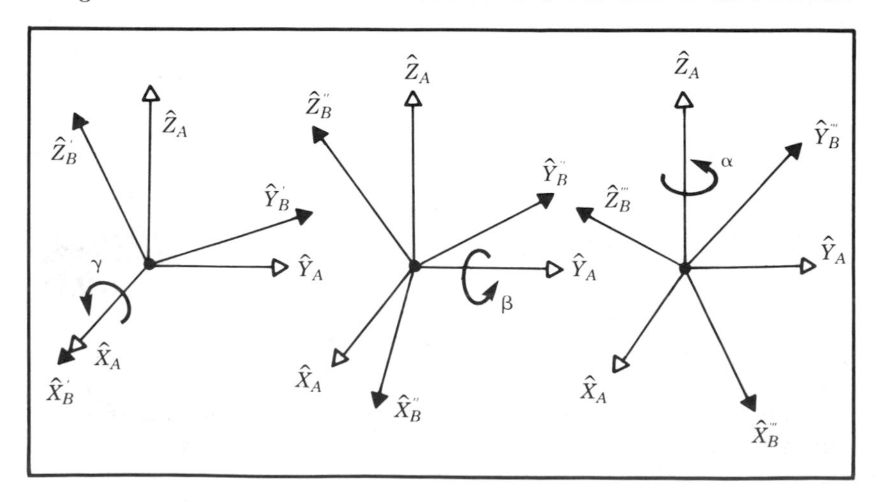

FIGURE 2.17 Roll, pitch, and yaw about fixed axes.

frame:

$$ {}^A_B R_{rpy}(\gamma,\beta,\alpha) = ROT({}^A\hat{Z}_A,\alpha)\ \ ROT({}^A\hat{Y}_A,\beta)\ \ ROT({}^A\hat{X}_A,\gamma) $$

$$ = \begin{bmatrix} c\alpha & -s\alpha & 0 \\ s\alpha & c\alpha & 0 \\ 0 & 0 & 1 \end{bmatrix} \begin{bmatrix} c\beta & 0 & s\beta \\ 0 & 1 & 0 \\ -s\beta & 0 & c\beta \end{bmatrix} \begin{bmatrix} 1 & 0 & 0 \\ 0 & c\gamma & -s\gamma \\ 0 & s\gamma & c\gamma \end{bmatrix}. \quad (2.56) $$

Where $c\alpha$ is shorthand for $\cos\alpha$ and $s\alpha$ for $\sin\alpha$, etc. Note the order of the rotations is (from the right) roll, pitch, and then yaw. Multiplying (2.56) out, we obtain

$$ {}^A_B R_{rpy}(\gamma,\beta,\alpha) = \begin{bmatrix} c\alpha c\beta & c\alpha s\beta s\gamma - s\alpha c\gamma & c\alpha s\beta c\gamma + s\alpha s\gamma \\ s\alpha c\beta & s\alpha s\beta s\gamma + c\alpha c\gamma & s\alpha s\beta c\gamma - c\alpha s\gamma \\ -s\beta & c\beta s\gamma & c\beta c\gamma \end{bmatrix}. \quad (2.57) $$

Keep in mind that the definition given here of roll, pitch, and yaw angles includes the order in which they are performed, as it must. Equation (2.57) is correct only for rotations performed in the order: roll, pitch, yaw.

The inverse problem, that of extracting equivalent roll, pitch, and yaw angles from a rotation matrix is often of interest. The solution depends on solving a set of transcendental equations: there are nine equations and three unknowns if (2.57) is equated to a given rotation matrix. Amongst the nine equations are six dependencies, so essentially we have three equations and three unknowns. Let:

$$ {}^A_B R = \begin{bmatrix} r_{11} & r_{12} & r_{13} \\ r_{21} & r_{22} & r_{23} \\ r_{31} & r_{32} & r_{33} \end{bmatrix}. \quad (2.58) $$

We note from (2.57) that as long as $c\beta \neq 0$ we can solve for α by taking the arc tangent of r_{21} over r_{11}, and we can solve for γ by taking the arc tangent of r_{32} over r_{33}. Finally, β is found by taking the square root of the sum of the squares of r_{11} and r_{21} in order to compute $\cos\beta$. Then, we can solve for β with the arc tangent of $-r_{31}$ over the computed cosine. In summary:

$$ \begin{aligned} \gamma &= \mathrm{Atan2}(r_{32}, r_{33}), \\ \beta &= \mathrm{Atan2}\left(-r_{31}, \sqrt{r_{11}^2 + r_{21}^2}\right), \\ \alpha &= \mathrm{Atan2}(r_{21}, r_{11}), \end{aligned} \quad (2.59) $$

where $\mathrm{Atan2}(y,x)$ is a two-argument arc tangent function.*

* $\mathrm{Atan2}(y,x)$ computes $\tan^{-1}(\frac{y}{x})$ but uses the signs of both x and y to determine the quadrant in which the resulting angle lies. For example, $\mathrm{Atan2}(-2.0,-2.0) = -135°$; whereas $\mathrm{Atan2}(2.0,2.0) = 45°$, a distinction which would be lost with a single argument arc tangent function. As we are frequently computing angles which can range over a full 360°, we will make use of the Atan2 function regularly. Note that Atan2 becomes undefined when both arguments are zero. It is sometimes called a "4-quadrant arc tangent," and some programming language libraries have it predefined.

Although a second solution exists, by using the positive square root in the formula for β, we always compute the single solution for which $-90.0 <= \beta <= 90.0$. This is usually a good practice, since we can then define one-to-one mapping functions between various representations of orientation. However, in some cases, calculating all solutions is important (more on this in Chapter 4). If $\beta = \pm 90.0$ degrees, the solution of (2.59) degenerates. In those cases, only the sum or the difference of α and γ may be computed. One possible convention is to choose $\alpha = 0.0$ in these cases, which has the results given below.

If $\beta = 90.0$ degrees, then

$$\begin{aligned} \gamma &= \mathrm{Atan2}(r_{12}, r_{22}), \\ \beta &= 90.0, \\ \alpha &= 0.0. \end{aligned} \tag{2.60}$$

If $\beta = -90.0$ degrees, then

$$\begin{aligned} \gamma &= -\mathrm{Atan2}(r_{12}, r_{22}), \\ \beta &= -90.0, \\ \alpha &= 0.0. \end{aligned} \tag{2.61}$$

Z-Y-X Euler angles

Another possible description of a frame $\{B\}$ is as follows:

> Start with the frame coincident with a known frame $\{A\}$. First rotate $\{B\}$ about $\hat{Z}_B$ by an angle α, then rotate about $\hat{Y}_B$ by an angle β, and then rotate about $\hat{X}_B$ by an angle γ.

Note that in this representation, each rotation is performed about an axis of $\{B\}$, rather than the fixed reference, $\{A\}$. Such a set of three rotations are called **Euler angles**. Note that each rotation takes place about an axis whose location depends upon the preceeding rotations. Because the three rotations occur about the axes $\hat{Z}$, $\hat{Y}$, and $\hat{X}$, we will call this representation **Z-Y-X Euler angles**.

Figure 2.18 shows the axes of $\{B\}$ after each Euler angle rotation is applied. Rotation α about $\hat{Z}$ causes $\hat{X}$ to rotate into $\hat{X}'$, and $\hat{Y}$ to rotate into $\hat{Y}'$, and so on. An additional "prime" gets added to each axis with each rotation.

Suppose we wish to find a rotation matrix as a function of the Z-Y-X Euler angles, ${}^A_B R_{zyx}(\alpha, \beta, \gamma)$. Note that each rotation is performed about a unit vector of $\{B\}$. These axes, when described in terms of $\{A\}$, are

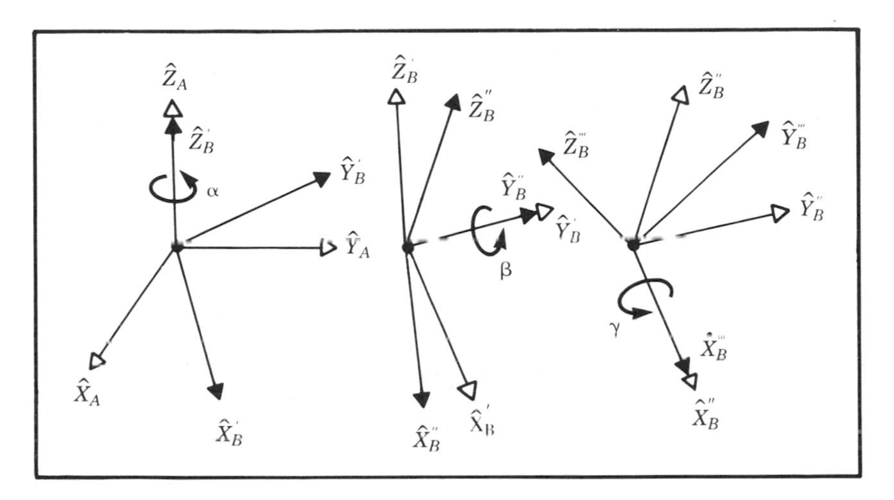

FIGURE 2.18 Z-Y-X Euler angles.

a function of the preceding rotations. If we look at the problem in the opposite sense, that is, consider the orientation of $\{A\}$ relative to $\{B\}$, things will be simpler because all rotations occur about unit directions of the "reference" frame. Consider a vector in $\{A\}$, ${}^{A}P$, and how its description relative to $\{B\}$ changes as we go through the three rotations. First, when we rotate $\{B\}$ by α we can view this as rotating $\{A\}$ by $-\alpha$. Therefore, our vector ${}^{A}P$ has undergone a rotation of $-\alpha$ about ${}^{B}\hat{Z}$. Likewise, we take this newly transformed vector and rotate it $-\beta$ about ${}^{B}\hat{Y}$, and finally we rotate by $-\gamma$ about ${}^{B}\hat{X}$. The final description of any vector ${}^{A}P$ is thus seen to be

$$\begin{aligned}{}^{B}P &= {}^{B}_{A}R_{zyx}(\alpha,\beta,\gamma) \quad {}^{A}P \\ &= ROT({}^{B}\hat{X},-\gamma) \quad ROT({}^{B}\hat{Y},-\beta) \quad ROT({}^{B}\hat{Z},-\alpha) \quad {}^{A}P.\end{aligned} \tag{2.62}$$

Now, because ${}^{B}_{A}R^{-1} = {}^{A}_{B}R$ and $ROT^{-1}(\hat{K},-\theta) = ROT(\hat{K},\theta)$, we can compute ${}^{A}_{B}R$ as

$$\begin{aligned}{}^{A}_{B}R_{zyx}(\alpha,\beta,\gamma) &= {}^{B}_{A}R^{-1}(\alpha,\beta,\gamma) \\ &= ROT({}^{B}\hat{Z},\alpha) \quad ROT({}^{B}\hat{Y},\beta) \quad ROT({}^{B}\hat{X},\gamma) \\ &= \begin{bmatrix} c\alpha & -s\alpha & 0 \\ s\alpha & c\alpha & 0 \\ 0 & 0 & 1 \end{bmatrix} \begin{bmatrix} c\beta & 0 & s\beta \\ 0 & 1 & 0 \\ -s\beta & 0 & c\beta \end{bmatrix} \begin{bmatrix} 1 & 0 & 0 \\ 0 & c\gamma & -s\gamma \\ 0 & s\gamma & c\gamma \end{bmatrix},\end{aligned} \tag{2.63}$$

where $c\alpha = \cos\alpha$ and $s\alpha = \sin\alpha$, etc. Multiplying out, we obtain

$${}^{A}_{B}R_{zyx}(\alpha,\beta,\gamma) = \begin{bmatrix} c\alpha c\beta & c\alpha s\beta s\gamma - s\alpha c\gamma & c\alpha s\beta c\gamma + s\alpha s\gamma \\ s\alpha c\beta & s\alpha s\beta s\gamma + c\alpha c\gamma & s\alpha s\beta c\gamma - c\alpha s\gamma \\ -s\beta & c\beta s\gamma & c\beta c\gamma \end{bmatrix}. \tag{2.64}$$

Note that the result is exactly the same as that obtained for the same three rotations taken in the opposite order about fixed axes!

The solution for extracting Z-Y-X Euler angles from a rotation matrix is the same as stated in the preceding section for rotations about fixed axes; only the order and interpretation of the angles is different. It is restated here for reference:

Given:

$$ {}^{A}_{B}R = \begin{bmatrix} r_{11} & r_{12} & r_{13} \\ r_{21} & r_{22} & r_{23} \\ r_{31} & r_{32} & r_{33} \end{bmatrix}. \tag{2.65} $$

Then:

$$ \begin{aligned} \alpha &= \text{Atan2}(r_{21}, r_{11}), \\ \beta &= \text{Atan2}(-r_{31}, \sqrt{r_{11}^2 + r_{21}^2}), \\ \gamma &= \text{Atan2}(r_{32}, r_{33}). \end{aligned} \tag{2.66} $$

Although a second solution exists, by using the positive square root in the formula for β, we always compute the single solution for which $-90.0 <= \beta <= 90.0$. This is usually a good practice, since we can then define one-to-one mapping functions between various representations of orientation. If $\beta = \pm 90.0$ degrees, the solution of (2.66) degenerates. In those cases, only the sum or the difference of α and γ may be computed. One possible convention is to choose $\alpha = 0.0$ in these cases, which has the results given below.

If $\beta = 90.0$ degrees, then

$$ \begin{aligned} \alpha &= 0.0, \\ \beta &= 90.0, \\ \gamma &= \text{Atan2}(r_{12}, r_{22}). \end{aligned} \tag{2.67} $$

If $\beta = -90.0$ degrees, then

$$ \begin{aligned} \alpha &= 0.0, \\ \beta &= -90.0, \\ \gamma &= -\text{Atan2}(r_{12}, r_{22}). \end{aligned} \tag{2.68} $$

Z-Y-Z Euler angles

Another possible description of a frame $\{B\}$ is as follows:

> Start with the frame coincident with a known frame $\{A\}$. First rotate $\{B\}$ about $\hat{Z}_B$ by an angle α, then rotate about $\hat{Y}_B$ by an angle β, and then rotate about $\hat{Z}_B$ by an angle γ.

Note that since rotations are described relative to the frame we are moving, $\{B\}$, this is an Euler angle description. Because the three rotations occur about the axes $\hat{Z}$, $\hat{Y}$, and $\hat{Z}$, we will call this representation **Z-Y-Z Euler angles.**

Following a development exactly as in the last section we arrive at the following equivalent rotation matrix:

$$
{}^{A}_{B}R_{zyz}(\alpha,\beta,\gamma) = \begin{bmatrix} c\alpha c\beta c\gamma - s\alpha s\gamma & -c\alpha c\beta s\gamma - s\alpha c\gamma & c\alpha s\beta \\ s\alpha c\beta c\gamma + c\alpha s\gamma & -s\alpha c\beta s\gamma + c\alpha c\gamma & s\alpha s\beta \\ -s\beta c\gamma & s\beta s\gamma & c\beta \end{bmatrix}. \qquad \textbf{(2.69)}
$$

The solution for extracting Z-Y-Z Euler angles from a rotation matrix is stated below.

Given:

$$
{}^{A}_{B}R = \begin{bmatrix} r_{11} & r_{12} & r_{13} \\ r_{21} & r_{22} & r_{23} \\ r_{31} & r_{32} & r_{33} \end{bmatrix}. \qquad \textbf{(2.70)}
$$

If $\sin\beta \neq 0$, then

$$
\begin{aligned}
\alpha &= \mathrm{Atan2}(r_{23}, r_{13}), \\
\beta &= \mathrm{Atan2}(\sqrt{r_{31}^2 + r_{32}^2}, r_{33}), \\
\gamma &= \mathrm{Atan2}(r_{32}, -r_{31}).
\end{aligned} \qquad \textbf{(2.71)}
$$

Although a second solution exists, by using the positive square root in the formula for β, we always compute the single solution for which $0.0 <= \beta <= 180.0$. If $\beta = 0.0$ or 180.0 degrees, the solution of (2.71) degenerates. In those cases, only the sum or the difference of α and γ may be computed. One possible convention is to choose $\alpha = 0.0$ in these cases, which has the results given below.

If $\beta = 0.0$ degrees, then

$$
\begin{aligned}
\alpha &= 0.0, \\
\beta &= 0.0, \\
\gamma &= \mathrm{Atan2}(-r_{12}, r_{11}).
\end{aligned} \qquad \textbf{(2.72)}
$$

If $\beta = 180.0$ degrees, then

$$\begin{aligned} \alpha &= 0.0, \\ \beta &= 180.0, \\ \gamma &= \mathrm{Atan2}(r_{12}, -r_{11}). \end{aligned} \tag{2.73}$$

Equivalent angle-axis

With the notation $ROT(\hat{X}, 30.0)$ we give the description of an orientation by giving an axis, $\hat{X}$, and an angle, 30.0 degrees. This is an example of an **equivalent angle-axis** representation. If the axis is a *general* direction (rather than one of the unit directions) any orientation may be obtained through proper axis and angle selection. Consider the following description of a frame $\{B\}$:

> Start with the frame coincident with a known frame $\{A\}$. Then rotate $\{B\}$ about the vector ${}^{A}\hat{K}$ by an angle θ according to the right hand rule.

Vector $\hat{K}$ is sometimes called the equivalent axis of a finite rotation. A general orientation of $\{B\}$ relative to $\{A\}$ may be written as ${}^{A}_{B}R({}^{A}\hat{K}, \theta)$ or $ROT({}^{A}\hat{K}, \theta)$ and will be called the equivalent angle-axis representation. The specification of the vector ${}^{A}\hat{K}$ requires only two parameters because its length is always taken to be one. The angle specifies a third parameter. Often we will multiply the unit direction, $\hat{K}$, with the amount of rotation, θ, to form a compact 3×1 vector description of orientation, denoted by K (no "hat"). See Fig. 2.19.

When the axis of rotation is chosen as one of the principle axes of $\{A\}$, then the equivalent rotation matrix takes on the familiar form of planar rotations:

$$ROT({}^{A}\hat{X}, \theta) = \begin{bmatrix} 1 & 0 & 0 \\ 0 & \cos\theta & -\sin\theta \\ 0 & \sin\theta & \cos\theta \end{bmatrix}, \tag{2.74}$$

$$ROT({}^{A}\hat{Y}, \theta) = \begin{bmatrix} \cos\theta & 0 & \sin\theta \\ 0 & 1 & 0 \\ -\sin\theta & 0 & \cos\theta \end{bmatrix}, \tag{2.75}$$

$$ROT({}^{A}\hat{Z}, \theta) = \begin{bmatrix} \cos\theta & -\sin\theta & 0 \\ \sin\theta & \cos\theta & 0 \\ 0 & 0 & 1 \end{bmatrix}. \tag{2.76}$$

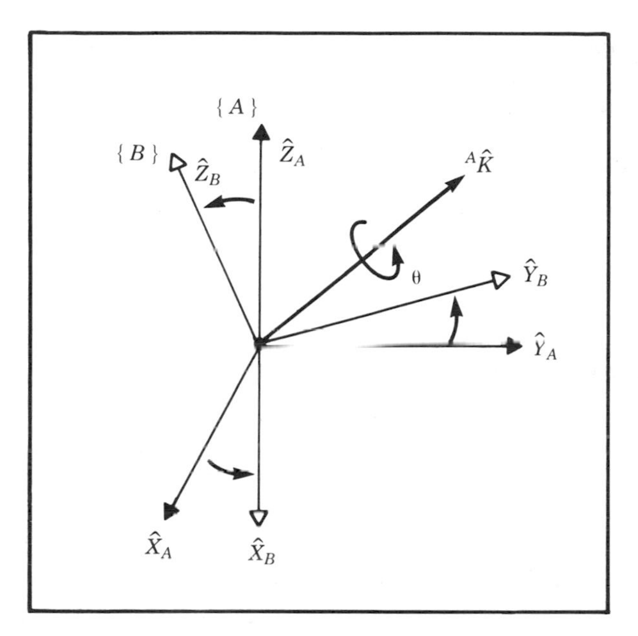

FIGURE 2.19 Equivalent angle-axis representation.

If the axis of rotation is a general axis, it can be shown (see Exercise 2.6) that the equivalent rotation matrix is:

$$ROT({}^A\hat{K},\theta) = \begin{bmatrix} k_x k_x v\theta + c\theta & k_x k_y v\theta - k_z s\theta & k_x k_z v\theta + k_y s\theta \\ k_x k_y v\theta + k_z s\theta & k_y k_y v\theta + c\theta & k_y k_z v\theta - k_x s\theta \\ k_x k_z v\theta - k_y s\theta & k_y k_z v\theta + k_x s\theta & k_z k_z v\theta + c\theta \end{bmatrix}. \quad (2.77)$$

Where $c\theta = \cos\theta$, $s\theta = \sin\theta$, $v\theta = 1 - \cos\theta$, and ${}^A\hat{K} = [k_x \; k_y \; k_z]^T$. The sign of θ is determined by the right hand rule with the thumb pointing along the positive sense of ${}^A\hat{K}$.

Equation (2.77) converts from angle-axis representation to rotation matrix representation. Note that given any axis of rotation and any angular amount, we can easily construct an equivalent rotation matrix.

The inverse problem, namely that of determining $\hat{K}$ and θ from a given rotation matrix, is left as an exercise (Exercises 2.6, 2.7). A partial result is given below [3]. If

$${}^A_BR = \begin{bmatrix} r_{11} & r_{12} & r_{13} \\ r_{21} & r_{22} & r_{23} \\ r_{31} & r_{32} & r_{33} \end{bmatrix}, \quad (2.78)$$

then

$$\theta = \text{Acos}\left(\frac{r_{11} + r_{22} + r_{33} - 1}{2}\right)$$

$$\hat{K} = \frac{1}{2\sin\theta}\begin{bmatrix} r_{32} - r_{23} \\ r_{13} - r_{31} \\ r_{21} - r_{12} \end{bmatrix}. \quad (2.79)$$

This solution always computes a value of θ between 0 and 180 degrees. For any axis-angle pair $({}^A\hat{K}, \theta)$ there is another pair, namely $(-{}^A\hat{K}, -\theta)$, which results in the same orientation in space, with the same rotation matrix describing it. Therefore in converting from a rotation matrix into angle-axis representation, we are faced with choosing between solutions. A more serious problem is that for small angular rotations, the axis becomes ill-defined. Clearly, if the amount of rotation goes to zero, the axis of rotation becomes completely undefined. The solution given by (2.79) fails if $\theta = 0°$ or $\theta = 180°$. See Exercise 2.7.

EXAMPLE 2.8

A frame $\{B\}$ is described as follows: initially coincident with $\{A\}$ we rotate $\{B\}$ about the vector ${}^A\hat{K} = [0.707\ 0.707\ 0.0]^T$ (passing through the origin) by an amount $\theta = 30$ degrees. Give the frame description of $\{B\}$.

Substituting into (2.77) yields the rotation matrix part of the frame description. Since there was no translation of the origin the position vector is $[0\ 0\ 0]^T$. So:

$$ {}^A_BT = \begin{bmatrix} 0.933 & 0.067 & 0.354 & 0.0 \\ 0.067 & 0.933 & -0.354 & 0.0 \\ -0.354 & 0.354 & 0.866 & 0.0 \\ 0.0 & 0.0 & 0.0 & 1.0 \end{bmatrix}. \quad \blacksquare \tag{2.80} $$

Up to this point, all rotations we have discussed have been about axes which pass through the origin of the reference system. If we encounter a problem for which this is not true, we may reduce the problem to the "axis through the origin" case by defining additional frames whose origins lie on the axis, and then solving a transform equation.

EXAMPLE 2.9

A frame $\{B\}$ is described as follows: initially coincident with $\{A\}$ we rotate $\{B\}$ about the vector ${}^A\hat{K} = [0.707\ 0.707\ 0.0]^T$, passing through the point ${}^AP = [1.0\ 2.0\ 3.0]$, by an amount $\theta = 30$ degrees. Give the frame description of $\{B\}$.

Before performing the rotation, $\{A\}$ and $\{B\}$ are coincident. We define two new frames, $\{A'\}$ and $\{B'\}$, which are coincident with one another and have the same orientation as $\{A\}$ and $\{B\}$ respectively, but are translated relative to $\{A\}$ by an offset which places their origins on the axis of rotation. We will choose

$$ {}^A_{A'}T = \begin{bmatrix} 1.0 & 0.0 & 0.0 & 1.0 \\ 0.0 & 1.0 & 0.0 & 2.0 \\ 0.0 & 0.0 & 1.0 & 3.0 \\ 0.0 & 0.0 & 0.0 & 1.0 \end{bmatrix}. \tag{2.81} $$

Similarly the description of $\{B\}$ in terms of $\{B'\}$ is

$$ {}^{B'}_{B}T = \begin{bmatrix} 1.0 & 0.0 & 0.0 & -1.0 \\ 0.0 & 1.0 & 0.0 & -2.0 \\ 0.0 & 0.0 & 1.0 & -3.0 \\ 0.0 & 0.0 & 0.0 & 1.0 \end{bmatrix}. \tag{2.82} $$

Now, keeping other relationships fixed, we can rotate $\{B'\}$ relative to $\{A'\}$. This is a rotation about an axis which passes through the origin, so we may use (2.77) to compute $\{B'\}$ relative to $\{A'\}$. Substituting into (2.77) yields the rotation matrix part of the frame description. Since there was no translation of the origin, the position vector is $[0\ 0\ 0]^T$. So we have

$$ {}^{A'}_{B'}T = \begin{bmatrix} 0.933 & 0.067 & 0.354 & 0.0 \\ 0.067 & 0.933 & -0.354 & 0.0 \\ -0.354 & 0.354 & 0.866 & 0.0 \\ 0.0 & 0.0 & 0.0 & 1.0 \end{bmatrix}. \tag{2.83} $$

Finally, we can write a transform equation to compute the desired frame,

$$ {}^{A}_{B}T = {}^{A}_{A'}T\ {}^{A'}_{B'}T\ {}^{B'}_{B}T, \tag{2.84} $$

which evaluates to

$$ {}^{A}_{B}T = \begin{bmatrix} 0.933 & 0.067 & 0.354 & 1.12 \\ 0.067 & 0.933 & -0.354 & -1.13 \\ -0.354 & 0.354 & 0.866 & -0.04 \\ 0.000 & 0.000 & 0.000 & 1.00 \end{bmatrix}. \tag{2.85} $$

A rotation about an axis which does not pass through the origin causes a change in position, plus the same final orientation as if the axis had passed through the origin. Note that we could have used any definition of $\{A'\}$ and $\{B'\}$ such that their origins were on the axis of rotation. Our particular choice of orientation was arbitrary, and our choice of the position of the origin was one of an infinity of possible choices lying along the axis of rotation. Also, see Exercise 2.14. ∎

Taught and predefined orientations

In many robot systems it will be possible to "teach" positions and orientations using the robot itself. The manipulator is moved to a desired location and this position is recorded. A frame taught in this manner need not necessarily be one to which the robot will be commanded to return; it could be a part location or a fixture location. In other words, the robot is used as a measuring tool having six degrees of freedom. Teaching an orientation like this completely obviates the need for the human programmer to deal with orientation representation at all. In the computer the taught point is stored as a rotation matrix, or whatever, but the user never has to see or understand it. Robot systems which allow teaching of frames using the robot are thus highly recommended.

Besides teaching frames some systems might have a set of predefined orientations like "pointing down" or "pointing left." These specifications are very easy for humans to deal with. However, if this were the only means of describing and specifying orientation, the system would be very limited.

2.9 Transformation of free vectors

We have been concerned mostly with position vectors in this chapter. In later chapters we will discuss velocity and force vectors as well. These vectors will transform differently because they are a different *type* of vector.

In mechanics one makes a distinction between the equality and the equivalence of vectors. *Two vectors are equal if they have the same dimensions, magnitude, and direction.* Two vectors which are considered *equal* may have different lines of actions, for example, the three equal vectors in Fig 2.20. These velocity vectors have the same dimensions, magnitude, and direction, and so are equal according to our definition.

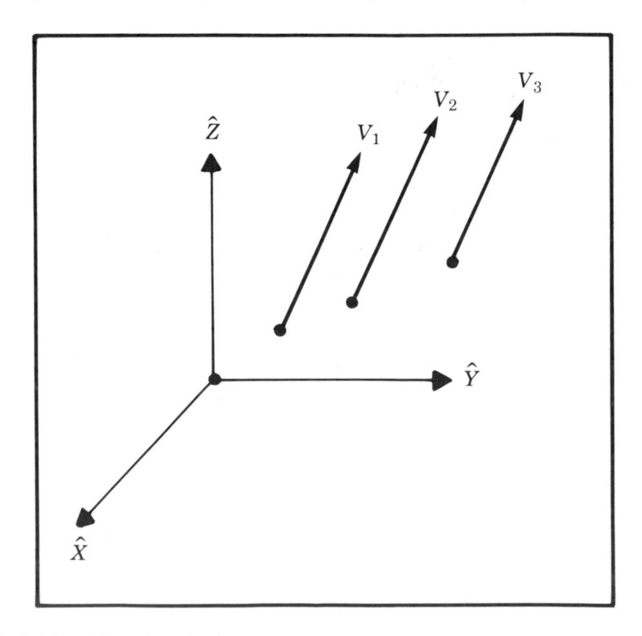

FIGURE 2.20 Equal velocity vectors.

Two vectors are equivalent in a certain capacity if each produces the very same effect in this capacity. Thus, if the criterion in Fig. 2.20 is distance traveled, all three vectors give the same result and are thus equivalent in this capacity. If the criterion is height above the xy plane, then the vectors are not equivalent despite their equality. Thus, relationships between vectors and notions of equivalence *depend entirely on the situation at hand.* Furthermore, vectors which are not equal may cause equivalent effects in certain cases.

We will define two basic classes of vector quantities which may be helpful.

A **line vector** refers to a vector which, along with direction and magnitude, is also dependent on its **line of action** as far as determining its effects is concerned. Often the effects of a force vector depend upon its line of action (or point of application), and so it would be considered a line vector.

A **free vector** refers to a vector which may be positioned anywhere in space without loss or change of meaning provided that magnitude and direction are preserved.

For example, a pure moment vector is always a free vector. If we have a moment vector, ${}^{B}N$, which is known in terms of $\{B\}$, then we calculate the same moment in terms of frame $\{A\}$ as

$$ {}^{A}N = {}^{A}_{B}R \, {}^{B}N. \tag{2.86}$$

That is, since all that counts is the magnitude and direction (in the case of a free vector), only the rotation matrix relating the two systems is used in transforming. The relative locations of the origins does not enter into the calculation.

Likewise, a velocity vector written in $\{B\}$, ${}^{B}V$, is written in $\{A\}$ as

$$ {}^{A}V = {}^{A}_{B}R \, {}^{B}V. \tag{2.87}$$

The velocity of a point is a free vector, so all that is important is its direction and magnitude. The operation of rotation (as in (2.87)) does not affect the magnitude, and accomplishes the rotation which changes the description of the vector from $\{B\}$ to $\{A\}$. Note that, ${}^{A}P_{BORG}$ which would appear in a position vector transformation, does not appear in a velocity transform. For example, in Fig. 2.21, if ${}^{B}V = 5\hat{X}$, then ${}^{A}V = 5\hat{Y}$.

Velocity vectors and force and moment vectors will be more fully introduced in Chapter 5.

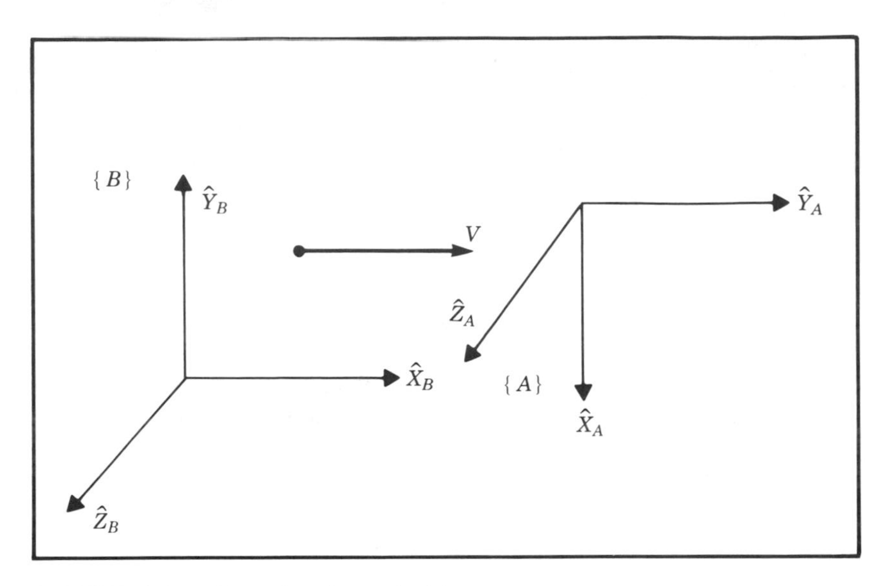

FIGURE 2.21 Transforming velocities.

2.10 Computational considerations

The availability of inexpensive computing power is largely responsible for the growth of the robotics industry; yet for some time to come, efficient computation will remain an important issue in the design of a manipulation system.

While the homogeneous representation is useful as a conceptual entity, typical transformation software used in industrial manipulation systems does not make use of them directly since the time spent multiplying by zeros and ones is wasteful. Usually, the computations shown in (2.36) and (2.40) are performed, rather than the direct multiplication or inversion of 4×4 matrices.

The *order* in which transformations are applied can make a large difference in the amount of computation required to compute the same quantity. Consider performing multiple rotations of a vector, as in

$$^{A}P = {}^{A}_{B}R \, {}^{B}_{C}R \, {}^{C}_{D}R \, {}^{D}P. \qquad \textbf{(2.88)}$$

One choice is to first multiply the three rotation matrices together, to form ${}^{A}_{D}R$ in the expression

$$^{A}P = {}^{A}_{D}R \, {}^{D}P. \qquad \textbf{(2.89)}$$

Forming ${}^{A}_{D}R$ from its three constituents requires 54 multiplications and 36 additions. Performing the final matrix-vector multiplication of (2.89) requires an additional 9 multiplications and 6 additions, bringing the total to 63 multiplications, 42 additions.

If instead we transform the vector through the matrices one at a time, i.e.,

$$\begin{aligned}
{}^{A}P &= {}^{A}_{B}R\, {}^{B}_{C}R\, {}^{C}_{D}R\, {}^{D}P \\
{}^{A}P &= {}^{A}_{B}R\, {}^{B}_{C}R\, {}^{C}P \\
{}^{A}P &= {}^{A}_{B}R\, {}^{B}P \\
{}^{A}P &= {}^{A}P,
\end{aligned} \tag{2.90}$$

the total computation requires only 27 multiplications and 18 additions, less than half the computations required by the other method.

Of course, in some cases, the relationships ${}^{A}_{B}R$, ${}^{B}_{C}R$, and ${}^{C}_{D}R$ may be constant, and there may be many ${}^{D}P_i$ which need to be transformed into ${}^{A}P_i$. In this case, it is more efficient to calculate ${}^{A}_{D}R$ once, and then use it for all future mappings. See also Exercise 2.16.

EXAMPLE 2.10

Give a method of computing the product of two rotation matrices, ${}^{A}_{B}R\, {}^{B}_{C}R$, using less than 27 multiplications and 18 additions .

Where $\hat{L}_i$ are the columns of ${}^{B}_{C}R$, and $\hat{C}_i$ are the three columns of the result, compute:

$$\begin{aligned}
\hat{C}_1 &= {}^{A}_{B}R\, \hat{L}_1, \\
\hat{C}_2 &= {}^{A}_{B}R\, \hat{L}_2, \\
\hat{C}_3 &= \hat{C}_1 \times \hat{C}_2,
\end{aligned} \tag{2.91}$$

which requires 24 multiplications and 15 additions. ∎

References

[1] B. Noble, "Applied Linear Algebra," Prentice Hall, 1969.
[2] D. Ballard, and C. Brown, "Computer Vision," Prentice-Hall, 1982.
[3] O. Bottema, and B. Roth, "Theoretical Kinematics," North Holland, Amsterdam, 1979.
[4] R.P. Paul, "Robot Manipulators," MIT Press, Cambridge, Mass., 1981.
[5] I. Shames, "Engineering Mechanics," Second Edition, Prentice-Hall, 1967.
[6] Symon, "Mechanics," Third Edition, Addison-Wesley, 1971.
[7] B. Gorla, and M. Renaud, "Robots Manipulateurs," Cepadues-Editions, Toulouse, 1984.

Exercises

2.1 [15] A vector ${}^{A}P$ is rotated about $\hat{Z}_A$ by θ degrees and is subsequently rotated about $\hat{X}_A$ by ϕ degrees. Give the rotation matrix which accomplishes these rotations in the given order.

2.2 [15] A vector ${}^{A}P$ is rotated about $\hat{Y}_A$ by 30 degrees and is subsequently rotated about $\hat{X}_A$ by 45 degrees. Give the rotation matrix which accomplishes these rotations in the given order.

2.3 [16] A frame $\{B\}$ is located as follows: initially coincident with a frame $\{A\}$ we rotate $\{B\}$ about $\hat{Z}_B$ by θ degrees and then we rotate the resulting frame about $\hat{X}_B$ by ϕ degrees. Give the rotation matrix which will change the description of vectors from ${}^{B}P$ to ${}^{A}P$.

2.4 [16] A frame $\{B\}$ is located as follows: initially coincident with a frame $\{A\}$ we rotate $\{B\}$ about $\hat{Z}_B$ by 30 degrees and then we rotate the resulting frame about $\hat{X}_B$ by 45 degrees. Give the rotation matrix which will change the description of vectors from ${}^{B}P$ to ${}^{A}P$.

2.5 [13] ${}^{A}_{B}R$ is a 3×3 matrix with eigenvalues 1, $e^{+\alpha i}$, and $e^{-\alpha i}$, where $i = \sqrt{-1}$. What is the physical meaning of the eigenvector of ${}^{A}_{B}R$ associated with the eigenvalue 1?

2.6 [21] Derive equation (2.77).

2.7 [24] Describe (or program) an algorithm which extracts the equivalent angle and axis of a rotation matrix. Equation (2.79) is a good start, but make sure your algorithm handles the special cases of $\theta = 0°$ and $\theta = 180°$.

2.8 [29] Write a subroutine which changes representation of orientation from rotation matrix form to equivalent angle-axis form. A Pascal-style procedure declaration would begin:

```
Procedure RMTOAA(VAR R:mat33; VAR K:vec3; VAR theta: real);
```

Write another subroutine which changes from equivalent angle-axis representation to rotation matrix representation:

```
Procedure AATORM(VAR K:vec3; VAR theta: real: VAR R:mat33);
```

Run these procedures on several cases of test data back-to-back and verify that you get back what you put in. Include some of the difficult cases!

2.9 [27] Do Exercise 2.8 for roll, pitch, yaw angles about fixed axes.

2.10 [27] Do Exercise 2.8 for Z - Y - Z Euler angles.

2.11 [10] Under what condition do two rotation matrices representing finite rotations commute? A proof is not required.

2.12 [14] A velocity vector is given by

$$ {}^{B}V = \begin{bmatrix} 10.0 \\ 20.0 \\ 30.0 \end{bmatrix}. $$

Given:

$$
{}^{A}_{B}T = \begin{bmatrix} 0.866 & -0.500 & 0.000 & 11.0 \\ 0.500 & 0.866 & 0.000 & -3.0 \\ 0.000 & 0.000 & 1.000 & 9.0 \\ 0 & 0 & 0 & 1 \end{bmatrix}.
$$

Compute ${}^{A}V$.

2.13 [21] The following frame definitions are given as known. Draw a frame diagram (like that of Fig. 2.15) which qualitatively shows their arrangement. Solve for ${}^{B}_{C}T$.

$$
{}^{U}_{A}T = \begin{bmatrix} 0.866 & -0.500 & 0.000 & 11.0 \\ 0.500 & 0.866 & 0.000 & -1.0 \\ 0.000 & 0.000 & 1.000 & 8.0 \\ 0 & 0 & 0 & 1 \end{bmatrix},
$$

$$
{}^{B}_{A}T = \begin{bmatrix} 1.000 & 0.000 & 0.000 & 0.0 \\ 0.000 & 0.866 & -0.500 & 10.0 \\ 0.000 & 0.500 & 0.866 & -20.0 \\ 0 & 0 & 0 & 1 \end{bmatrix},
$$

$$
{}^{C}_{U}T = \begin{bmatrix} 0.866 & -0.500 & 0.000 & -3.0 \\ 0.433 & 0.750 & -0.500 & -3.0 \\ 0.250 & 0.433 & 0.866 & 3.0 \\ 0 & 0 & 0 & 1 \end{bmatrix}.
$$

2.14 [31] Develop a general formula to obtain ${}^{A}_{B}T$, where, starting from initial coincidence, $\{B\}$ is rotated by θ about $\hat{K}$ where $\hat{K}$ passes through the point ${}^{A}P$ (not through the origin of $\{A\}$ in general).

2.15 [34] $\{A\}$ and $\{B\}$ are frames differing only in orientation. $\{B\}$ is attained as follows: starting coincident with $\{A\}$, $\{B\}$ is rotated by θ radians about unit vector $\hat{K}$. That is,

$$
{}^{A}_{B}R = {}^{A}_{B}R({}^{A}\hat{K}, \theta).
$$

Show that:

$$
{}^{A}_{B}R = e^{K\theta},
$$

where

$$
K = \begin{bmatrix} 0 & -k_z & k_y \\ k_z & 0 & -k_x \\ -k_y & k_x & 0 \end{bmatrix}.
$$

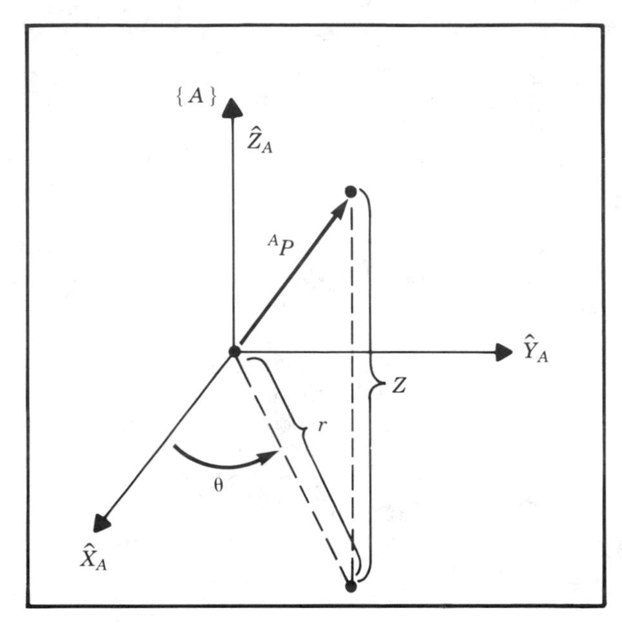

FIGURE 2.22 Cylindrical coordinates.

2.16 [22] A vector must be mapped through three rotation matrices:

$$^AP = {}^A_BR\, {}^B_CR\, {}^C_DR\, {}^DP.$$

One choice is to first multiply the three rotation matrices together, to form ${}^A_DR$ in the expression:

$$^AP = {}^A_DR\, {}^DP.$$

Another choice is to transform the vector through the matrices one at a time, that is,

$$^AP = {}^A_BR\, {}^B_CR\, {}^C_DR\, {}^DP$$

$$^AP = {}^A_BR\, {}^B_CR\, {}^CP$$

$$^AP = {}^A_BR\, {}^BP$$

$$^AP = {}^AP.$$

Because ${}^DP$ is changing at 100 Hz., we must recalculate ${}^AP$ at this rate. However, the three rotation matrices are also changing as determined by a vision system which gives us new values for ${}^A_BR$, ${}^B_CR$, and ${}^C_DR$ at 30 Hz. What is the best way to organize the computation to minimize the calculation effort (multiplications and additions)?

2.17 [16] Another familiar set of three coordinates which can be used to describe a point in space is cylindrical coordinates. The three coordinates are defined as illustrated in Fig. 2.22. The coordinate θ gives a direction in the xy plane along which to radially translate by an amount r. Finally, z is given to specify the height above the xy-plane. Determine the Cartesian coordinates of the point ${}^AP$ in terms of the cylindrical coordinates θ, r, and z.

2.18 [18] Another set of three coordinates which can be used to describe a point in space are spherical coordinates. The three coordinates are defined as illustrated in Fig. 2.23. The angles α and β can be thought of as describing azimuth and elevation of a ray projecting into space. The third coordinate, r, is the radial distance along that ray to the point being described. Determine the Cartesian coordinates of the point ${}^AP$ in terms of the spherical coordinates α, β, and r.

2.19 [24] An object is rotated about its $\hat{X}$ axis by an amount ϕ, and then it is rotated about its *new* $\hat{Y}$ axis by an amount ψ. From our study of Euler angles, we know that the resulting orientation is given by:

$$ROT(\hat{X}, \phi) ROT(\hat{Y}, \psi).$$

Whereas if the two rotations had occurred about axes of the fixed reference frame, the result would be

$$ROT(\hat{Y}, \psi) ROT(\hat{X}, \phi).$$

It appears that the order of multiplication depends upon whether rotations are described relative to fixed axes, or those of the frame being

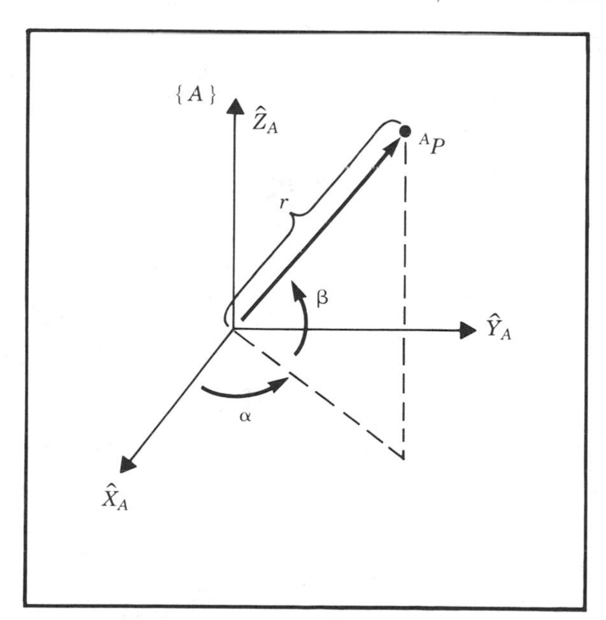

FIGURE 2.23 Spherical coordinates.

moved. It is more appropriate, however, to realize that in the case of specifying a rotation about an axis of the frame being moved, we are specifying a rotation in the fixed system given by (for this example)

$$ROT(\hat{X},\phi)ROT(\hat{Y},\psi)ROT^{-1}(\hat{X},\phi).$$

This *similarity transform* [1], multiplying the original $ROT(\hat{X},\phi)$ on the left reduces to the resulting expression in which *it looks as if* the order of matrix multiplication has been reversed. Taking this viewpoint, give a derivation for the form of the rotation matrix which is equivalent to the Z-Y-Z Euler angle set (α,β,γ) (the result is given by (2.69)).

2.20 [20] Imagine rotating a vector Q about a vector $\hat{K}$ by an amount θ to form a new vector, Q'. That is

$$Q' = ROT(\hat{K},\theta)Q.$$

Use (2.77) to derive **Rodriques' formula**, which is

$$Q' = Q\cos\theta + \sin\theta(\hat{K}\times Q) + (1-\cos\theta)(\hat{K}\cdot\hat{Q})\hat{K}.$$

Programming Exercise (Part 2)

1. If your function library does not include an Atan2 function subroutine, write one.

2. To make a friendly user interface, we wish to describe orientations in the planar world with a single angle, θ, instead of a 2×2 rotation matrix. The user will always communicate in terms of angle θ, but internally we will need the rotation matrix form. For the position vector part of a frame, the user will specify an x and a y value. So, we want to allow the user to specify a *frame* as a 3-tuple: $(x,\ y,\ \theta)$. Internally, we wish to use a 2×1 position vector and a 2×2 rotation matrix, so we need conversion routines. Write a subroutine whose Pascal definition would begin:

```
Procedure UTOI(VAR uform: vec3; VAR iform: frame);
```

Where "UTOI" stands for "User form TO Internal form." The first argument is the 3-tuple $(x,\ y,\ \theta)$, and the second argument is of type frame. The type "frame" consists of a (2x1) position vector and a (2x2) rotation matrix. If desired, you may represent the frame with a (3x3) homogeneous transform in which the third row is [0 0 1]. The inverse routine will also be necessary:

```
Procedure ITOU(VAR iform: frame; VAR uform: vec3);
```

3. Write a subroutine to multiply two transforms together. Use the following procedure heading:

```
Procedure TMULT(VAR brela, crelb, crela: frame);
```

The first two arguments are inputs, and the third is an output. Note that the names of the arguments document what the program does (brela $= {}^{A}_{B}T$).

4. Write a subroutine to invert a transform. Use the following procedure heading:

```
Procedure TINVERT(VAR brela, arelb: frame);
```

The first argument is the input, the second the output. Note that the names of the arguments document what the program does (brela $= {}^{A}_{B}T$).

5. The following frame definitions are given as known. These frames are input in the user representation of [x y θ] (where θ is in degrees). Draw a frame diagram (like Fig. 2.15, only in 2-D) which qualitatively shows their arrangement. Write a program which calls TMULT and TINVERT (defined in 3 and 4 above) as many times as needed to solve for ${}^{B}_{C}T$.

$$ {}^{U}_{A}T = [x \quad y \quad \theta] = [11.0 \quad -1.0 \quad 30.0], $$

$$ {}^{B}_{A}T = [x \quad y \quad \theta] = [0.0 \quad 7.0 \quad 45.0], $$

$$ {}^{C}_{U}T = [x \quad y \quad \theta] = [-3.0 \quad -3.0 \quad -30.0]. $$

Print out ${}^{B}_{C}T$ in both internal and user representation.

3

MANIPULATOR KINEMATICS

3.1 Introduction

Kinematics is the science of motion which treats motion without regard to the forces which cause it. Within the science of kinematics one studies the position, velocity, acceleration, and all higher order derivatives of the position variables (with respect to time or any other variable(s)). Hence, the study of the kinematics of manipulators refers to all the geometrical and time based properties of the motion. The relationships between these motions and the forces and torques which cause them is the problem of dynamics, and is the subject of Chapter 6.

In this chapter, we consider position and orientation of the manipulator linkages in static situations. In Chapters 5 and 6 we will consider the kinematics when velocities and accelerations are involved.

In order to deal with the complex geometry of a manipulator we will affix frames to the various parts of the mechanism and then describe the relationship between these frames. The study of manipulator kinematics involves, among other things, how the locations of these frames change as the mechanism articulates. The central topic of this chapter is a method to compute the position and orientation of the manipulator's end-effector relative to the base of the manipulator as a function of the joint variables.

3.2 Link description

A manipulator may be thought of as a set of bodies connected in a chain by joints. These bodies are called links. Each joint usually exhibits one degree of freedom. Most manipulators have joints which are like hinges, called revolute joints, or have sliding joints called prismatic joints. In the rare case that a mechanism is built with a joint having n degrees of freedom, it can be modeled as n joints of one degree of freedom connected with $n-1$ links of zero length. Therefore, without loss of generality, we will consider only manipulators which have joints with a single degree of freedom.

The links are numbered starting from the immobile base of the arm, which might be called link 0. The first moving body is link 1, and so on, out to the free end of the arm, which is link n. In order to position an end-effector generally in 3-space, a minimum of 6 joints are required.* Typical manipulators have 5 or 6 joints. Some robots may actually not be as simple as a single kinematic chain—they may have parallelogram linkages or other closed kinematic structures. We will consider one such manipulator later in this chapter.

A single link of a typical robot has many attributes which a mechanical designer had to consider during its design. These include the type of material used, the strength and stiffness of the link, the location and type of the joint bearings, the external shape, the weight and inertia, etc. However, for the purposes of obtaining the kinematic equations of the mechanism, *a link is considered only as a rigid body which defines the relationship between two neighboring joint axes of a manipulator.* Joint axes are defined by lines in space. Joint axis i is defined by a line in space, or a vector direction, about which link i rotates relative to link $i-1$. It turns out that for kinematic purposes, a link can be specified with two numbers which define the relative location of the two axes in space.

For any two axes in 3-space there exists a well-defined measure of distance between them. This distance is measured along a line which is mutually perpendicular to both axes. This mutual perpendicular always exists and is unique except when both axes are parallel, in which case there are many mutual perpendiculars of equal length. Figure 3.1 shows link $i-1$ and the mutually perpendicular line along which the **link length**,

* This makes good intuitive sense as the description of an object in space requires six parameters—three for position and three for orientation.

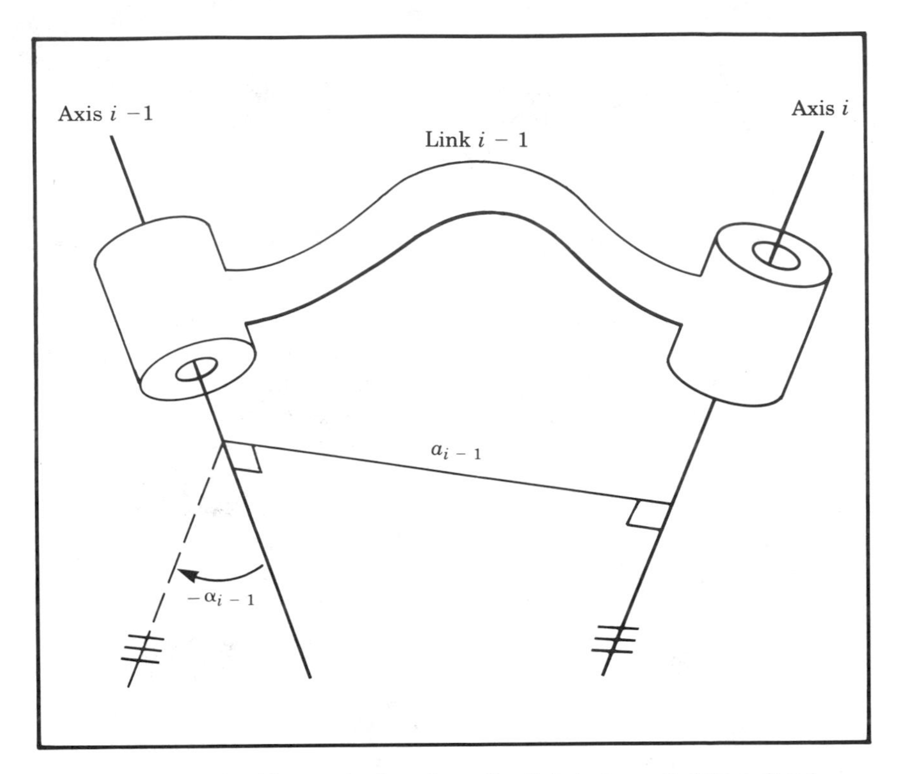

FIGURE 3.1 The kinematic function of a link is to maintain a fixed relationship between the two joint axes it supports. This relationship can be described with two parameters, the link length, a, and the link twist, α.

a_{i-1}, is measured. The second parameter needed to define the relative location of the two axes is called the **link twist**. If we imagine a plane whose normal is the mutually perpendicular line just constructed, we can project both axes $i-1$ and i onto this plane and measure the angle between them. This angle is measured from axis $i-1$ to axis i in the right hand sense about a_{i-1}.* We will use this definition of the twist of link $i-1$, α_{i-1}. In Fig. 3.1, α_{i-1} is indicated as the angle between axis $i-1$ and axis i (the lines with the triple hash marks are parallel). In the case of intersecting axes, twist is measured in the plane containing both axes, but the sense of α_{i-1} is lost. In this special case, one is free to assign the sign of α_{i-1} arbitrarily.

You should convince yourself that these two parameters, length and twist, as defined above, can be used to define the relationship between any two lines (in this case axes) in space.

* In this case a_{i-1} is given a direction as pointing from axis $i-1$ to axis i.

EXAMPLE 3.1

Figure 3.2 shows the mechanical drawings of a robot link. If this link is used in a robot with bearing "A" used for the lower numbered joint, give the length and twist of this link. Assume that holes are centered in each bearing.

By inspection, the common perpendicular lies right down the middle of the metal bar connecting the bearings, so the link length is 7 inches. The end view actually shows a projection of the bearings onto the plane whose normal is the mutual perpendicular. Link twist is measured in the right hand sense about the common perpendicular from axis i 1 to axis i, so in this example, it is clearly +45 degrees. ■

3.3 Link connection description

The problem of connecting the links of a robot together is again one filled with many questions for the mechanical designer to resolve. These include the strength of the joint, lubrication, bearing and gearing mounting, etc. However, for the investigation of kinematics, we need only worry about two quantities which will completely specify the way in which links are connected together.

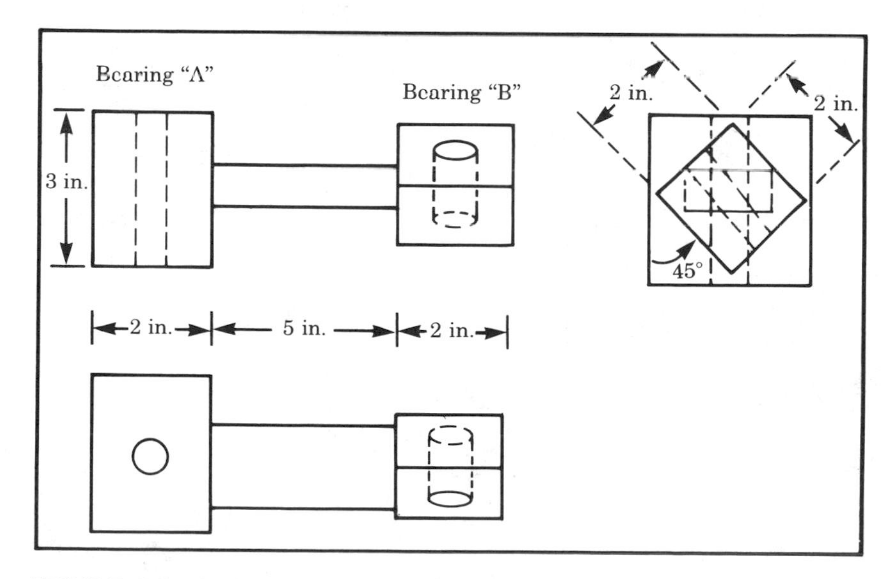

FIGURE 3.2 A simple link which supports two revolute axes.

Intermediate links in the chain

Neighboring links have a common joint axis between them. One parameter of interconnection has to do with the distance along this common axis from one link to the next. This parameter is called the **link offset**. The offset at joint axis i is called d_i. The second parameter describes the amount of rotation about this common axis between one link and its neighbor. This is called the **joint angle**, θ_i.

Figure 3.3 shows the interconnection of link $i-1$ and link i. Recall that a_{i-1} is the mutual perpendicular between the two axes of link $i-1$. Likewise a_i is the mutual perpendicular defined for link i. The first parameter of interconnection is the link offset, d_i, which is the signed distance measured along the axis of joint i from the point where a_{i-1} intersects the axis to the point where a_i intersects the axis. The offset d_i is indicated in Fig. 3.3. The link offset d_i is variable if joint i is prismatic. The second parameter of interconnection is the angle made between an extension of a_{i-1} and a_i measured about the axis of joint i. This is indicated in Fig. 3.3, where the lines with the double hash marks are parallel. This parameter is named θ_i, and is variable for a revolute joint.

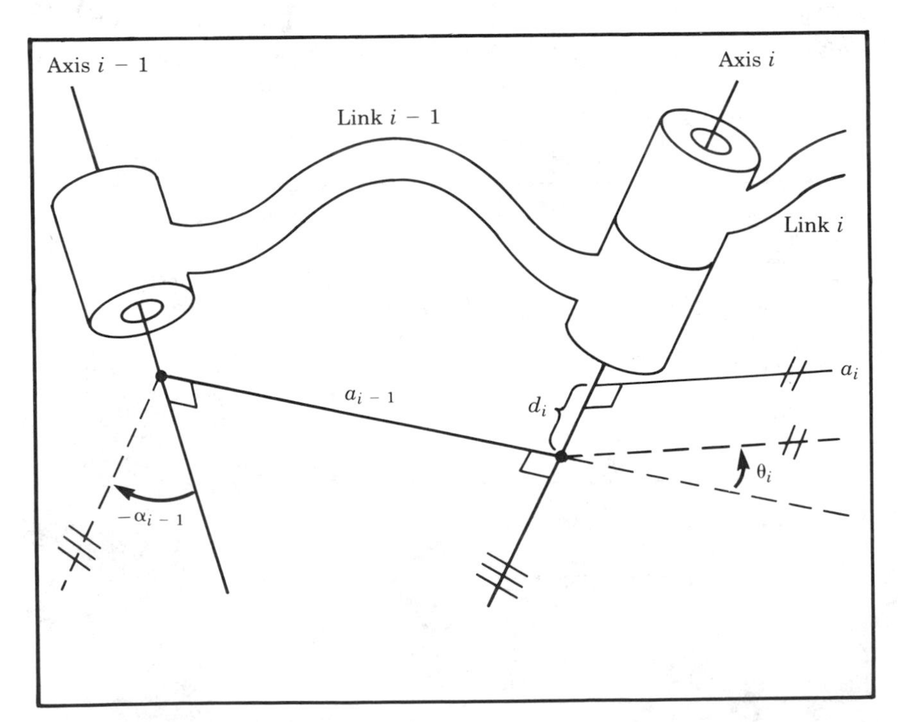

FIGURE 3.3 The link offset, d, and the joint angle, θ, are two parameters which may be used to describe the nature of the connection between neighboring links.

First and last links in the chain

Link length, a_i, and link twist, α_i, depend on joint axes i and $i+1$. Hence a_1 through a_{n-1} and α_1 through α_{n-1} are defined as discussed above in Section 3.3. At the ends of the chain, it will be our convention to assign zero to these quantities. That is, $a_0 = a_n = 0.0$ and $\alpha_0 = \alpha_n = 0.0$.* Link offset, d_i, and joint angle, θ_i, are well defined for joints 2 through $n-1$ according to the conventions discussed above in Section 3.3. If joint 1 is revolute, the zero position for θ_1 may be chosen arbitrarily and $d_1 = 0.0$ will be our convention. Similarly, if joint 1 is prismatic, the zero position of d_1 may be chosen arbitrarily, and $\theta_1 = 0.0$ will be our convention. Exactly the same statements apply to joint n.

These conventions have been chosen so that in case where a quantity could be assigned arbitrarily, a zero value is assigned so that later calculations will be as simple as possible.

Link parameters

Hence any robot can be described kinematically by giving the values of four quantities for each link. Two describe the link itself, and two describe the link's connection to a neighboring link. In the usual case of a revolute joint, θ_i is called the **joint variable**, and the other three quantities would be fixed **link parameters**. For prismatic joints, d_i is the joint variable and the other three quantities are fixed link parameters. The definition of mechanisms by means of these quantities is a convention usually called the **Denavit-Hartenberg notation** [1].† Other methods of describing mechanisms are available but are not presented here.

At this point we could inspect any mechanism and determine the Denavit-Hartenberg parameters which describe it. For a 6-jointed robot 18 numbers would be required to completely describe the fixed portion of its kinematics. In the case of a 6-jointed robot with all revolute joints, the 18 numbers are in the form of 6 sets of (a_i, α_i, d_i).

EXAMPLE 3.2

Two links, as described in Fig. 3.2, are connected as links 1 and 2 of a robot. Joint 2 is composed of a "B" bearing of link 1 and an "A" bearing of link 2 arranged so that the flat surfaces of the "A" and "B" bearings lie flush against each other. What is d_2?

---

* In fact, a_n and α_n do not need to be defined at all.

† Note that many related conventions go by the name of Denavit-Hartenberg, but differ in a few details. For example, the version used in this book differs from much of the robotic literature in the manner of frame numbering. Unlike some other conventions, in this book frame $\{i\}$ is attached to link i and has its origin lying on joint axis i.

The link offset d_2 is the offset at joint 2, which is the distance, measured along the joint 2 axis, between the mutual perpendicular of link 1 and that of link 2. From the drawings in Fig. 3.2, this is 2.5 inches. ■

Before introducing more examples we will define a convention for attaching a frame to each link of the manipulator.

3.4 Convention for affixing frames to links

In order to describe the location of each link relative to its neighbors we define a frame attached to each link. The link frames are named by number according to the link to which they are attached. That is, frame $\{i\}$ is attached rigidly to link i.

Intermediate links in the chain

The convention we will use to locate frames on the links is as follows: The $\hat{Z}$-axis of frame $\{i\}$, called $\hat{Z}_i$, is coincident with the joint axis i. The origin of frame $\{i\}$ is located where the a_i perpendicular intersects the joint i axis. $\hat{X}_i$ points along a_i in the direction from joint i to joint $i+1$.

In the special case of $a_i = 0$, $\hat{X}_i$ is chosen normal to the plane of $\hat{Z}_i$ and $\hat{Z}_{i+1}$. We define α_i as being measured in the right hand sense about $\hat{X}_i$, and so we see that the freedom of choosing the sign of α_i in this case corresponds to two choices for the direction of $\hat{X}_i$. $\hat{Y}_i$ is formed by the right hand rule to complete the ith frame. Figure 3.4 shows the location of frames $\{i-1\}$ and $\{i\}$ for a general manipulator.

First and last links in the chain

We attach a frame to the base of the robot, or link 0, called frame $\{0\}$. This frame does not move, and for the problem of arm kinematics can be considered the reference frame. We may describe the position of all other link frames in terms of this frame.

Since frame $\{0\}$ is arbitrary, it always simplifies matters to choose $\hat{Z}_0$ along axis 1 and to locate frame $\{0\}$ so that it coincides with frame $\{1\}$ when joint variable 1 is zero. Using this convention we will always have $a_0 = 0.0$, $\alpha_0 = 0.0$. Additionally, this insures that $d_1 = 0.0$ if joint 1 is revolute, or $\theta_1 = 0.0$ if joint 1 is prismatic.

For joint n revolute, the direction of $\hat{X}_N$ is chosen so that it aligns with $\hat{X}_{N-1}$ when $\theta_n = 0.0$, and the origin of frame $\{N\}$ is chosen so that $d_n = 0.0$. For joint n prismatic, the direction of $\hat{X}_N$ is chosen so

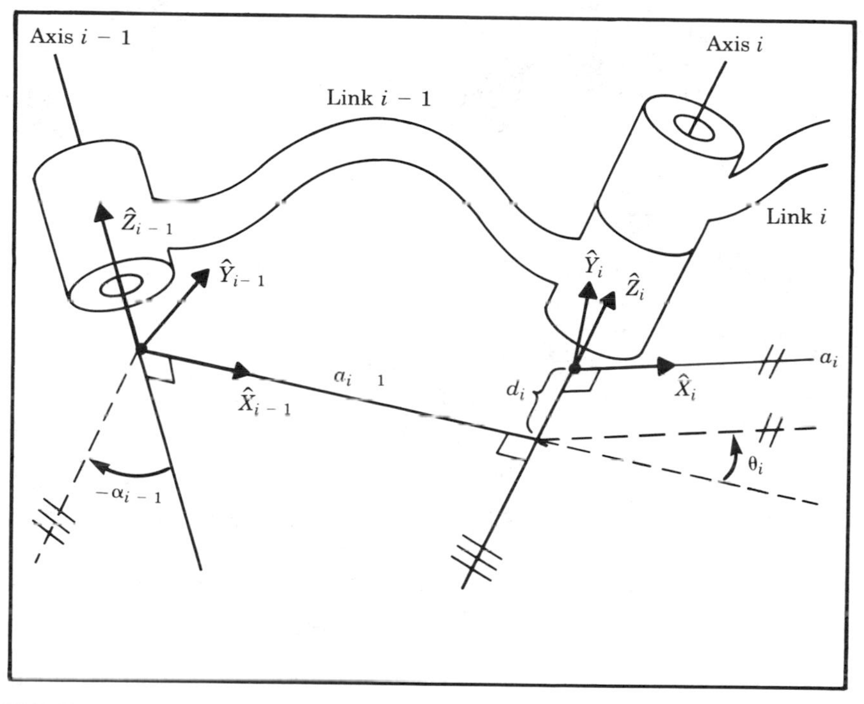

FIGURE 3.4 Link frames are attached so that frame $\{i\}$ is attached rigidly to link i.

that $\theta_n = 0.0$, and the origin of frame $\{N\}$ is chosen at the intersection of $\hat{X}_{N-1}$ and joint axis n when $d_n = 0.0$.

Summary of the link parameters in terms of the link frames

If the link frames have been attached to the links according to our convention, the following definitions of the link parameters are valid:

a_i = the distance from $\hat{Z}_i$ to $\hat{Z}_{i+1}$ measured along $\hat{X}_i$;
α_i = the angle between $\hat{Z}_i$ and $\hat{Z}_{i+1}$ measured about $\hat{X}_i$;
d_i = the distance from $\hat{X}_{i-1}$ to $\hat{X}_i$ measured along $\hat{Z}_i$; and
θ_i = the angle between $\hat{X}_{i-1}$ and $\hat{X}_i$ measured about $\hat{Z}_i$.

We usually choose $a_i > 0$ since it corresponds to a distance; however, α_i, d_i, and θ_i are signed quantities.

A final note on uniqueness. The convention outlined above does not result in a unique attachment of frames to links. First of all, when we first align the $\hat{Z}_i$ axis with joint axis i, there are two choices of direction in which to point $\hat{Z}_i$. Furthermore, in the case of intersecting joint axes (i.e., $a_i = 0$), there are two choices for the direction of $\hat{X}_i$, corresponding to the choice of signs for the normal to the plane containing $\hat{Z}_i$ and $\hat{Z}_{i+1}$. Also when prismatic joints are present there is quite a bit of freedom in frame assignment. (See also Example 3.5.)

EXAMPLE 3.3

Figure 3.5 shows a 3-link planar arm with revolute joints. Assign link frames to the mechanism and give the Denavit-Hartenberg parameters.

We start by defining the reference frame, frame {0}. It is fixed to the base and aligns with frame {1} when the first joint variable (θ_1) is zero. Therefore we position frame {0} as shown in Fig. 3.6 with $\hat{Z}_0$ aligned with the joint 1 axis. For this arm, all joint axes are oriented perpendicular to the plane of the arm. Since the arm lies in a plane with all $\hat{Z}$ axes parallel, there are no link offsets (all d_i are zero). Since all joints are rotational, when they are at zero degrees, all $\hat{X}$ axes must align.

With these comments in mind it is easy to find the frame assignments shown in Fig. 3.6. The corresponding link parameters are shown in Fig. 3.7.

Note that since the joint axes are all parallel and all the $\hat{Z}$ axes are taken to point out of the paper, all α_i are zero. This is obviously a very simple mechanism. Note that our kinematic analysis always ends at a frame whose origin lies on the last joint axis, therefore l_3 does not

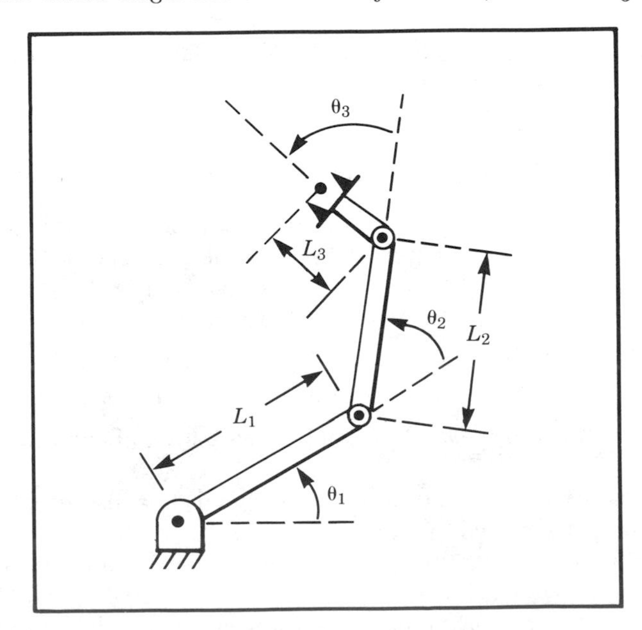

FIGURE 3.5 A 3-link planar arm.

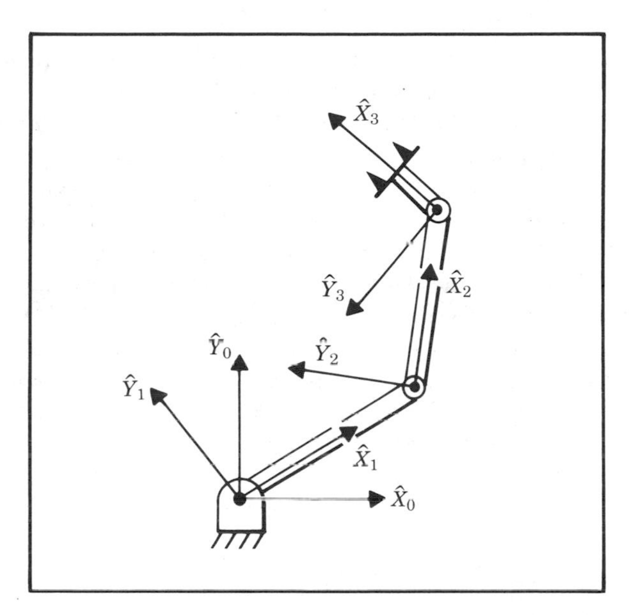

FIGURE 3.6 Link frame assignments.

i	α_{i-1}	a_{i-1}	d_i	θ_i
1	0	0	0	θ_1
2	0	L_1	0	θ_2
3	0	L_2	0	θ_3

FIGURE 3.7 Link parameters of the 3-link planar manipulator.

appear in the link parameters. Such final offsets to the hand are dealt with separately later. ■

EXAMPLE 3.4

Figure 3.8 shows a robot having three degrees of freedom and one prismatic joint. It is a "cylindrical" robot whose first two joints are analogous to polar coordinates when viewed from above. The last joint (joint 3) provides "roll" for the hand. On the right in Fig. 3.8 the arm is shown with the prismatic joint at minimum extension.

Figure 3.9 shows the assignment of link frames for this simple robot. Note that frame $\{0\}$ and frame $\{1\}$ are shown as exactly coincident in this figure because the robot is drawn for the position $\theta_1 = 0$. Note that frame $\{0\}$, although not at the bottom of the flanged base of the robot, is nonetheless rigidly affixed to link 0, the nonmoving part of the robot. Just as our link frames are not used to describe the kinematics all the way out to the hand, they need not be attached all the way back to the lowest part of the base of the robot. It is sufficient that frame $\{0\}$ be attached anywhere to the nonmoving link 0, and that frame $\{N\}$, the final frame, be attached anywhere to the last link of the manipulator. Other offsets can be handled later in a general way.

Note that while rotational joints rotate about the $\hat{Z}$ axis of the associated frame, prismatic joints slide along $\hat{Z}$. In the case where joint

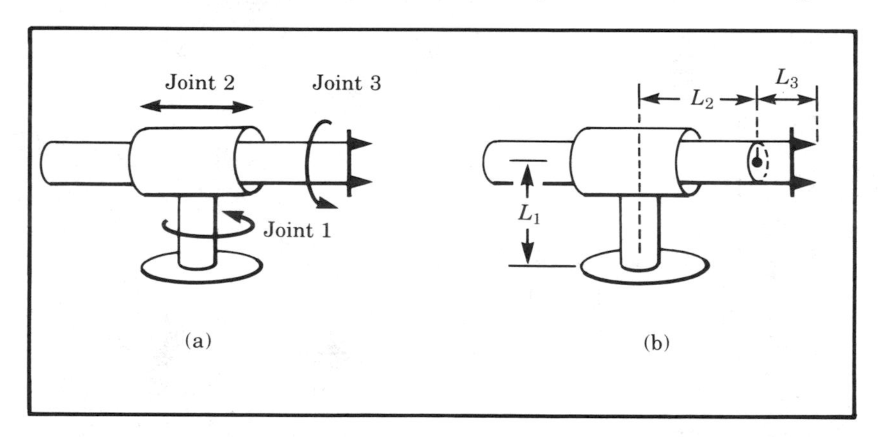

FIGURE 3.8 Manipulator having three degrees of freedom and one prismatic joint.

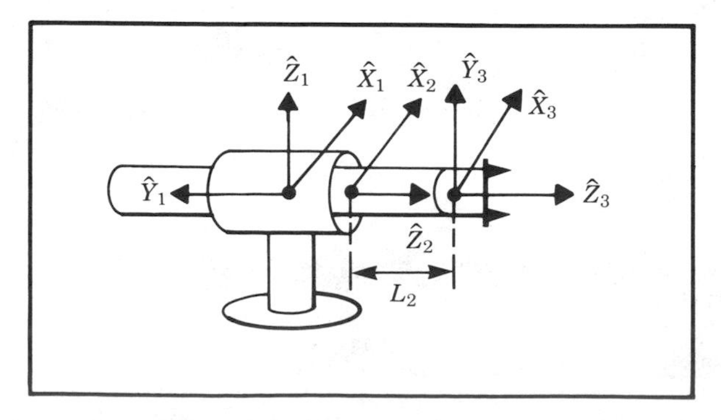

FIGURE 3.9 Link frame assignments.

i is prismatic, θ_i is a fixed constant and d_i is the variable. If d_i is zero at minimum extension of the link, then frame {2} should be attached where shown so that d_2 gives the true offset. The link parameters are shown in Fig. 3.10.

Note that θ_2 is zero for this robot and d_2 is a variable. Axes 1 and 2 intersect, so a_1 is zero. Angle α_1 must be 90 degrees in order to rotate $\hat{Z}_1$ so as to align with $\hat{Z}_2$ (about $\hat{X}_1$). ■

EXAMPLE 3.5

Figure 3.11 shows a three-link arm for which joint axes 1 and 2 intersect, and axes 2 and 3 are parallel. Demonstrate the nonuniqueness of frame assignments and of the Denavit-Hartenberg parameters by showing several possible correct assignments of frames {1} and {2}.

i	α_{i-1}	a_{i-1}	d_i	θ_i
1	0	0	0	θ_1
2	90°	0	d_2	0
3	0	0	L_2	θ_3

FIGURE 3.10 Link parameters for the manipulator having three degrees of freedom and one prismatic joint.

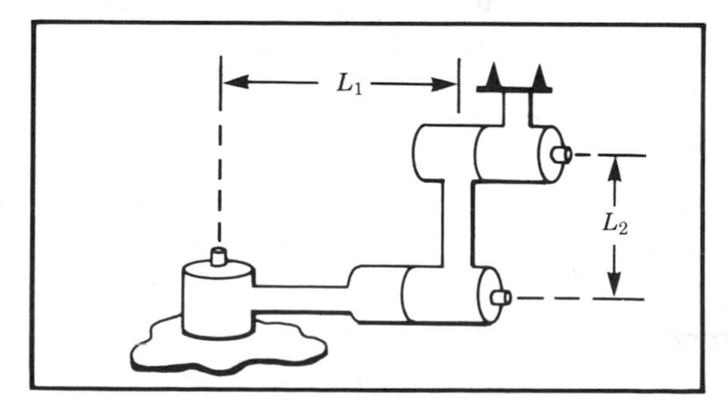

FIGURE 3.11 Three-link, nonplanar manipulator.

Figure 3.12 shows two possible frame assignments and corresponding parameters for the two possible choices of direction of $\hat{Z}_2$.

In general when $\hat{Z}_i$ and $\hat{Z}_{i+1}$ intersect, there are two choices for $\hat{X}_i$. In this example joint axes 1 and 2 intersect, so there are two choices for the direction of $\hat{X}_1$. Figure 3.13 shows two more possible frame assignments corresponding to the second choice of $\hat{X}_1$.

In fact, there are four more possibilities corresponding to the above four choices but with $\hat{Z}_1$ pointing downward. ■

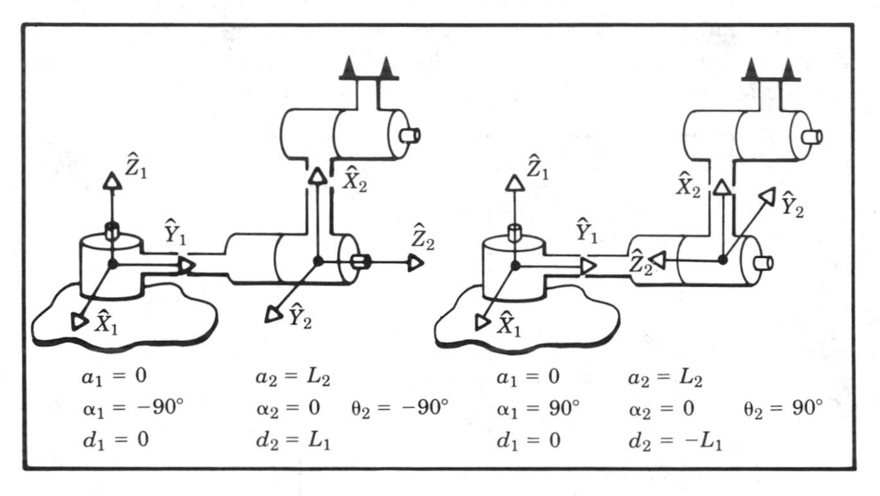

FIGURE 3.12 Two possible frame assignments.

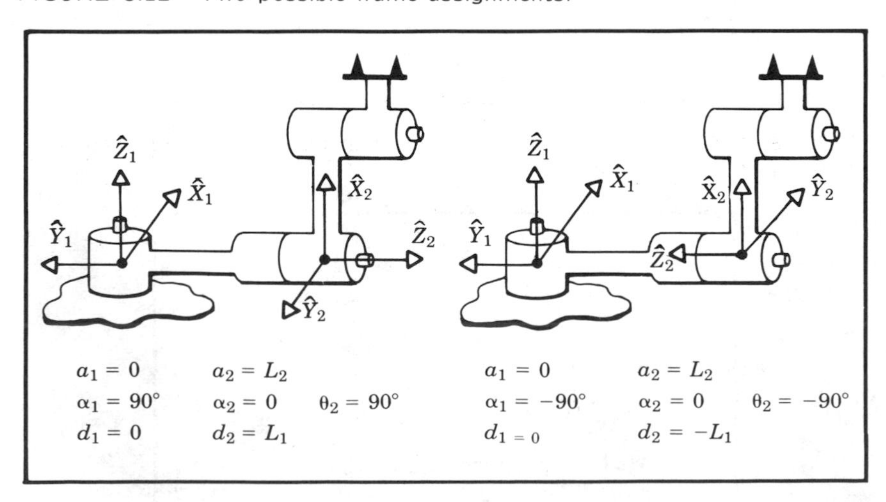

FIGURE 3.13 Two more possible frame assignments.

3.5 Manipulator kinematics

In this section we derive the general form of the transformation which relates the frames attached to neighboring links. We then concatenate these individual transformations to solve for the position and orientation of link n relative to link 0.

Derivation of link transformations

We wish to determine the transform which defines frame $\{i\}$ relative to the frame $\{i-1\}$. In general, this transformation will be a function of the four link parameters. For any *given* robot, this transformation will be a function of only one variable, the other three parameters being fixed by mechanical design. By defining a frame for each link we have broken the kinematics problem into n subproblems. In order to solve each of these subproblems, namely ${}^{i-1}_{i}T$, we will further break the problem into four sub-subproblems. *Each of these four transformations will be a function of one link parameter only, and will be simple enough that we can write down its form by inspection.* We begin by defining three intermediate frames for each link, namely: $\{P\}$, $\{Q\}$, and $\{R\}$.

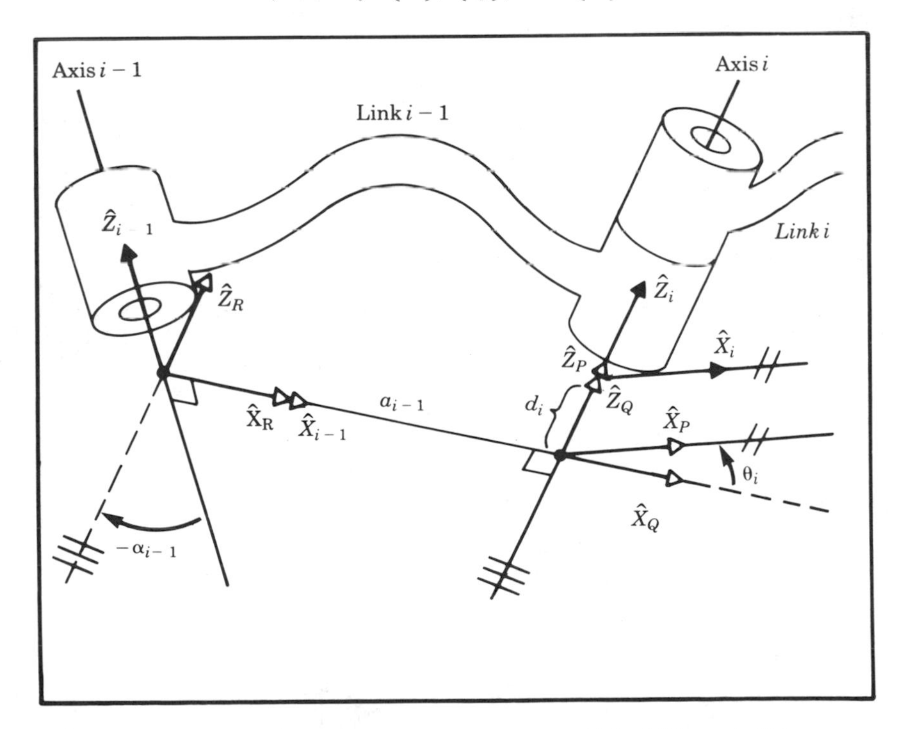

FIGURE 3.14 Location of intermediate frames $\{P\}$, $\{Q\}$, and $\{R\}$.

Figure 3.14 shows the same pair of joints as before with frames $\{P\}$, $\{Q\}$, and $\{R\}$ defined. Note that only the $\hat{X}$ and $\hat{Z}$ axes are shown for each frame to make the drawing clearer. Frame $\{R\}$ differs from frame $\{i-1\}$ only by a rotation of α_{i-1}. Frame $\{Q\}$ differs from $\{R\}$ by a translation a_{i-1}. Frame $\{P\}$ differs from $\{Q\}$ by a rotation θ_i, and frame $\{i\}$ differs from $\{P\}$ by a translation d_i. If we wish to write the transformation which transforms vectors defined in $\{i\}$ to their description in $\{i-1\}$ we may write

$$^{i-1}P = {}^{i-1}_{R}T \; {}^{R}_{Q}T \; {}^{Q}_{P}T \; {}^{P}_{i}T \; {}^{i}P, \tag{3.1}$$

or

$$^{i-1}P = {}^{i-1}_{i}T \; {}^{i}P, \tag{3.2}$$

where

$${}^{i-1}_{i}T = {}^{i-1}_{R}T \; {}^{R}_{Q}T \; {}^{Q}_{P}T \; {}^{P}_{i}T. \tag{3.3}$$

Considering each of these transformations, we see that (3.3) may be written:

$${}^{i-1}_{i}T = \mathrm{Rot}(\hat{X}_i, \alpha_{i-1}) \; \mathrm{Trans}(\hat{X}_i, a_{i-1}) \; \mathrm{Rot}(\hat{Z}_i, \theta_i) \; \mathrm{Trans}(\hat{Z}_i, d_i), \tag{3.4}$$

or

$${}^{i-1}_{i}T = \mathrm{Screw}(\hat{X}_i, a_{i-1}, \alpha_{i-1}) \; \mathrm{Screw}(\hat{Z}_i, d_i, \theta_i), \tag{3.5}$$

where $\mathrm{Screw}(\hat{Q}, r, \phi)$ stands for a translation along an axis $\hat{Q}$ by a distance r, and a rotation about the same axis by an angle ϕ. Multiplying out (3.4) we obtain the general form of ${}^{i-1}_{i}T$:

$${}^{i-1}_{i}T = \begin{bmatrix} c\theta_i & -s\theta_i & 0 & a_{i-1} \\ s\theta_i c\alpha_{i-1} & c\theta_i c\alpha_{i-1} & -s\alpha_{i-1} & -s\alpha_{i-1} d_i \\ s\theta_i s\alpha_{i-1} & c\theta_i s\alpha_{i-1} & c\alpha_{i-1} & c\alpha_{i-1} d_i \\ 0 & 0 & 0 & 1 \end{bmatrix}. \tag{3.6}$$

EXAMPLE 3.6

Using the link parameters shown in Fig. 3.10 for the robot of Fig. 3.8, compute the individual transformations for each link.

Substituting the parameters into (3.6) we obtain:

$$\begin{aligned} {}^{0}_{1}T &= \begin{bmatrix} c\theta_1 & -s\theta_1 & 0 & 0 \\ s\theta_1 & c\theta_1 & 0 & 0 \\ 0 & 0 & 1 & 0 \\ 0 & 0 & 0 & 1 \end{bmatrix}, \\ {}^{1}_{2}T &= \begin{bmatrix} 1 & 0 & 0 & 0 \\ 0 & 0 & -1 & -d_2 \\ 0 & 1 & 0 & 0 \\ 0 & 0 & 0 & 1 \end{bmatrix}, \\ {}^{2}_{3}T &= \begin{bmatrix} c\theta_3 & -s\theta_3 & 0 & 0 \\ s\theta_3 & c\theta_3 & 0 & 0 \\ 0 & 0 & 1 & l_2 \\ 0 & 0 & 0 & 1 \end{bmatrix}. \end{aligned} \tag{3.7}$$

Having derived these link transformations, it is a good idea to check them using common sense. For example the elements of the fourth column of each transform should give the coordinates of the origin of the next higher frame. ■

Concatenating link transformations

Once the link frames have been defined and the corresponding link parameters found, developing the kinematic equations is straightforward. Using the values of the link parameters the individual link transformation matrices can be computed. Then, the link transformations can be multiplied together to find the single transformation that relates frame $\{N\}$ to frame $\{0\}$:

$$ {}^{0}_{N}T = {}^{0}_{1}T \, {}^{1}_{2}T \, {}^{2}_{3}T \ldots {}^{N-1}_{N}T. \tag{3.8} $$

This transformation, ${}^{0}_{N}T$, will be a function of all n joint variables. If the robot's joint position sensors are queried, the Cartesian position and orientation of the last link may be computed by ${}^{0}_{N}T$.

3.6 Actuator space, joint space, and Cartesian space

The position of all the links of a manipulator of n degrees of freedom can be specified with a set of n joint variables. This set of variables is often referred to as the $n \times 1$ **joint vector**. The space of all such joint vectors is referred to as **joint space**. Thus far in this chapter we have been concerned with computing the **Cartesian space** description from knowledge of the joint space description. We use the term *Cartesian space* when position is measured along orthogonal axes, and orientation is measured according to any of the conventions outlined in Chapter 2. Sometimes the terms **task oriented space** or **operational space** are used for what we will call Cartesian space.

So far we have implicitly assumed that each kinematic joint is actuated directly with some sort of actuator. However, in the case of many industrial robots, this is not so. For example, sometimes two actuators work together in a differential pair to move a single joint, or sometimes a linear actuator is used to rotate a revolute joint through the use of a four-bar linkage. In these cases it is helpful to consider the notion of *actuator positions*. Since the sensors which measure the position of the manipulator are often located at the actuators, some computations must be performed to compute the joint vector as a function of a set of actuator values, or **actuator vector**.

As indicated in Fig. 3.15, there are three representations of a manipulator's position and orientation: descriptions in **actuator space**, **joint space**, and **Cartesian space**. In this chapter we are concerned with the mappings between representations as indicated by the solid arrows in Fig. 3.15. In Chapter 4, we will consider the inverse mappings indicated by the dashed arrows.

The manner in which actuators might be connected to move a joint is quite varied and, although they might be catalogued, we will not do so here. For each robot we design or seek to analyze, the correspondence between actuator positions and joint positions must be solved. In the following section we will solve an example problem for an industrial robot.

3.7 Examples: kinematics of two industrial robots

In this section we work out the kinematics of two industrial robots. First we consider the Unimation PUMA 560, a rotary joint manipulator with six degrees of freedom. We will solve for the kinematic equations as functions of the joint angles. For this example we will skip the additional problem of the relationship between actuator space and joint space. Second, we consider the Yasukawa Motoman L-3, a robot with five degrees of freedom and rotary joints. This example is done in detail, including the actuator-to-joint transformations. This example may be skipped on first reading of the book.

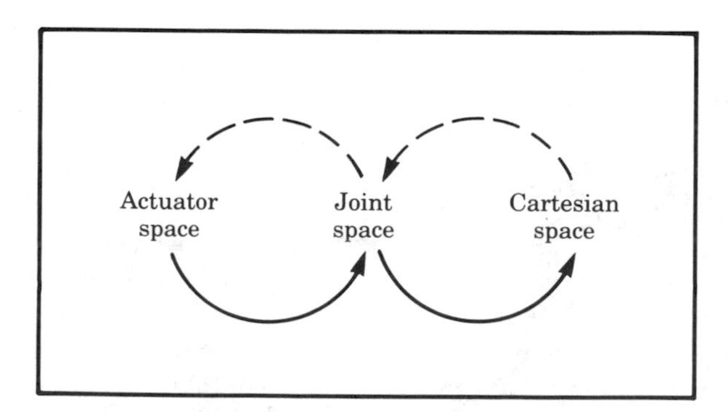

FIGURE 3.15 Mappings between kinematic descriptions.

The PUMA 560

The Unimation PUMA 560 (Fig. 3.16) is a robot with six degrees of freedom and all rotational joints. It is shown in Fig. 3.17 with link frame assignments in the position corresponding to all joint angles equal to zero.* Figure 3.18 shows detail of the forearm of the robot.

Note that the frame {0} (not shown) is coincident with frame {1} when θ_1 is zero. Note also that for this robot, as with many industrial robots, the joint axes of joints 4, 5, and 6 all intersect at a common point, and this point of intersection coincides with the origin of frames {4}, {5} and {6}. The link parameters corresponding to this placement

FIGURE 3.16 The Unimation PUMA 560.
Courtesy of Unimation Incorporated, Shelter Rock Lane, Danbury, Conn.

* Unimation has used a slightly different assignment of zero location of the joints, such that $\theta_3^* = \theta_3 - 180°$ where θ_3^* is the position of joint 3 using Unimation's convention.

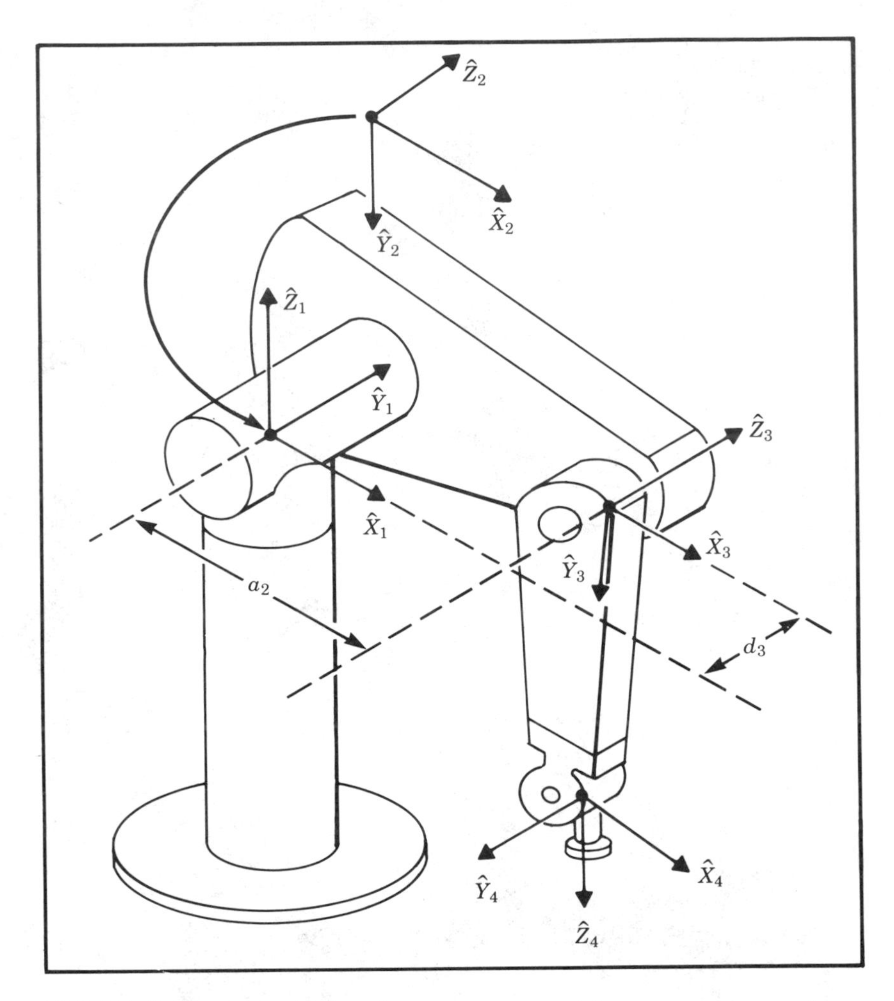

FIGURE 3.17 Some kinematic parameters and frame assignments for the PUMA 560 manipulator.

of link frames are shown in Fig. 3.19. In the case of the PUMA 560 a gearing arrangement in the wrist of the manipulator couples together the motions of joints 4, 5, and 6. What this means is that for these three joint we must make a distinction between joint space and actuator space and solve the complete kinematics in two steps. However, in this example, we will consider only the kinematics from joint space to Cartesian space.

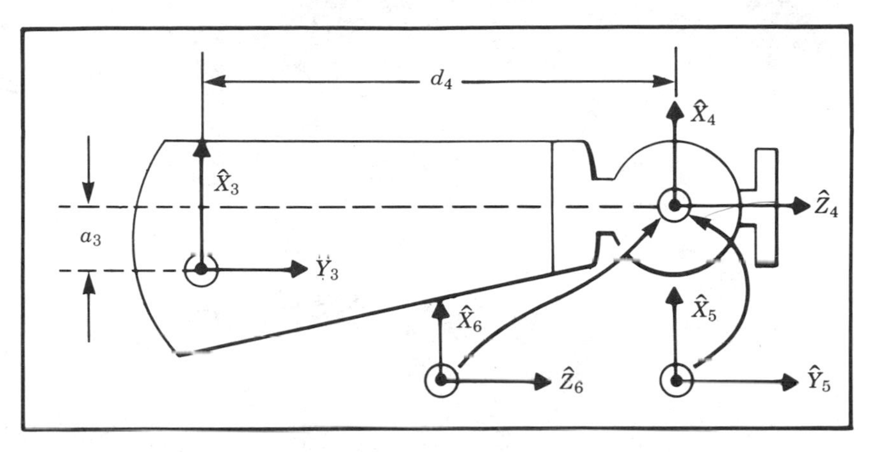

FIGURE 3.18 Kinematic parameters and frame assignments for the forearm of the PUMA 560 manipulator.

i	α_{i-1}	a_{i-1}	d_i	θ_i
1	0	0	0	θ_1
2	$-90°$	0	0	θ_2
3	0	a_2	d_3	θ_3
4	$-90°$	a_3	d_4	θ_4
5	$90°$	0	0	θ_5
6	$-90°$	0	0	θ_6

FIGURE 3.19 Link parameters of the PUMA 560.

Using (3.6) we compute each of the link transformations:

$$
\begin{aligned}
{}^0_1T &= \begin{bmatrix} c\theta_1 & -s\theta_1 & 0 & 0 \\ s\theta_1 & c\theta_1 & 0 & 0 \\ 0 & 0 & 1 & 0 \\ 0 & 0 & 0 & 1 \end{bmatrix}, \\
{}^1_2T &= \begin{bmatrix} c\theta_2 & -s\theta_2 & 0 & 0 \\ 0 & 0 & 1 & 0 \\ -s\theta_2 & -c\theta_2 & 0 & 0 \\ 0 & 0 & 0 & 1 \end{bmatrix}, \\
{}^2_3T &= \begin{bmatrix} c\theta_3 & -s\theta_3 & 0 & a_2 \\ s\theta_3 & c\theta_3 & 0 & 0 \\ 0 & 0 & 1 & d_3 \\ 0 & 0 & 0 & 1 \end{bmatrix}, \\
{}^3_4T &= \begin{bmatrix} c\theta_4 & -s\theta_4 & 0 & a_3 \\ 0 & 0 & 1 & d_4 \\ -s\theta_4 & -c\theta_4 & 0 & 0 \\ 0 & 0 & 0 & 1 \end{bmatrix}, \\
{}^4_5T &= \begin{bmatrix} c\theta_5 & -s\theta_5 & 0 & 0 \\ 0 & 0 & -1 & 0 \\ s\theta_5 & c\theta_5 & 0 & 0 \\ 0 & 0 & 0 & 1 \end{bmatrix}, \\
{}^5_6T &= \begin{bmatrix} c\theta_6 & -s\theta_6 & 0 & 0 \\ 0 & 0 & 1 & 0 \\ -s\theta_6 & -c\theta_6 & 0 & 0 \\ 0 & 0 & 0 & 1 \end{bmatrix}.
\end{aligned}
\qquad (3.9)
$$

We now form ${}^0_6T$ by matrix multiplication of the individual link matrices. While forming this product we will derive some sub-results which will be useful when solving the inverse kinematic problem in Chapter 4. We start by multiplying ${}^4_5T$ and ${}^5_6T$:

$$
{}^4_6T = {}^4_5T\,{}^5_6T = \begin{bmatrix} c_5c_6 & -c_5s_6 & -s_5 & 0 \\ s_6 & c_6 & 0 & 0 \\ s_5c_6 & -s_5s_6 & c_5 & 0 \\ 0 & 0 & 0 & 1 \end{bmatrix}, \qquad (3.10)
$$

where c_5 is shorthand for $\cos\theta_5$, s_5 for $\sin\theta_5$, and so on.* Then we have

$$
{}^3_6T = {}^3_4T\,{}^4_6T = \begin{bmatrix} c_4c_5c_6 - s_4s_6 & -c_4c_5s_6 - s_4c_6 & -c_4s_5 & a_3 \\ s_5c_6 & -s_5s_6 & c_5 & d_4 \\ -s_4c_5c_6 - c_4s_6 & s_4c_5s_6 - c_4c_6 & s_4s_5 & 0 \\ 0 & 0 & 0 & 1 \end{bmatrix}. \qquad (3.11)
$$

Because joints 2 and 3 are always parallel, multiplying ${}^1_2T$ and ${}^2_3T$ first and applying sum of angle formulas will yield a somewhat simpler final

* Depending on the amount of space available to show expressions, we use any of the three forms: $\cos\theta_5$, $c\theta_5$, or c_5.

expression. This can be done whenever two rotational joints have parallel axes, and we have

$$ {}^{1}_{3}T = {}^{1}_{2}T \, {}^{2}_{3}T = \begin{bmatrix} c_{23} & -s_{23} & 0 & a_2 c_2 \\ 0 & 0 & 1 & d_3 \\ -s_{23} & -c_{23} & 0 & -a_2 s_2 \\ 0 & 0 & 0 & 1 \end{bmatrix}, \tag{3.12} $$

where we have used the sum of angles formulas:

$$ c_{23} = c_2 c_3 - s_2 s_3, $$

$$ s_{23} = c_2 s_3 + s_2 c_3. $$

Then we have

$$ {}^{1}_{6}T = {}^{1}_{3}T \, {}^{3}_{6}T = \begin{bmatrix} {}^{1}r_{11} & {}^{1}r_{12} & {}^{1}r_{13} & {}^{1}p_x \\ {}^{1}r_{21} & {}^{1}r_{22} & {}^{1}r_{23} & {}^{1}p_y \\ {}^{1}r_{31} & {}^{1}r_{32} & {}^{1}r_{33} & {}^{1}p_z \\ 0 & 0 & 0 & 1 \end{bmatrix}, $$

where

$$ \begin{aligned} {}^{1}r_{11} &= c_{23}\left[c_4 c_5 c_6 - s_4 s_6\right] - s_{23} s_5 c_6, \\ {}^{1}r_{21} &= \quad s_4 c_5 c_6 \quad c_4 s_6, \\ {}^{1}r_{31} &= -s_{23}\left[c_4 c_5 c_6 - s_4 s_6\right] - c_{23} s_5 c_6, \\ \\ {}^{1}r_{12} &= -c_{23}\left[c_4 c_5 s_6 + s_4 c_6\right] + s_{23} s_5 s_6, \\ {}^{1}r_{22} &= s_4 c_5 s_6 - c_4 c_6, \\ {}^{1}r_{32} &= s_{23}\left[c_4 c_5 s_6 + s_4 c_6\right] + c_{23} s_5 s_6, \\ \\ {}^{1}r_{13} &= -c_{23} c_4 s_5 - s_{23} c_5, \\ {}^{1}r_{23} &= s_4 s_5, \\ {}^{1}r_{33} &= s_{23} c_4 s_5 - c_{23} c_5, \\ \\ {}^{1}p_x &= a_2 c_2 + a_3 c_{23} - d_4 s_{23}, \\ {}^{1}p_y &= d_3, \\ {}^{1}p_z &= -a_3 s_{23} - a_2 s_2 - d_4 c_{23}. \end{aligned} \tag{3.13} $$

Finally, we obtain the product of all six link transforms,

$$ {}^{0}_{6}T = {}^{0}_{1}T \, {}^{1}_{6}T = \begin{bmatrix} r_{11} & r_{12} & r_{13} & p_x \\ r_{21} & r_{22} & r_{23} & p_y \\ r_{31} & r_{32} & r_{33} & p_z \\ 0 & 0 & 0 & 1 \end{bmatrix}, $$

where

$$\begin{aligned}
r_{11} &= c_1\left[c_{23}(c_4c_5c_6 - s_4s_6) - s_{23}s_5c_6\right] + s_1(s_4c_5c_6 + c_4s_6),\\
r_{21} &= s_1\left[c_{23}(c_4c_5c_6 - s_4s_6) - s_{23}s_5c_6\right] - c_1(s_4c_5c_6 + c_4s_6),\\
r_{31} &= -s_{23}(c_4c_5c_6 - s_4s_6) - c_{23}s_5c_6,\\
\\
r_{12} &= c_1\left[c_{23}(-c_4c_5s_6 - s_4c_6) + s_{23}s_5s_6\right] + s_1(c_4c_6 - s_4c_5s_6),\\
r_{22} &= s_1\left[c_{23}(-c_4c_5s_6 - s_4c_6) + s_{23}s_5s_6\right] - c_1(c_4c_6 - s_4c_5s_6),\\
r_{32} &= -s_{23}(-c_4c_5s_6 - s_4c_6) + c_{23}s_5s_6,\\
\\
r_{13} &= -c_1(c_{23}c_4s_5 + s_{23}c_5) - s_1s_4s_5,\\
r_{23} &= -s_1(c_{23}c_4s_5 + s_{23}c_5) + c_1s_4s_5,\\
r_{33} &= s_{23}c_4s_5 - c_{23}c_5,\\
\\
p_x &= c_1\left[a_2c_2 + a_3c_{23} - d_4s_{23}\right] - d_3s_1,\\
p_y &= s_1\left[a_2c_2 + a_3c_{23} - d_4s_{23}\right] + d_3c_1,\\
p_z &= -a_3s_{23} - a_2s_2 - d_4c_{23}.
\end{aligned} \qquad \textbf{(3.14)}$$

Equations (3.14) constitute the kinematics of the PUMA 560. They specify how to compute the position and orientation of frame $\{6\}$ relative to frame $\{0\}$ of the robot. These are the basic equations for all kinematic analysis of this manipulator.

The Yasukawa Motoman L-3

The Yasukawa Motoman L-3 is a popular industrial manipulator with five degrees of freedom (Fig. 3.20). Unlike the examples we have seen thus far, the Motoman is not a simple open kinematic chain, but rather makes use of two linear actuators coupled to links 2 and 3 with four-bar linkages. Also, through a chain drive, joints 4 and 5 are operated by two actuators in a differential arrangement.

In this example we will solve the kinematics in two stages. First we will solve for joint angles from actuator positions; and second, we will solve for Cartesian position and orientation of the last link from joint angles.

Figure 3.21 shows the linkage mechanism which connects actuator number 2 to links 2 and 3 of the robot. The actuator is a linear one which directly controls the length of the segment labeled DC. Triangle ABC is fixed, as is the length BD. Joint 2 pivots about point B, and the

FIGURE 3.20 The Yasukawa Motoman L-3.
Courtesy of Machine Intelligence Corporation.

actuator pivots slightly about point C as the linkage moves. We give the following names to the constants (lengths and angles) associated with actuator 2:

$$\gamma_2 = AB, \quad \phi_2 = AC, \quad \alpha_2 = BC,$$
$$\beta_2 = BD, \quad \Omega_2 = \angle JBD, \quad l_2 = BJ,$$

and the following names to the variables:

$$\theta_2 = -\angle JBQ, \quad \psi_2 = \angle CBD, \quad g_2 = DC.$$

Figure 3.22 shows the linkage mechanism which connects actuator number 3 to links 2 and 3 of the robot. The actuator is a linear one which directly controls the length of the segment labeled HG. Triangle EFG is fixed, as is the length FH. Joint 3 pivots about point J, and the

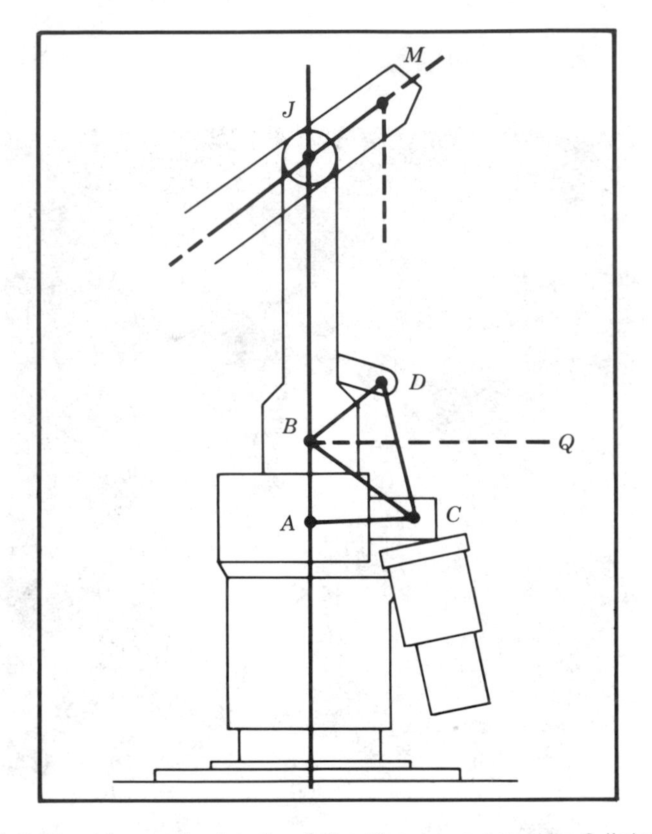

FIGURE 3.21 Kinematic details of the Yasukawa actuator 2 linkage.

actuator pivots slightly about point G as the linkage moves. We give the following names to the constants (lengths and angles) associated with actuator 3:

$$\gamma_3 = EF, \quad \phi_3 = EG, \quad \alpha_3 = GF,$$
$$\beta_3 = HF, \quad l_3 = JK,$$

and the following names to the variables:

$$\theta_3 = \angle PJK, \quad \psi_3 = \angle GFH, \quad g_3 = GH.$$

This arrangment of actuators and linkages has the following functional effect. Actuator 2 is used to position joint 2, and while doing so, link 3 remains in the same orientation relative to the base of the robot. Actuator 3 is used to adjust the orientation of link 3 relative to the base of the robot (rather than relative to the preceeding link as in a serial kinematic chain robot). One purpose of such a linkage arrangement is to increase the structural rigidity of the main linkages of the robot.

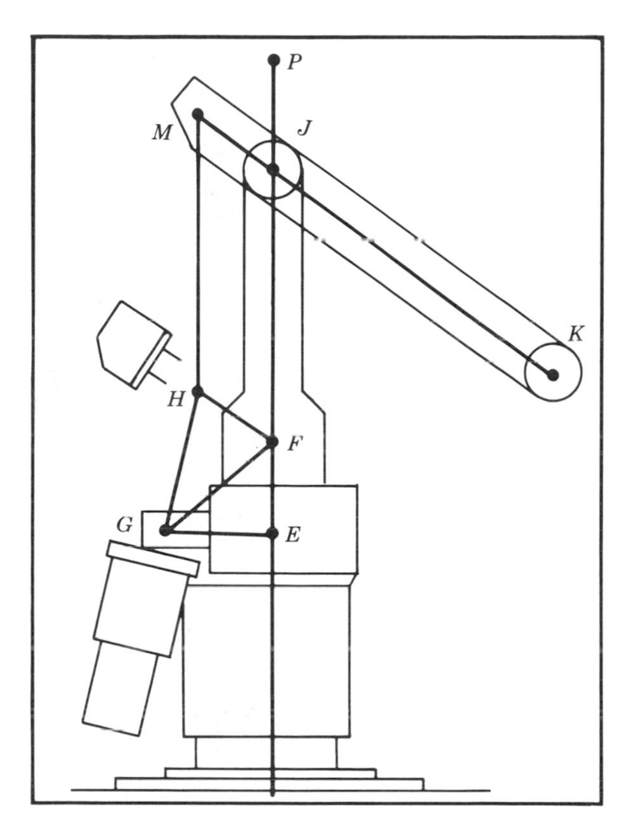

FIGURE 3.22 Kinematic details of the Yasukawa actuator 3 linkage.

This often pays off in terms of an increased ability to position the robot precisely.

The actuators for joints 4 and 5 are attached to link 1 of the robot with their axes aligned with that of joint 2 (points B and F in Figs. 3.21 and 3.22). They operate the wrist joints through two sets of chains—one set located interior to link 2, and the second set located interior to link 3. The effect of this transmission system along with its interaction with the actuation of links 2 and 3 is described functionally as follows: Actuator 4 is used to position joint 4 relative to the base of the robot, rather than relative to the preceeding link 3. This means that holding actuator 4 constant will keep link 4 at a constant orientation relative to the base of the robot, regardless of the positions of joints 2 and 3. Finally, actuator 5 behaves as if directly connected to joint 5.

We now state the equations which map a set of actuator values (A_i) to the equivalent set of joint values (θ_i). In this case, these equations were derived by straightforward plane geometry—mostly just application of the "law of cosines."* Appearing in these equations are scale (k_i) and offset (λ_i) constants for each actuator. For example, actuator 1 is directly connected to joint axis 1, and so the conversion is simple; it is just a matter of a scale factor plus an offset. Thus

$$\theta_1 = k_1 A_1 + \lambda_1,$$

$$\theta_2 = \cos^{-1}\left(\frac{(k_2 A_2 + \lambda_2)^2 - \alpha_2^2 - \beta_2^2}{-2\alpha_2\beta_2}\right) + \tan^{-1}\left(\frac{\phi_2}{\gamma_2}\right) + \Omega_2 - 270^\circ,$$

$$\theta_3 = \cos^{-1}\left(\frac{(k_3 A_3 + \lambda_3)^2 - \alpha_3^2 - \beta_3^2}{-2\alpha_3\beta_3}\right) - \theta_2 + \tan^{-1}\left(\frac{\phi_3}{\gamma_3}\right) - 90^\circ, \qquad (3.15)$$

$$\theta_4 = -k_4 A_4 - \theta_2 - \theta_3 + \lambda_4 + 180^\circ,$$

$$\theta_5 = -k_5 A_5 + \lambda_5.$$

Figure 3.23 shows the attachment of the link frames. In this figure the manipulator is shown in a position corresponding to the joint vector $\Theta = (0, -90^\circ, 90^\circ, 90^\circ, 0)$. Figure 3.24 shows the link parameters for this manipulator.

The resulting link transformation matrices are

$$
\begin{aligned}
{}^0_1T &= \begin{bmatrix} c\theta_1 & -s\theta_1 & 0 & 0 \\ s\theta_1 & c\theta_1 & 0 & 0 \\ 0 & 0 & 1 & 0 \\ 0 & 0 & 0 & 1 \end{bmatrix}, \\
{}^1_2T &= \begin{bmatrix} c\theta_2 & -s\theta_2 & 0 & 0 \\ 0 & 0 & 1 & 0 \\ -s\theta_2 & -c\theta_2 & 0 & 0 \\ 0 & 0 & 0 & 1 \end{bmatrix}, \\
{}^2_3T &= \begin{bmatrix} c\theta_3 & -s\theta_3 & 0 & l_2 \\ s\theta_3 & c\theta_3 & 0 & 0 \\ 0 & 0 & 1 & 0 \\ 0 & 0 & 0 & 1 \end{bmatrix}, \qquad (3.16) \\
{}^3_4T &= \begin{bmatrix} c\theta_4 & -s\theta_4 & 0 & l_3 \\ s\theta_4 & c\theta_4 & 0 & 0 \\ 0 & 0 & 1 & 0 \\ 0 & 0 & 0 & 1 \end{bmatrix}, \\
{}^4_5T &= \begin{bmatrix} c\theta_5 & -s\theta_5 & 0 & 0 \\ 0 & 0 & -1 & 0 \\ s\theta_5 & c\theta_5 & 0 & 0 \\ 0 & 0 & 0 & 1 \end{bmatrix}.
\end{aligned}
$$

* If a triangle's angles are labelled a, b, and c, where angle a is opposite side A, and so on, then $A^2 = B^2 + C^2 - 2BC\cos a$.

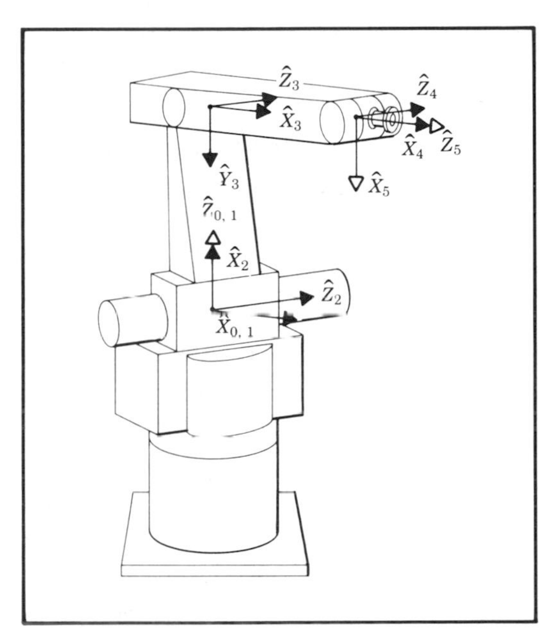

FIGURE 3.23 Assignment of link frames for the Yasukawa L-3.

i	α_{i-1}	a_{i-1}	d_i	θ_i
1	0	0	0	θ_1
2	-90°	0	0	θ_2
3	0	L_2	0	θ_3
4	0	L_3	0	θ_4
5	90°	0	0	θ_5

FIGURE 3.24 Link parameters of the Yasukawa L-3 Manipulator.

Forming the product to obtain ${}^0_5T$ we obtain

$$ {}^0_5T = \begin{bmatrix} r_{11} & r_{12} & r_{13} & p_x \\ r_{21} & r_{22} & r_{23} & p_y \\ r_{31} & r_{32} & r_{33} & p_z \\ 0 & 0 & 0 & 1 \end{bmatrix}, $$

where

$$ \begin{aligned} r_{11} &= c_1 c_{234} c_5 - s_1 s_5, \\ r_{21} &= s_1 c_{234} c_5 + c_1 s_5, \\ r_{31} &= -s_{234} c_5, \\[1em] r_{12} &= -c_1 c_{234} s_5 - s_1 c_5, \\ r_{22} &= -s_1 c_{234} s_5 + c_1 c_5, \\ r_{32} &= s_{234} s_5, \\[1em] r_{13} &= c_1 s_{234}, \\ r_{23} &= s_1 s_{234}, \\ r_{33} &= c_{234}, \\[1em] p_x &= c_1 \left(l_2 c_2 + l_3 c_{23} \right), \\ p_y &= s_1 \left(l_2 c_2 + l_3 c_{23} \right), \\ p_z &= -l_2 s_2 - l_3 s_{23}. \end{aligned} \tag{3.17} $$

We have developed the kinematic equations for the Yasukawa Motoman in two steps. In the first step we compute a joint vector from an actuator vector, and in the second step we compute a position and orientation of the wrist frame from the joint vector. If we wish to compute only Cartesian position and not joint angles, it is possible to derive equations which map directly from actuator space to Cartesian space which are somewhat simpler computationally than the two-step approach (see Exercise 3.10).

3.8 Frames with standard names

As a matter of convention it will be helpful if we assign specific names and locations to certain "standard" frames associated with a robot and its workspace. Figure 3.25 shows a typical situation in which a robot has grasped some sort of tool and wishes to position the tool tip to a user-defined location. The five frames indicated in Fig. 3.25 are so often referred to that we will define names for them. The naming and subsequent use of these five frames in a robot programming and control system facilitates providing general capabilities in an easily understandable way. All robot motions will be described in terms of these frames.

Brief definitions of the frames shown in Fig. 3.25 are listed below.

The base frame, $\{B\}$

$\{B\}$ is located at the base of the manipulator. It is merely another name for frame $\{0\}$. It is affixed to a nonmoving part of the robot, sometimes called link 0.

The station frame, $\{S\}$

$\{S\}$ is located in a task relevant location. In Fig. 3.26, it is at the corner of a table upon which the robot is to work. As far as the user of this robot system is concerned, $\{S\}$ is the universe frame and all actions of the robot are made relative to it. It is sometimes called the task frame, the world frame, or the universe frame. The station frame is always specified with respect to the base frame, that is, ${}^B_S T$.

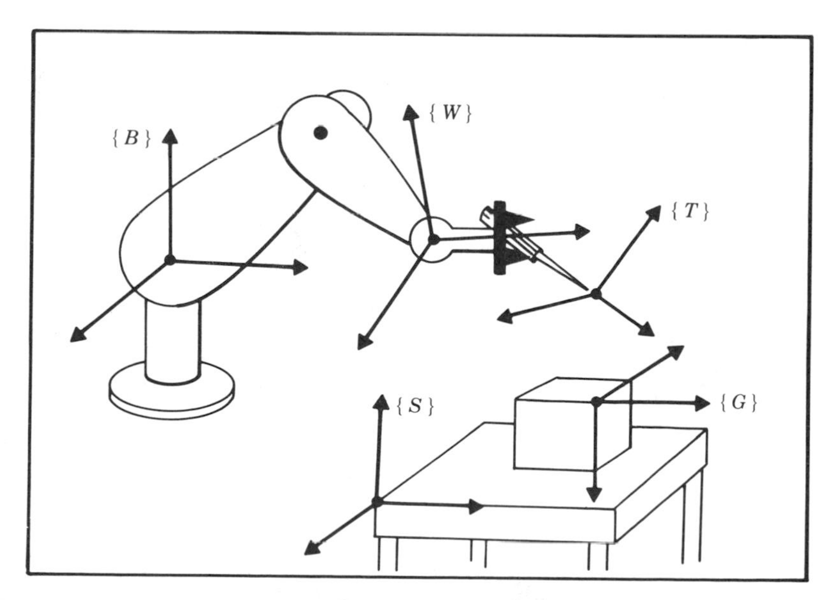

FIGURE 3.25 The standard frames.

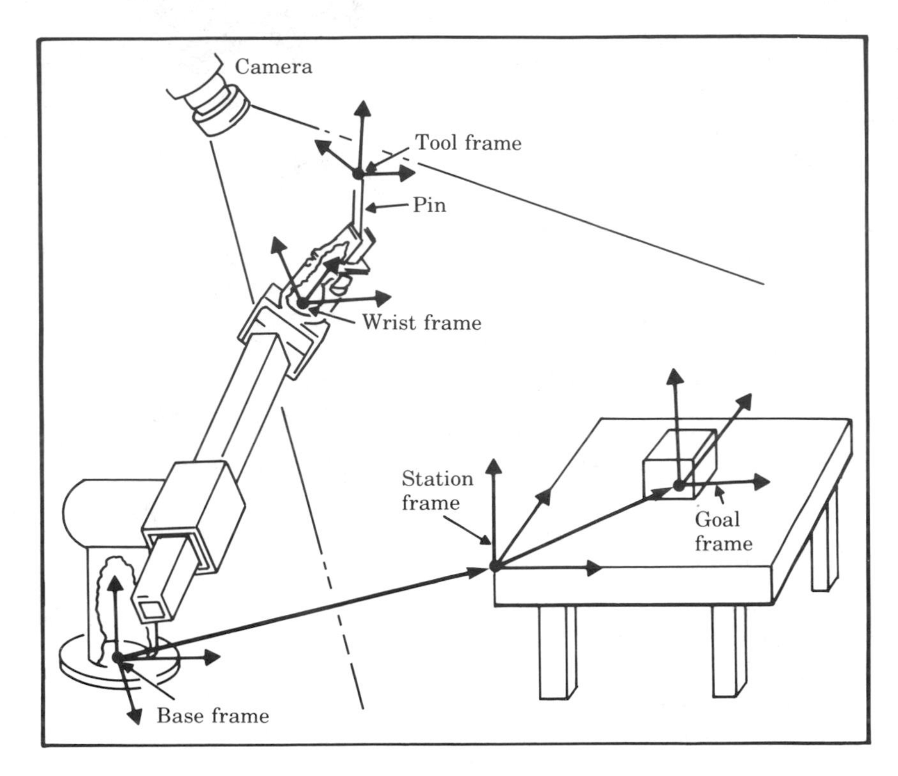

FIGURE 3.26 Example of the assignment of standard frames.

The wrist frame, $\{W\}$

$\{W\}$ is affixed to the last link of the manipulator. It is another name for frame $\{N\}$, the link frame attached to the last link of the robot. Very often $\{W\}$ has its origin fixed at a point called the wrist of the manipulator, and $\{W\}$ moves with the last link of the manipulator. It is defined relative to the base frame. That is, $\{W\} = {}^B_W T = {}^0_N T$.

The tool frame, $\{T\}$

$\{T\}$ is affixed to the end of any tool the robot happens to be holding. When the hand is empty, $\{T\}$ is usually located with its origin between the finger-tips of the robot. The tool frame is always specified with respect to the wrist frame. In Fig. 3.26 the tool frame is defined with its origin at the tip of a pin that the robot is holding.

The goal frame, $\{G\}$

$\{G\}$ is a description of the location to which the robot is to move the tool. Specifically this means that at the end of the motion, the tool frame should be brought to coincidence with the goal frame. $\{G\}$ is always specified relative to the station frame. In Fig. 3.26 the goal is located at a hole into which we want the pin to be inserted.

All robot motions may be described in terms of these frames without loss of generality. Their use helps to give us a standard language for talking about robot tasks.

3.9 WHERE Is the tool?

One of the first capabilities a robot must have is to be able to calculate the position and orientation of the tool it is holding (or of its empty hand) with respect to a convenient coordinate system. That is, we wish to calculate the value of the tool frame, $\{T\}$, relative to the station, $\{S\}$. Once ${}^{B}_{W}T$ has been computed using the kinematic equations we can use Cartesian transforms, as studied in Chapter 2, to calculate $\{T\}$ relative to $\{S\}$. Solving a simple transform equation leads to:

$$ {}^{S}_{T}T = {}^{B}_{S}T^{-1}\, {}^{B}_{W}T\, {}^{W}_{T}T. \quad (\mathbf{3.18}) $$

Equation (3.18) implements what is called the `WHERE` function in some robot systems. It computes "where" the arm is. For the situation in Fig. 3.26, the output of `WHERE` would be the position and orientation of the pin relative to the table top.

Equation (3.18) can be thought of as *generalizing* the kinematics. ${}^{S}_{T}T$ computes the kinematics due to the geometry of the linkages along with a general transform (which might be considered a fixed link) at the base end (${}^{B}_{S}T$) and another at the end-effector (${}^{W}_{T}T$). These extra transforms allow us to include tools with offsets and twists, and to operate with respect to an arbitrary station frame.

3.10 Computational considerations

In many practical manipulator systems, the time required to perform kinematic calculations is a consideration. In this section we briefly discuss various issues involved in computing manipulator kinematics as exemplified by (3.14) for the case of the PUMA 560.

One choice to be made is the use of fixed- or floating-point representation of the quantities involved. Many implementations use floating point for ease of software development, since the programmer does not have to be concerned with scaling operations due to the relative magnitude of the variables. However, when speed is crucial, fixed-point representation is quite possible because the variables do not have a large dynamic range, and these ranges are fairly well known. Rough estimations of the number of bits needed in fixed-point representation seem to indicate that 24 are sufficient [2].

By factoring equations such as (3.14), it is possible to reduce the number of multiplications and additions at the cost of creating local variables, which is usually a good trade off. The point is to avoid computing common terms over and over throughout the computation. There has been some application of computer assisted automatic factorization of such equations [3].

The major expense in calculating kinematics is often the calculation of the transcendental functions, i.e., sine and cosine. When these functions are available as part of a standard library, they are often computed from a series expansion at the cost of many multiply times. At the expense of the required memory, many manipulation systems employ table lookup implementations of the transcendental functions. Depending on the scheme, this reduces the amount of time required to calculate a sine or cosine to 2 or 3 multiply times or less [4].

The computation of the kinematics as in (3.14) is redundant in that nine quantities are calculated to represent orientation. One means which usually reduces computation is to calculate only two columns of the rotation matrix and then compute a cross product (requiring only 6 multiplications and 3 adds) to compute the third column. Obviously, one chooses the two least complicated columns to compute.

References

[1] J. Denavit and R.S. Hartenberg, "A Kinematic Notation for Lower-Pair Mechanisms Based on Matrices," *J. Applied Mechanics*, pp. 215-221, June 1955.

[2] T. Turner, J. Craig, W. Gruver, "A Microprocessor Architecture for Advanced Robot Control," 14th ISIR, Stockholm, Sweden, October 1984.

[3] W. Schiehlen, "Computer Generation of Equations of Motion" in "Computer Aided Analysis and Optimization of Mechanical System Dynamics," E.J. Haug, editor, Springer-Verlag, 1984.

[4] C. Ruoff, "Fast Trigonometric Functions for Robot Control," *Robotics Age*, November 1981.

Exercises

3.1 [15] Compute the kinematics of the planar arm from Example 3.3.

3.2 [37] Imagine an arm like the PUMA 560 except that joint 3 is replaced with a prismatic joint. Assume the prismatic joint slides along the direction of $\hat{X}_1$ in Fig. 3.17; however, there is still an offset equivalent to d_3 to be accounted for. Make any additional assumptions needed. Derive the kinematic equations.

3.3 [25] The arm with three degrees of freedom shown in Fig. 3.27 is like the one in Example 3.3 except that joint 1's axis is not parallel to the other two. Instead, there is a twist of 90 degrees in magnitude between axes 1 and 2. Derive link parameters and the kinematic equations for ${}^{B}_{W}T$. Note that no l_3 need be defined.

3.4 [22] The arm with three degrees of freedom shown in Fig. 3.28 has joints 1 and 2 perpendicular, and joints 2 and 3 parallel. As pictured, all joints are at their zero location. Note that the positive sense of the joint angle is indicated. Assign link frames {0} through {3} for this arm—that is, sketch the arm showing the attachment of the frames. Then derive the transformation matrices ${}^{0}_{1}T$, ${}^{1}_{2}T$, and ${}^{2}_{3}T$.

3.5 [26] Write a subroutine to compute the kinematics of a PUMA 560. Code for speed, trying to minimize the number of multiplications as much as possible. Use the procedure heading

```
Procedure KIN(VAR theta: vec6; VAR wrelb: frame);
```

Count a sine or cosine evaluation as costing 5 multiply times. Count additions as costing 0.333 multiply times, and assignment statements as 0.2 multiply times. Count a square root computation as 4 multiply times. How many multiply times do you need?

3.6 [20] Write a subroutine to compute the kinematics of the cylindrical arm in Example 3.4. Use the procedure heading

```
Procedure KIN(VAR jointvar: vec3; VAR wrelb: frame);
```

Count a sine or cosine evaluation as costing 5 multiply times. Count additions as costing 0.333 multiply times, and assignment statements

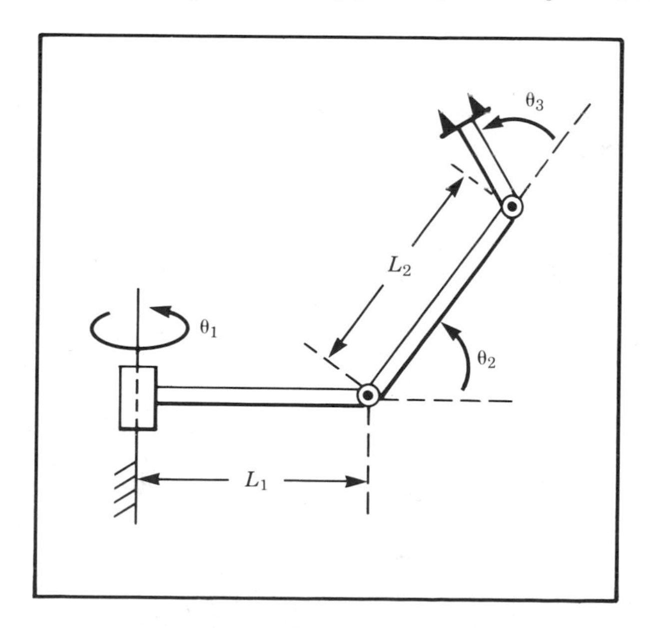

FIGURE 3.27 The 3-link non-planar arm of Exercise 3.

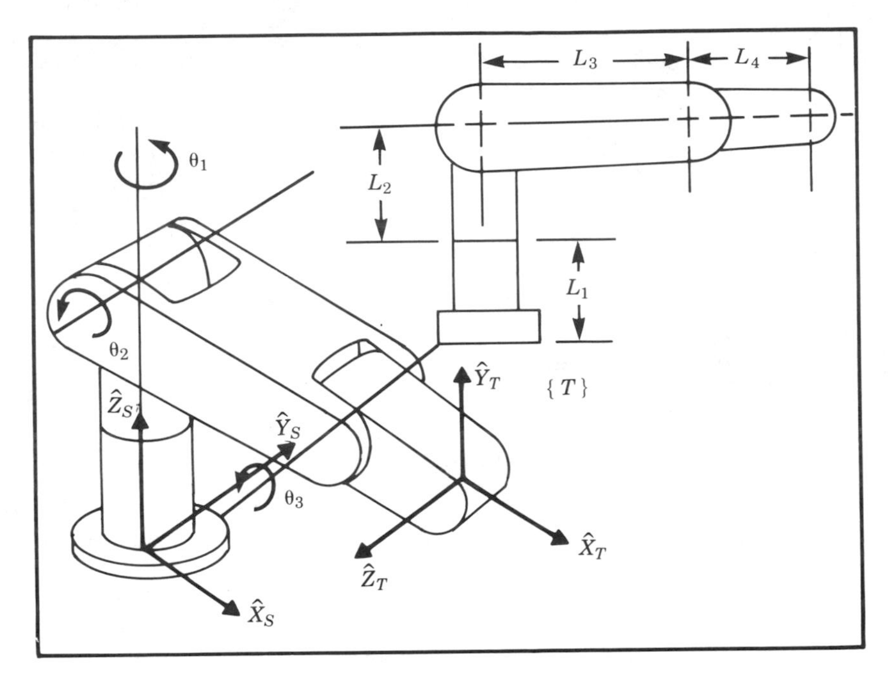

FIGURE 3.28 Two views of the 3-link of Exercise 4.

as 0.2 multiply times. Count a square root computation as 4 multiply times. How many multiply times do you need?

3.7 [22] Write a subroutine to compute the kinematics of the arm in Exercise 3.3. Use the procedure heading

```
Procedure KIN(VAR theta: vec3; VAR wrelb: frame);
```

Count a sine or cosine evaluation as costing 5 multiply times. Count additions as costing 0.333 multiply times, and assignment statements as 0.2 multiply times. Count a square root computation as 4 multiply times. How many multiply times do you need?

3.8 [13] In Fig. 3.29 the location of the tool, ${}^{W}_{T}T$, is not accurately known. Using force control, the robot feels around with the tool tip until it inserts it into the socket (or Goal) at location ${}^{S}_{G}T$. Once in this "calibration" configuration (in which $\{G\}$ and $\{T\}$ are coincident), the position of the robot, ${}^{B}_{W}T$, is determined by reading the joint angle sensors and computing the kinematics. Assuming ${}^{B}_{S}T$ and ${}^{S}_{G}T$ are known, give the transform equation to compute the unknown tool frame, ${}^{W}_{T}T$.

3.9 [11] For the 2-link manipulator shown in Fig. 3.30a, the link transformation matrices, ${}^{0}_{1}T$ and ${}^{1}_{2}T$, were determined. Their product is:

$$
{}^{0}_{2}T = \begin{bmatrix} c\theta_1 c\theta_2 & -c\theta_1 s\theta_2 & s\theta_1 & l_1 c\theta_1 \\ s\theta_1 c\theta_2 & -s\theta_1 s\theta_2 & -c\theta_1 & l_1 s\theta_1 \\ s\theta_2 & c\theta_2 & 0 & 0 \\ 0 & 0 & 0 & 1 \end{bmatrix}.
$$

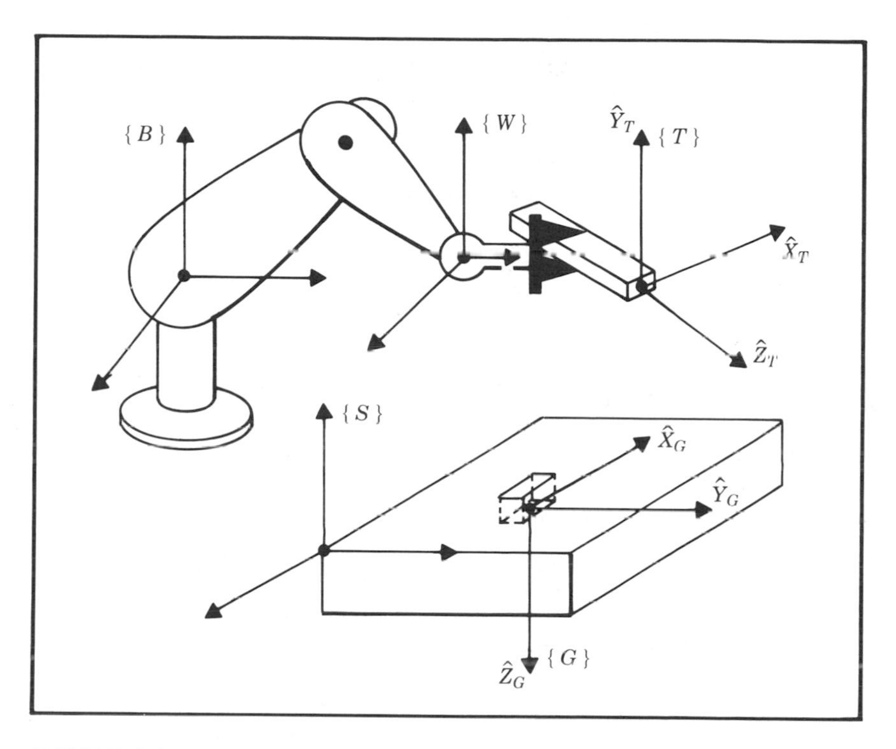

FIGURE 3.29 Determination of the tool frame of Exercise 3.8

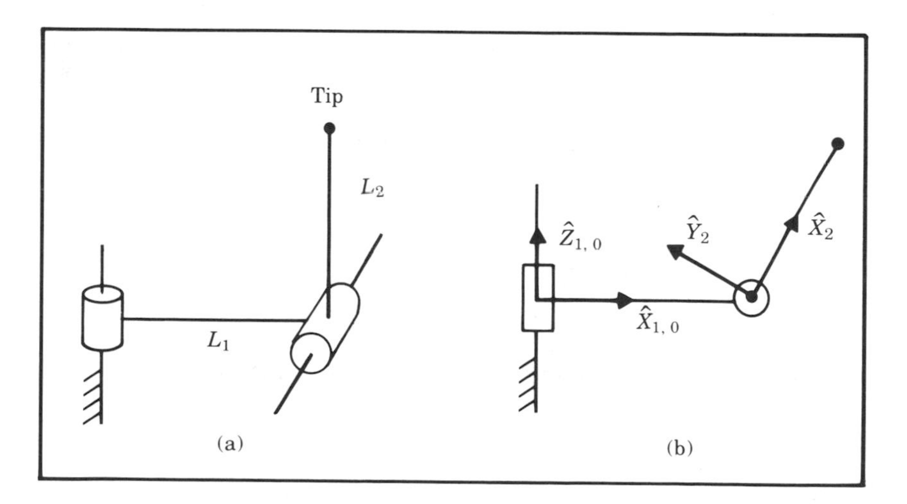

FIGURE 3.30 Two-link arm with frame assignments (Exercise 3.9).

The link frame assignments used are indicated in Fig. 3.30b. Note that frame {0} is coincident with frame {1} when $\theta_1 = 0$. The length of the second link is l_2. Find an expression for the vector ${}^0P_{tip}$ which locates the tip of the arm relative to the {0} frame.

3.10 [39] Derive kinematic equations for the Yasukawa Motoman robot (see Section 3.7) which compute the position and orientation of the wrist frame directly from actuator values, rather than first computing the joint angles. A solution is possible which requires only 33 multiplications, 2 square roots, and 6 sine or cosine evaluations.

Programming Exercise (Part 3)

1. Write a subroutine to compute the kinematics of the planar arm in Example 3.3. That is, a routine with the joint angles' values as input, and a frame (the wrist frame relative to the base frame) as output. Use the procedure heading

```
Procedure KIN(VAR theta: vec3; VAR wrelb: frame);
```

where "wrelb" is the wrist frame relative to the base frame, ${}^B_W T$. The type "frame" consists of a 2×2 rotation matrix and a 2×1 position vector. If desired, you may represent the frame with a 3×3 homogeneous transform in which the third row is $[\ 0\ 0\ 1\ ]$. (The manipulator data are $l_1 = l_2 = 0.5$ meters.)

2. Write a routine which calculates where the tool is relative to the station frame. The input to the routine is a vector of joint angles:

```
Procedure WHERE(VAR theta: vec3; VAR trels: frame);
```

Obviously, WHERE must make use of descriptions of the tool frame and the robot base frame in order to compute the location of the tool relative to the station. The value of ${}^W_T T$ and ${}^B_S T$ should be stored in global memory (or, as a second choice, you may pass them as arguments in WHERE).

3. A tool frame and a station frame are defined by the user for a certain task as below:

$$ {}^W_T T = [x \quad y \quad \theta] = [0.1 \quad 0.2 \quad 30.0], $$

$$ {}^B_S T = [x \quad y \quad \theta] = [-0.1 \quad 0.3 \quad 0.0]. $$

Calculate the position and orientation of the tool relative to the station frame for the following three configurations (in units of degrees) of the arm:

$$ [\theta_1 \quad \theta_2 \quad \theta_3] = [0.0 \quad 90.0 \quad -90.0], $$

$$ [\theta_1 \quad \theta_2 \quad \theta_3] = [-23.6 \quad -30.3 \quad 48.0], $$

$$ [\theta_1 \quad \theta_2 \quad \theta_3] = [130.0 \quad 40.0 \quad 12.0]. $$

4

INVERSE MANIPULATOR KINEMATICS

4.1 Introduction

In the last chapter we considered the problem of computing the position and orientation of the tool relative to the user's work station given the joint angles of the manipulator. In this chapter we investigate the more difficult problem: Given the desired position and orientation of the tool relative to the station, how do we compute the set of joint angles which will achieve this desired result? Whereas in Chapter 3 we studied the **direct kinematics** of manipulators, here we study the **inverse kinematics** of manipulators.

Solving the problem of finding the required joint angles to place the tool frame, $\{T\}$, relative to the station frame, $\{S\}$, is split into two parts. First, frame transformations are performed to find the wrist frame,

$\{W\}$, relative to the base frame, $\{B\}$, and then the inverse kinematics are used to solve for the joint angles.

4.2 Solvability

The problem of solving the kinematic equations of a manipulator is a nonlinear one. Given the numerical value of ${}^0_N T$ we attempt to find values of $\theta_1, \theta_2, \ldots, \theta_n$. Consider the equations given in (3.14). In the case of the PUMA 560 manipulator, the precise statement of our current problem is: Given ${}^0_6 T$ as sixteen numeric values (four of which are trivial), solve (3.14) for the six joint angles θ_1 through θ_6.

For the case of an arm with six degrees of freedom (like the one corresponding to the equations in (3.14)) we have twelve equations and six unknowns. However, among the nine equations arising from the rotation matrix portion of ${}^0_6 T$, only three equations are independent. These added with the three equations from the position vector portion of ${}^0_6 T$ give six equations with six unknowns. These equations are nonlinear, transcendental equations which can be quite difficult to solve. The equations of (3.14) are those of a robot which had very simple link parameters—many of the α_i were 0 or ± 90 degrees. Many link offsets and lengths were zero. It is easy to imagine that for the case of a general mechanism with six degrees of freedom (with all link parameters nonzero) the kinematic equations would be much more complex than those of (3.14). As with any nonlinear set of equations, we must concern ourselves with the existence of solutions, multiple solutions, and the method of solution.

Existence of solutions

The question of whether solutions exist or not raises the question of manipulator **workspace**. Roughly speaking, workspace is that volume of space which the end-effector of the manipulator can reach. For a solution to exist, the specified goal point must lie within the workspace. Sometimes it is useful to consider two definitions of workspace: **dextrous workspace** is that volume of space which the robot end-effector can reach with all orientations. That is, at each point in the dextrous workspace, the end-effector can be arbitrarily oriented. The **reachable workspace** is that volume of space which the robot can reach in at least one orientation. Clearly, the dextrous workspace is a subset of the reachable workspace.

Consider the workspace of the 2-link manipulator in Fig. 4.1. If $l_1 = l_2$ then the reachable workspace consists of a disc of radius $2l_1$.

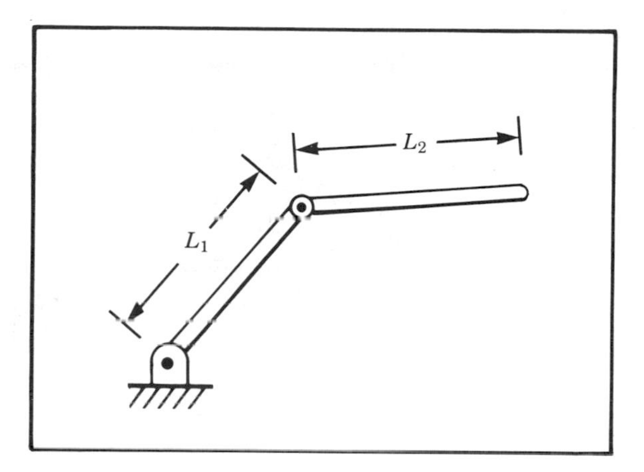

FIGURE 4.1 Two-link manipulator with link lengths l_1 and l_2.

The dextrous workspace consists of only a single point, the origin. If $l_1 \neq l_2$ then there is no dextrous workspace, and the reachable workspace becomes a ring of outer radius $l_1 + l_2$ and inner radius of $|l_1 - l_2|$. Inside the reachable workspace there are two possible orientations of the end-effector. On the boundaries of the workspace there is only one possible orientation.

These considerations of workspace for the 2-link manipulator have all assumed that the joints can rotate 360 degrees. This is rarely true for actual mechanisms. When joint limits are a subset of the full 360 degrees, then the workspace is obviously correspondingly reduced, either in extent, or in the number of possible orientations attainable. For example, if the arm in Fig. 4.1 has full 360 motion for θ_1, but only $0 <- \theta_2 <- 180$, then the reachable workspace has the same extent, but only one orientation is attainable at each point.

When a manipulator has less than six degrees of freedom, it cannot attain general goal positions and orientations in 3-space. Clearly, the planar manipulator in Fig. 4.1 cannot reach out of the plane, so any goal point with a nonzero Z-coordinate value can be quickly rejected as unreachable. In many realistic situations, manipulators with four or five degrees of freedom are employed which operate out of a plane, but clearly cannot reach general goals. Each such manipulator must be studied to understand its workspace. In general, the workspace of such a robot is a subset of a **subspace** which can be associated with any particular robot. Given a general goal frame specification, an interesting problem arises in connection with manipulators of less than six degrees of freedom: What is the nearest attainable goal frame?

Workspace also depends on the tool frame transformation, since it is usually the tool-tip which is discussed when we speak of reachable points in space. Generally, the tool transformation is performed independently of the manipulator kinematics and inverse kinematics, so we are often led to consider the workspace of the wrist frame, $\{W\}$. For a given end-effector, a tool frame, $\{T\}$, is defined; given a goal frame, $\{G\}$, the corresponding $\{W\}$ frame is calculated, and then we ask: Does this desired position and orientation of $\{W\}$ lie in the workspace? In this way the workspace which we must concern ourselves with (in a computational sense) is a different one than the one imagined by the user, who is concerned with the workspace of the end-effector (the $\{T\}$ frame).

If the desired position and orientation of the wrist frame is in the workspace, then at least one solution exists.

Multiple solutions

Another possible problem encountered in solving kinematic equations is that of multiple solutions. A planar arm with three revolute joints has a large dextrous workspace in the plane (given "good" link lengths and large joint ranges) since any position in the interior of its workspace could be reached with any orientation. Figure 4.2 shows a 3-link planar arm with it's end-effector at a certain position and orientation. The dashed lines indicate a second posible configuration in which the same end-effector position and orientation are achieved.

The fact that a manipulator has multiple solutions may cause problems because the system has to be able to choose one. The criteria upon which to base a decision vary, but a very reasonable choice would be the *closest* solution. For example, if the the manipulator is at point A as in Fig. 4.3, and we wish to move it to point B, a good choice would be the solution which minimizes the amount that each joint is required to move. Hence, in the absence of the obstacle, the upper, dashed configuration in Fig. 4.3 would be chosen. This suggests that one

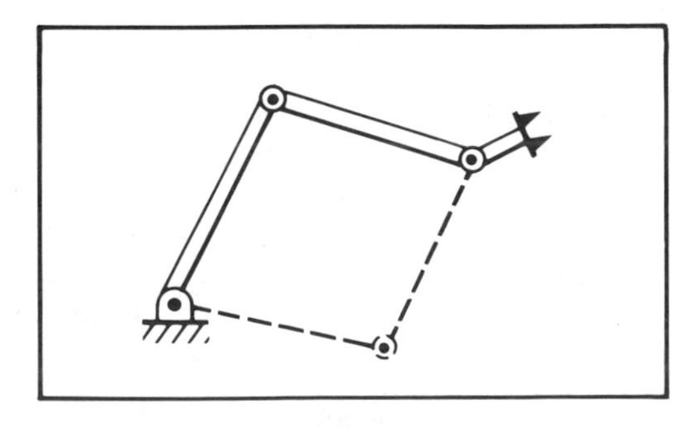

FIGURE 4.2 Three-link manipulator. Dashed lines indicate a second solution.

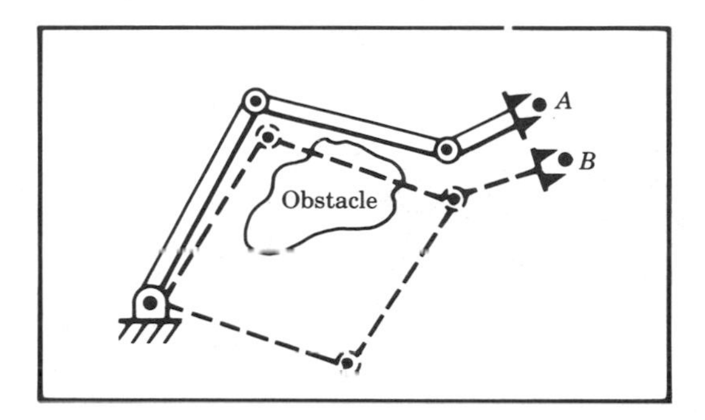

FIGURE 4.3 One of the two possible solutions to reach point *B* causes a collision.

input argument to our kinematic inverse procedure might be the present position of the manipulator. In this way, if there is a choice, our algorithm can choose the closest solution in joint-space. However, the notion of "close" might be defined in several ways. For example, typical robots may have three large links followed by three smaller, orienting links near the end-effector. In this case, weights might be applied in the calculation of which solution is "closer" so that the selection favors moving smaller joints rather than moving the large joints when a choice exists. The presence of obstacles may force the "farther" solution to be chosen in cases where the "closer" solution would cause a collision—in general, we need to be able to calculate all the possible solutions. Thus, in Fig. 4.3 the presence of the obstacle implies the lower dashed configuration is to be used to reach point B.

The number of solutions depends upon the number of joints in the manipulator, but is also a function of the link parameters (α_i, a_i, and d_i for a rotary joint manipulator) and the allowable range of motion of the joints. For example, the PUMA 560 can reach certain goals with eight different solutions. Figure 4.4 shows four solutions which all position the hand with the same position and orientation. For each solution pictured, there is another solution in which the last three joints "flip" to an alternate configuration according to the formulas:

$$\begin{aligned}\theta'_4 &= \theta_4 + 180^\circ \\ \theta'_5 &= -\theta_5 \\ \theta'_6 &= \theta_6 + 180^\circ\end{aligned} \tag{4.1}$$

So in total there can be eight solutions for a single goal. Due to limits on joint ranges, some of these eight may not be accesible.

In general, the more nonzero link parameters there are, the more ways there will be to reach a certain goal. For example, consider

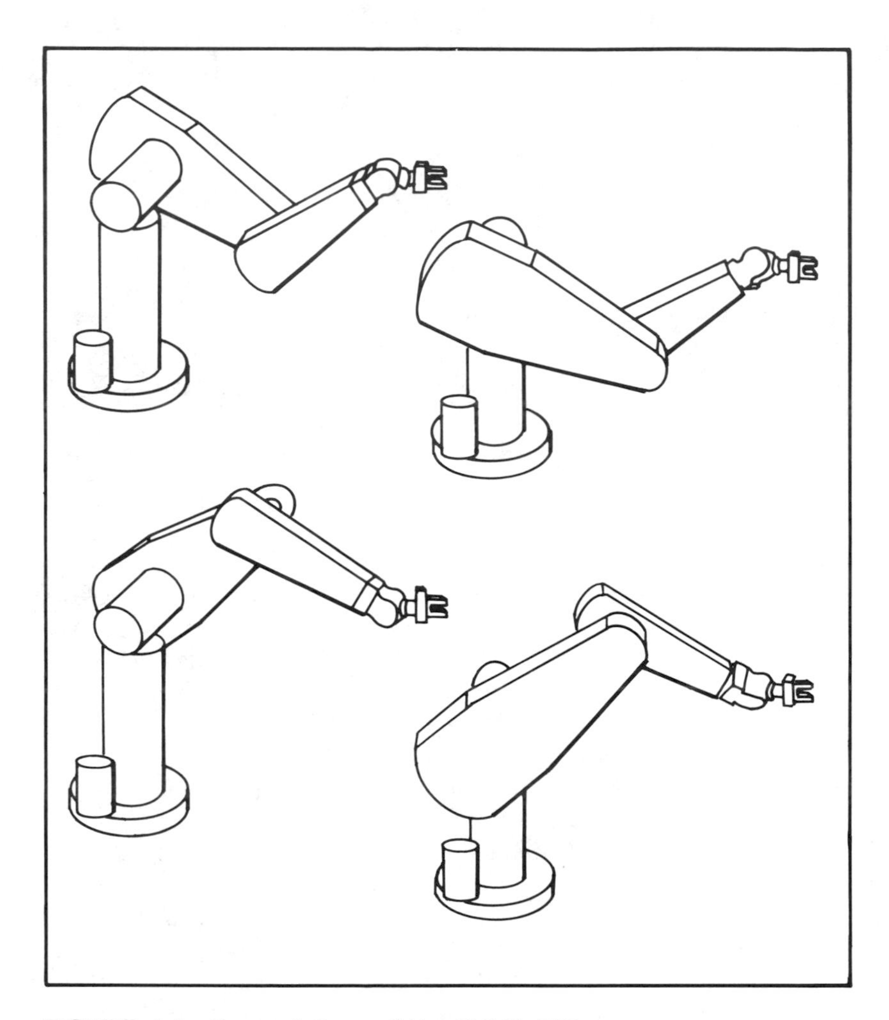

FIGURE 4.4 Four solutions of the PUMA 560.

a manipulator with six rotational joints. Figure 4.5 shows how the maximum number of solutions is related to how many of the link length parameters (the a_i) are zero. The more that are nonzero, the bigger the maximum number of solutions. For a completely general rotary-jointed manipulator with six degrees of freedom, there are up to 16 solutions possible [1], [6].

a_i	Number of solutions
$a_1 = a_3 = a_5 = 0$	$\leqslant 4$
$a_3 = a_5 = 0$	$\leqslant 8$
$a_3 = 0$	$\leqslant 16$
All $a_i = 0$	$\leqslant 16$

FIGURE 4.5 Number of solutions vs. nonzero a_i.

Method of solution

Unlike linear equations, there are no general algorithms which may be employed to solve a set of nonlinear equations. In considering methods of solution it will be wise to define what constitutes the "solution" of a given manipulator.

A manipulator will be considered **solvable** if the joint variables can be determined by an algorithm which allows one to determine *all* the sets of joint variables associated with a given position and orientation [2].

The main point of this definition is that we require, in the case of multiple solutions, that it be possible to calculate all solutions. Hence, we do not consider some numerical iterative procedures as solving the manipulator since some of these methods are not guaranteed to find all the solutions.

We will split all proposed manipulator solution strategies into two broad classes: **closed form solutions** and **numerical solutions**. Because of their iterative nature, numerical solutions generally are much slower than the corresponding closed form solution; in fact, so much so that for most uses, we are not interested in the numerical approach to solution of kinematics. Numerical solution to nonlinear equations is a whole field of study in itself (see [6]), and is beyond the scope of this text.

We will restrict our attention to closed form solution methods. In this context "closed form" means a solution method based on analytic expressions or on the solution of a polynomial of degree 4 or less, such that noniterative calculations suffice to arrive at a solution. Within the class of closed form solutions we distinguish two methods of obtaining the solution: **algebraic** and **geometric**. These distinctions are somewhat hazy, since any geometric methods brought to bear are applied by means of algebraic expressions, so both methods are similar. The methods differ perhaps in approach only.

A major recent result in kinematics is that, according to our definition of solvability, *all systems with revolute and prismatic joints having a total of six degrees of freedom in a single series chain are now solvable.* However, this general solution is a numerical one. Only in special cases may robots with six degrees of freedom be solved analytically. These robots for which an analytic or closed form solution exists are characterised by several intersecting joint axes, and/or many α_i equal to 0 or ± 90 degrees. Since numerical solutions are generally time consuming relative to evaluating analytic expressions it is considered very important to design a manipulator such that a closed form solution exists. Manipulator designers discovered this very soon and now virtually all industrial manipulators are designed sufficiently simply so that a closed form solution may be developed.

A sufficient condition that a manipulator with six revolute joints will have a closed form solution is that three neighboring joint axes intersect at a point. In Section 4.5 we will present this result [4]. Almost every manipulator with six degrees of freedom built today has three axes intersecting. For example, axes 4, 5, and 6 of the PUMA 560 intersect.

4.3 The notion of manipulator subspace when $n < 6$

The set of reachable goal frames for a given manipulator constitutes its reachable workspace. For a manipulator with n degrees of freedom where $n < 6$, this reachable workspace can be thought of as a portion of an n degree of freedom **subspace**. In the same manner in which the workspace of a six degree of freedom manipulator is a subset of space, the workspace of a simpler manipulator is a subset of its subspace. For example, the subspace of the two-link robot of Fig. 4.1 is a plane, but the workspace is a subset of this plane, namely a circle of radius $l_1 + l_2$ for the case that $l_1 = l_2$.

One way to specify the subspace of an n degree of freedom manipulator is to give an expression for its wrist or tool frame as a function of n variables which locate it. If we consider these n variables to be free, then as they take on all possible values, the subspace is generated.

EXAMPLE 4.1

Give a description of the subspace of ${}^B_W T$ for the three-link manipulator from Chapter 3, Fig. 3.5.

The subspace of ${}^B_W T$ is given by:

$$
{}^B_W T = \begin{bmatrix} c_\phi & -s_\phi & 0.0 & x \\ s_\phi & c_\phi & 0.0 & y \\ 0.0 & 0.0 & 1.0 & 0.0 \\ 0 & 0 & 0 & 1 \end{bmatrix}, \qquad \textbf{(4.2)}
$$

where x and y give the position of the wrist, and ϕ describes the orientation of the terminal link. As x, y, and ϕ are allowed to take on arbitrary values, the subspace is generated. Any wrist frame which does not have the structure of (4.2) lies outside the subspace (and therefore lies outside the workspace) of this manipulator. Link lengths and joint limits restrict the workspace of the manipulator to be a subset of this subspace. ■

EXAMPLE 4.2

Give a description of the subspace of ${}^0_2T$ for the polar manipulator with two degrees of freedom shown in Fig. 4.6. We have:

$$ {}^0P_{2ORG} = \begin{bmatrix} x \\ y \\ 0 \end{bmatrix}, \tag{4.3} $$

where x and y can take any values. The orientation is restricted because the ${}^0\hat{Z}_2$ axis must point in a direction which depends on x and y. The ${}^0\hat{Y}_2$ axis always points down, and the ${}^0\hat{X}_2$ axis can be computed as the cross product ${}^0\hat{Y}_2 \times {}^0\hat{Z}_2$. In terms of x and y we have

$$ {}^0\hat{Z}_2 = \begin{bmatrix} \frac{x}{\sqrt{x^2+y^2}} \\ \frac{y}{\sqrt{x^2+y^2}} \\ 0 \end{bmatrix}. \tag{4.4} $$

The subspace can therefore be given as

$$ {}^0_2T = \begin{bmatrix} \frac{y}{\sqrt{x^2+y^2}} & 0 & \frac{x}{\sqrt{x^2+y^2}} & x \\ \frac{-x}{\sqrt{x^2+y^2}} & 0 & \frac{y}{\sqrt{x^2+y^2}} & y \\ 0 & -1 & 0 & 0 \\ 0 & 0 & 0 & 1 \end{bmatrix}. \quad ■ \tag{4.5} $$

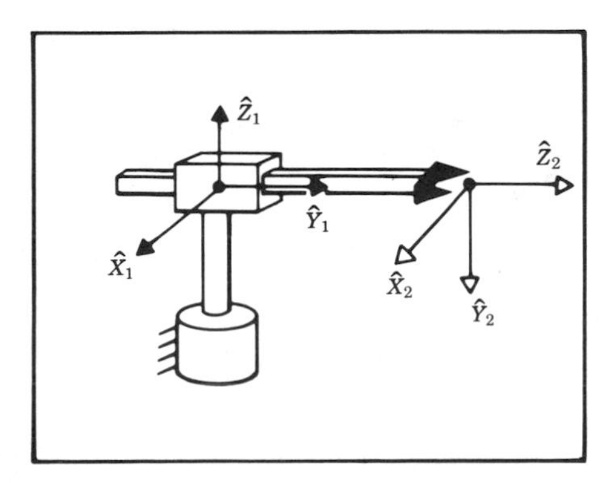

FIGURE 4.6 A polar 2-link manipulator.

Usually in defining a goal for a manipulator with n degrees of freedom we use n parameters to specify the goal. If, on the other hand, we give a specification of a full six degrees of freedom, we will not in general be able to reach the goal with an $n < 6$ manipulator. In this case, we may be interested in instead reaching a goal which lies in the manipulator's subspace and is as "near" as possible to the original desired goal.

Hence, when specifying *general* goals for a manipulator with less than six degrees of freedom, a solution strategy is:

1. Given a general goal frame, ${}^S_G T$, compute a modified goal frame, ${}^S_G T'$, such that ${}^S_G T'$ lies in the manipulator's subspace and is as "near" to ${}^S_G T$ as possible. A definition of "near" must be chosen.
2. Compute the inverse kinematics to find joint angles using ${}^S_G T'$ as the desired goal. Note that a solution still may not be possible if the goal point is not in the manipulator's workspace.

It generally makes sense to position the tool frame origin to the desired location, and then choose an attainable orientation which is near the desired orientation. As we saw in Examples 4.1 and 4.2, computation of the subspace is dependent on manipulator geometry. Each manipulator must be individually considered to arrive at a method of making this computation.

In Section 4.6 we will give an example of *projecting* a general goal into the subspace of a manipulator with five degrees of freedom in order to compute joint angles which will result in the manipulator reaching the nearest attainable frame to the desired.

4.4 Algebraic vs. geometric

As an introduction to solving kinematic equations, we will consider two different approaches to the solution of a simple planar 3-link manipulator.

Algebraic solution

Consider the 3-link planar manipulator introduced in Chapter 3. It is shown with its link parameters in Fig. 4.7.

Following the method of Chapter 3, we may use the link parameters

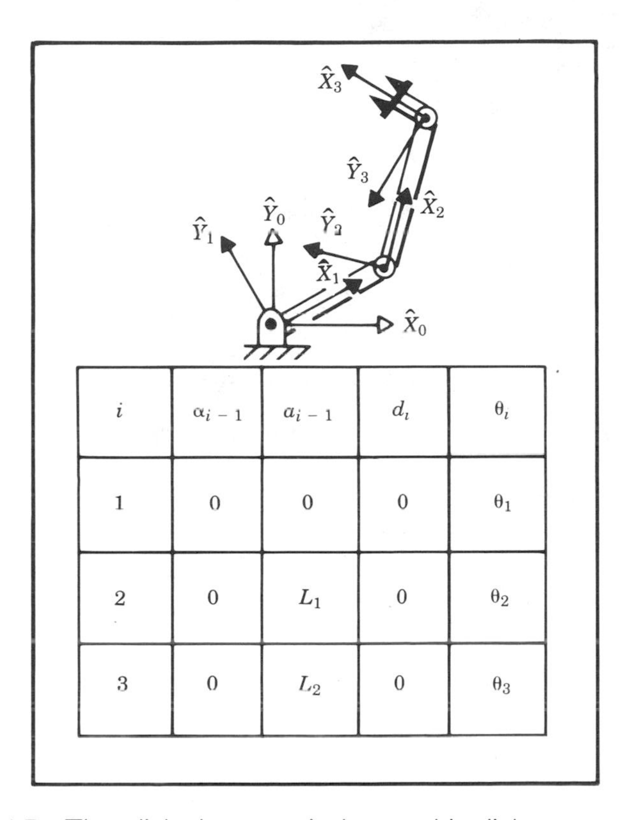

i	α_{i-1}	a_{i-1}	d_i	θ_i
1	0	0	0	θ_1
2	0	L_1	0	θ_2
3	0	L_2	0	θ_3

FIGURE 4.7 Three-link planar manipulator and its link parameters.

easily to find the kinematic equations of this arm:

$$ {}^{B}_{W}T = {}^{0}_{3}T = \begin{bmatrix} c_{123} & -s_{123} & 0.0 & l_1c_1 + l_2c_{12} \\ s_{123} & c_{123} & 0.0 & l_1s_1 + l_2s_{12} \\ 0.0 & 0.0 & 1.0 & 0.0 \\ 0 & 0 & 0 & 1 \end{bmatrix}. \qquad \textbf{(4.6)} $$

To focus our discussion on inverse kinematics, we will assume that the necessary transformations have been performed so that the goal point is a specification of the wrist frame relative to the base frame, that is, ${}^{B}_{W}T$. Because we are working with a planar manipulator, specification of these goal points can be most easily accomplished by specifying three numbers: x, y, and ϕ, where ϕ is the orientation of link 3 in the plane (relative to the $+\hat{X}$ axis). Hence, rather than giving a general ${}^{B}_{W}T$ as a goal specification, we will assume a transformation with the structure

$$ {}^{B}_{W}T = \begin{bmatrix} c_{\phi} & -s_{\phi} & 0.0 & x \\ s_{\phi} & c_{\phi} & 0.0 & y \\ 0.0 & 0.0 & 1.0 & 0.0 \\ 0 & 0 & 0 & 1 \end{bmatrix}. \qquad \textbf{(4.7)} $$

All attainable goals must lie in the subspace implied by the structure of equation (4.7). By equating (4.6) and (4.7) we arive at a set of four nonlinear equations which must be solved for θ_1, θ_2, and θ_3:

$$c_\phi = c_{123}, \tag{4.8}$$

$$s_\phi = s_{123}, \tag{4.9}$$

$$x = l_1 c_1 + l_2 c_{12}, \tag{4.10}$$

$$y = l_1 s_1 + l_2 s_{12}. \tag{4.11}$$

We now begin our algebraic solution of equations (4.8) through (4.11). If we square both (4.10) and (4.11) and add them, we obtain

$$x^2 + y^2 = l_1^2 + l_2^2 + 2 l_1 l_2 c_2, \tag{4.12}$$

where we have made use of

$$\begin{aligned} c_{12} &= c_1 c_2 - s_1 s_2, \\ s_{12} &= c_1 s_2 + s_1 c_2. \end{aligned} \tag{4.13}$$

Solving (4.12) for c_2 we obtain

$$c_2 = \frac{x^2 + y^2 - l_1^2 - l_2^2}{2 l_1 l_2}. \tag{4.14}$$

In order for a solution to exist, the right hand side of (4.14) must have a value between -1 and 1. In the solution algorithm, this constraint would be checked at this time to determine if a solution exists. Physically, if this constraint is not satisfied, then the goal point is too far away for the manipulator to reach.

Assuming the goal is in the workspace, we write an expression for s_2 as

$$s_2 = \pm\sqrt{1 - c_2^2}. \tag{4.15}$$

Finally, we compute θ_2 using the two-argument arctangent routine*

$$\theta_2 = \text{Atan2}(s_2, c_2). \tag{4.16}$$

The choice of signs in (4.15) corresponds to the multiple solution in which we can choose the "elbow-up" or the "elbow-down" solution. In determining θ_2 we have used one of the recurring methods for solving the type of kinematic relationships that often arise, namely to determine both the sine and cosine of the desired joint angle, and then apply the

---

* See Section 2.8.

two argument arctangent. This insures that we have found all solutions, and that the solved angle is in the proper quadrant.

Having found θ_2 we may solve (4.10) and (4.11) for θ_1. We write (4.10) and (4.11) in the form

$$x = k_1 c_1 - k_2 s_1, \tag{4.17}$$

$$y = k_1 s_1 + k_2 c_1, \tag{4.18}$$

where

$$\begin{aligned} k_1 &= l_1 + l_2 c_2, \\ k_2 &= l_2 s_2. \end{aligned} \tag{4.19}$$

In order to solve an equation of this form, we perform a change of variables. Actually, we are changing the way in which we write the constants k_1 and k_2.

If

$$r = +\sqrt{k_1^2 + k_2^2}$$

and

$$\gamma = \text{Atan2}(k_2, k_1), \tag{4.20}$$

then

$$\begin{aligned} k_1 &= r \cos\gamma, \\ k_2 &= r \sin\gamma. \end{aligned} \tag{4.21}$$

Equations (4.17) and (4.18) can now be written

$$\frac{x}{r} = \cos\gamma\cos\theta_1 - \sin\gamma\sin\theta_1, \tag{4.22}$$

$$\frac{y}{r} = \cos\gamma\sin\theta_1 + \sin\gamma\cos\theta_1, \tag{4.23}$$

or

$$\cos(\gamma + \theta_1) = \frac{x}{r}, \tag{4.24}$$

$$\sin(\gamma + \theta_1) = \frac{y}{r}. \tag{4.25}$$

Using the two-argument arctangent we get

$$\gamma + \theta_1 = \text{Atan2}\left(\frac{y}{r}, \frac{x}{r}\right) = \text{Atan2}\,(y, x)\,, \tag{4.26}$$

and so

$$\theta_1 = \text{Atan2}(y, x) - \text{Atan2}(k_2, k_1). \tag{4.27}$$

Note that when a choice of sign is made in the solution of θ_2 above, it will cause a sign change in k_2, thus affecting θ_1. The substitutions used, (4.20) and (4.21), constitute a method of solution of a frequently appearing form in kinematics, namely that of (4.10) or (4.11). Note also that if $x = y = 0$ then (4.27) becomes undefined—in this case θ_1 is arbitrary.

Finally, from (4.8) and (4.9) we can solve for the sum of θ_1 through θ_3:

$$\theta_1 + \theta_2 + \theta_3 = \text{Atan2}(s_\phi, c_\phi) = \phi, \tag{4.28}$$

from which we can solve for θ_3 since we know the first two angles. It is typical with manipulators that have two or more links moving in a plane that in the course of solution, expressions for sums of joint angles arise.

Geometric solution

In a geometric approach to finding a manipulator's solution, we try to decompose the spatial geometry of the arm into several plane geometry problems. For many manipulators (particularly when the $\alpha_i = 0$ or ± 90) this can be done quite easily. Joint angles can then be solved for using the tools of plane geometry. For the arm with three degrees of freedom shown in Fig. 4.7, since the arm is planar, we can apply plane geometry directly to find a solution.

Figure 4.8 shows the triangle formed by l_1, l_2, and the line joining the origin of frame {0} with the origin of frame {3}. The dashed lines represent the other possible configuration of the triangle which would lead to the same position of the frame {3}. Considering the solid triangle,

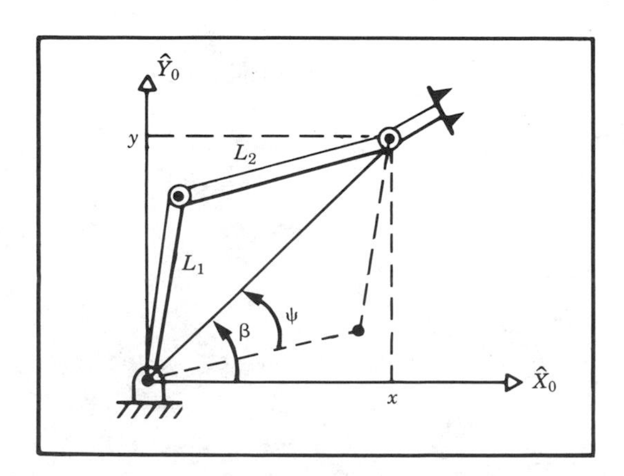

FIGURE 4.8 Plane geometry associated with a 3-link planar robot.

we can apply the "law of cosines" to solve for θ_2:

$$x^2 + y^2 = l_1^2 + l_2^2 - 2l_1 l_2 \cos(180 + \theta_2). \quad \textbf{(4.29)}$$

Since $\cos(180 + \theta_2) = -\cos(\theta_2)$, we have

$$c_2 = \frac{x^2 + y^2 - l_1^2 - l_2^2}{2l_1 l_2}. \quad \textbf{(4.30)}$$

In order for this triangle to exist, the distance to the goal point $\sqrt{x^2 + y^2}$ must be less than or equal to the sum of the link lengths, $l_1 + l_2$. This condition would be checked at this point in a computational algorithm to verify existence of solutions. This condition is not satisfied when the goal point is out of reach of the manipulator. Assuming a solution exists, this equation is solved for that value of θ_2 which lies between 0 and -180 degrees since only for these values does the triangle in Fig. 4.8 exist. The other possible solution (the one indicated by the dashed line triangle) may be found by symmetry to be $\theta_2' = -\theta_2$.

To solve for θ_1 we find expressions for angles ψ and β as indicated in Fig. 4.8. First, β may be in any quadrant depending on the signs of x and y. So we must use a two argument arctangent:

$$\beta = \text{Atan2}(y, x). \quad \textbf{(4.31)}$$

We again apply the law of cosines to find ψ:

$$\cos\psi = \frac{x^2 + y^2 + l_1^2 - l_2^2}{2l_1\sqrt{x^2 + y^2}}, \quad \textbf{(4.32)}$$

where the arccosine must be solved so that $0 <= \psi <= 180°$ in order that the geometry which lead to (4.32) is preserved. These considerations are typical when using a geometric approach—we must apply the formulas we derive only over a range of variables such that the geometry is preserved. Then we have

$$\theta_1 = \beta \pm \psi, \quad \textbf{(4.33)}$$

where the plus sign is used if $\theta_2 < 0$ and the minus sign if $\theta_2 > 0$.

We know that angles in a plane add, so the sum of the 3 joint angles must be the orientation of the last link:

$$\theta_1 + \theta_2 + \theta_3 = \phi, \quad \textbf{(4.34)}$$

which is solved for θ_3 to complete our solution.

4.5 Pieper's solution when three axes intersect

As mentioned earlier, although a completely general robot with six degrees of freedom does not have a closed form solution, certain important special cases can be solved. Pieper studied manipulators with six degrees of freedom in which three consecutive axes intersect at a point [4]. In this section, we outline the method he developed for the case of all six joints revolute, with the last three axes intersecting. His method applies to other configurations which include prismatic joints as well, and the interested reader should see [4]. Pieper's work applies to the majority of commercially available industrial robots.

When the last three axes intersect, the origins of link frames {4}, {5}, and {6} are all located at this point of intersection. This point is given in base coordinates as

$$ {}^{0}P_{4ORG} = {}^{0}_{1}T\ {}^{1}_{2}T\ {}^{2}_{3}T\ {}^{3}P_{4ORG} \tag{4.35} $$

or, using the fourth column of (3.6) for $i = 4$,

$$ {}^{0}P_{4ORG} = {}^{0}_{1}T\ {}^{1}_{2}T\ {}^{2}_{3}T \begin{bmatrix} a_3 \\ -d_4 s\alpha_3 \\ d_4 c\alpha_3 \\ 1 \end{bmatrix}; \tag{4.36} $$

or

$$ {}^{0}P_{4ORG} = {}^{0}_{1}T\ {}^{1}_{2}T \begin{bmatrix} f_1(\theta_3) \\ f_2(\theta_3) \\ f_3(\theta_3) \\ 1 \end{bmatrix}, \tag{4.37} $$

where

$$ \begin{bmatrix} f_1 \\ f_2 \\ f_3 \\ 1 \end{bmatrix} = {}^{2}_{3}T \begin{bmatrix} a_3 \\ -d_4 s\alpha_3 \\ d_4 c\alpha_3 \\ 1 \end{bmatrix}. \tag{4.38} $$

Using (3.6) for ${}^{2}_{3}T$ in (4.38) yields the following expressions for f_i:

$$ \begin{aligned} f_1 &= a_3 c_3 + d_4 s\alpha_3 s_3 + a_2, \\ f_2 &= a_3 c\alpha_2 s_3 - d_4 s\alpha_3 c\alpha_2 c_3 - d_4 s\alpha_2 c\alpha_3 - d_3 s\alpha_2, \\ f_3 &= a_3 s\alpha_2 s_3 - d_4 s\alpha_3 s\alpha_2 c_3 + d_4 c\alpha_2 c\alpha_3 + d_3 c\alpha_2. \end{aligned} \tag{4.39} $$

Using (3.6) for ${}^{0}_{1}T$ and ${}^{1}_{2}T$ in (4.37) we obtain

$$ {}^{0}P_{4ORG} = \begin{bmatrix} c_1 g_1 - s_1 g_2 \\ s_1 g_1 + c_1 g_2 \\ g_3 \\ 1 \end{bmatrix} \tag{4.40} $$

where

$$
\begin{aligned}
g_1 &= c_2 f_1 - s_2 f_2 + a_1, \\
g_2 &= s_2 c\alpha_1 f_1 + c_2 c\alpha_1 f_2 - s\alpha_1 f_3 - d_2 s\alpha_1, \\
g_3 &= s_2 s\alpha_1 f_1 + c_2 s\alpha_1 f_2 + c\alpha_1 f_3 + d_2 c\alpha_1.
\end{aligned} \qquad (4.41)
$$

We now write an expression for the magnitude squared of ${}^0P_{4ORG}$, which is seen from (4.40) to be

$$ r = g_1^2 + g_2^2 + g_3^2 \qquad (4.42) $$

or, using (4.41) for the g_i, we have

$$ r = f_1^2 + f_2^2 + f_3^2 + a_1^2 + d_2^2 + 2d_2 f_3 + 2a_1(c_2 f_1 - s_2 f_2). \qquad (4.43) $$

We now write this equation, along with the Z component equation from (4.40), as a system of two equations in the form

$$
\begin{aligned}
r &= (k_1 c_2 + k_2 s_2) 2a_1 + k_3, \\
z &= (k_1 s_2 - k_2 c_2) s\alpha_1 + k_4,
\end{aligned} \qquad (4.44)
$$

where

$$
\begin{aligned}
k_1 &= f_1, \\
k_2 &= -f_2, \\
k_3 &= f_1^2 + f_2^2 + f_3^2 + a_1^2 + d_2^2 + 2d_2 f_3, \\
k_4 &= f_3 c\alpha_1 + d_2 c\alpha_1.
\end{aligned} \qquad (4.45)
$$

Equation (4.44) is useful because dependence on θ_1 has been eliminated, and dependence on θ_2 takes a simple form.

At this point we introduce a very important geometric substitution used often in solving kinematic equations. Transcendental equations are difficult to solve because, although there may be just one variable, say θ, it generally appears as $\sin\theta$ and $\cos\theta$. Making the following substitutions, however, yields an expression in terms of a single variable, $\tan\frac{\theta}{2}$:

$$
\begin{aligned}
\cos\theta &= \frac{1 - \tan^2\frac{\theta}{2}}{1 + \tan^2\frac{\theta}{2}}, \\
\sin\theta &= \frac{2\tan\frac{\theta}{2}}{1 + \tan^2\frac{\theta}{2}}.
\end{aligned} \qquad (4.46)
$$

Now let us consider the solution of (4.44) for θ_3. We distinguish three cases:

1. If $a_1 = 0$ then we have $r = k_3$ where r is known. The right hand side (k_3) is a function of θ_3 only. After making the substitution (4.46), a quadratic equation in $\tan\frac{\theta_3}{2}$ may be solved for θ_3.

2. If $s\alpha_1 = 0$ then we have $z = k_4$ where z is known. Again, after substituting (4.46) a quadratic equation arises which may be solved for θ_3.

3. Otherwise, eliminate s_2 and c_2 from (4.44) to obtain

$$\frac{(r-k_3)^2}{4a_1^2} + \frac{(z-k_4)^2}{s^2\alpha_1} = k_1^2 + k_2^2. \qquad \textbf{(4.47)}$$

This equation, after the (4.46) substitution for θ_3, results in an equation of degree four, which may be solved for θ_3.*

Having solved for θ_3, we may solve (4.44) for θ_2, and (4.40) for θ_1.

To complete our solution, we need to solve for θ_4, θ_5, and θ_6. Since these axes intersect, these joint angles affect the orientation of only the last link. We can compute them based only upon the rotation portion of the specified goal, ${}^0_6R$. Having obtained θ_1, θ_2, and θ_3, we can compute ${}^0_3R$, and since the problem was specified given ${}^0_6R$, we can compute

$${}^3_6R = {}^0_3R^{-1}\,{}^0_6R. \qquad \textbf{(4.48)}$$

For many manipulators, these last three angles can be solved for by using exactly the Z-Y-Z Euler angle solution given in Chapter 2 applied to ${}^3_6R$. For any manipulator (with intersecting axes 4, 5, and 6), the last three joint angles can be solved for as a set of appropriately defined Euler angles. Since there are always two solutions for these last three joints, the total number of solutions for the manipulator will be twice the number found for the first three joints.

4.6 Examples of inverse manipulator kinematics

In this section we work out the inverse kinematics of two industrial robots. One manipulator solution is done purely algebraically, while the second solution is partially algebraic and partially geometric. While the following solutions do not constitute a cook-book method of solving manipulator kinematics, they do show many of the common manipulations which are likely to appear in most kinematic solutions.

---

* It is helpful to note that $f_1^2 + f_2^2 + f_3^2 = a_3^2 + d_4^2 + d_3^2 + a_2^2 + 2d_4d_3c\alpha_3 + 2a_2a_3c_3 + 2a_2d_4s\alpha_3s_3$

The Unimation PUMA 560

As an example of the algebraic solution technique applied to a manipulator with six degrees of freedom, we will solve the kinematic equations of the PUMA 560 which were developed in Chapter 3. This solution is in the style of [5].

We wish to solve

$$
{}^{0}_{6}T = \begin{bmatrix} r_{11} & r_{12} & r_{13} & p_x \\ r_{21} & r_{22} & r_{23} & p_y \\ r_{31} & r_{32} & r_{33} & p_z \\ 0 & 0 & 0 & 1 \end{bmatrix} \tag{4.49}
$$

$$
= {}^{0}_{1}T(\theta_1)\ {}^{1}_{2}T(\theta_2)\ {}^{2}_{3}T(\theta_3)\ {}^{3}_{4}T(\theta_4)\ {}^{4}_{5}T(\theta_5)\ {}^{5}_{6}T(\theta_6)
$$

for θ_i when ${}^{0}_{6}T$ is given as numeric values.

A restatement of (4.49) which puts the dependence on θ_1 on the left hand side of the equation is

$$
\left[{}^{0}_{1}T(\theta_1)\right]^{-1}\ {}^{0}_{6}T = {}^{1}_{2}T(\theta_2)\, {}^{2}_{3}T(\theta_3)\, {}^{3}_{4}T(\theta_4)\, {}^{4}_{5}T(\theta_5)\, {}^{5}_{6}T(\theta_6). \tag{4.50}
$$

Inverting ${}^{0}_{1}T$ we write (4.50) as

$$
\begin{bmatrix} c_1 & s_1 & 0 & 0 \\ -s_1 & c_1 & 0 & 0 \\ 0 & 0 & 1 & 0 \\ 0 & 0 & 0 & 1 \end{bmatrix} \begin{bmatrix} r_{11} & r_{12} & r_{13} & p_x \\ r_{21} & r_{22} & r_{23} & p_y \\ r_{31} & r_{32} & r_{33} & p_z \\ 0 & 0 & 0 & 1 \end{bmatrix} = {}^{1}_{6}T, \tag{4.51}
$$

Where ${}^{1}_{6}T$ is given by equation (3.13) developed in Chapter 3. This simple technique of multiplying each side of a transform equation by an inverse is often used to advantage in separating out variables in search of a solvable equation.

Equating the $(2,4)$ elements from both sides of (4.51), we have

$$
-s_1 p_x + c_1 p_y = d_3. \tag{4.52}
$$

To solve an equation of this form, we make the trigonometric substitutions

$$
\begin{aligned} p_x &= \rho \cos\phi, \\ p_y &= \rho \sin\phi, \end{aligned} \tag{4.53}
$$

where

$$
\begin{aligned} \rho &= \sqrt{p_x^2 + p_y^2}, \\ \phi &= \operatorname{Atan2}(p_y, p_x). \end{aligned} \tag{4.54}
$$

Substituting (4.53) into (4.52), we obtain

$$c_1 s_\phi - s_1 c_\phi = \frac{d_3}{\rho}. \tag{4.55}$$

Using the difference of angles formula:

$$\sin(\phi - \theta_1) = \frac{d_3}{\rho}. \tag{4.56}$$

Hence

$$\cos(\phi - \theta_1) = \pm\sqrt{1 - \frac{d_3^2}{\rho^2}}, \tag{4.57}$$

and so

$$\phi - \theta_1 = \text{Atan2}\left(\frac{d_3}{\rho}, \pm\sqrt{1 - \frac{d_3^2}{\rho^2}}\right). \tag{4.58}$$

Finally, the solution for θ_1 may be written:

$$\theta_1 = \text{Atan2}\,(p_y, p_x) - \text{Atan2}\left(d_3, \pm\sqrt{p_x^2 + p_y^2 - d_3^2}\right). \tag{4.59}$$

Note that we have found two possible solutions for θ_1 corresponding to the plus-or-minus sign in (4.59). Now that θ_1 is known, the left hand side of (4.51) is known. If we equate the $(1, 4)$ elements from both sides of (4.51) and also the $(3, 4)$ elements, we obtain

$$\begin{aligned} c_1 p_x + s_1 p_y &= a_3 c_{23} - d_4 s_{23} + a_2 c_2, \\ -p_z &= a_3 s_{23} + d_4 c_{23} + a_2 s_2. \end{aligned} \tag{4.60}$$

If we square equations (4.60) and (4.52) and add the resulting equations, we obtain

$$a_3 c_3 - d_4 s_3 = K, \tag{4.61}$$

where

$$K = \frac{p_x^2 + p_y^2 + p_z^2 - a_2^2 - a_3^2 - d_3^2 - d_4^2}{2a_2}. \tag{4.62}$$

Note that dependence on θ_1 has been removed from (4.61). Equation (4.61) is of the same form as (4.52) and so may be solved by the same kind of trigonometric substitution to yield a solution for θ_3:

$$\theta_3 = \text{Atan2}\,(a_3, d_4) - \text{Atan2}\left(K, \pm\sqrt{a_3^2 + d_4^2 - K^2}\right). \tag{4.63}$$

The plus or minus sign in (4.63) leads to two different solutions for θ_3.

If we consider (4.49) again, we can now rewrite it so that all the left-hand side is a function of only knowns and θ_2:

$$\left[{}^0_3T(\theta_2)\right]^{-1}\ {}^0_6T = {}^3_4T(\theta_4)\ {}^4_5T(\theta_5)\ {}^5_6T(\theta_6), \tag{4.64}$$

or

$$\begin{bmatrix} c_1c_{23} & s_1c_{23} & -s_{23} & -a_2c_3 \\ -c_1s_{23} & -s_1s_{23} & -c_{23} & a_2s_3 \\ -s_1 & c_1 & 0 & -d_3 \\ 0 & 0 & 0 & 1 \end{bmatrix} \begin{bmatrix} r_{11} & r_{12} & r_{13} & p_x \\ r_{21} & r_{22} & r_{23} & p_y \\ r_{31} & r_{32} & r_{33} & p_z \\ 0 & 0 & 0 & 1 \end{bmatrix} = {}^3_6T, \tag{4.65}$$

where ${}^3_6T$ is given by equation (3.11) developed in Chapter 3. Equating the $(1,4)$ elements from both sides of (4.65), as well as the $(2,4)$ elements, we get

$$\begin{aligned} c_1c_{23}p_x + s_1c_{23}p_y - s_{23}p_z - a_2c_3 &= a_3, \\ -c_1s_{23}p_x - s_1s_{23}p_y - c_{23}p_z + a_2s_3 &= d_4, \end{aligned} \tag{4.66}$$

These equations may be solved simultaneously for s_{23} and c_{23}, resulting in

$$\begin{aligned} s_{23} &= \frac{(-a_3 - a_2c_3)p_z + (c_1p_x + s_1p_y)(a_2s_3 - d_4)}{p_z^2 + (c_1p_x + s_1p_y)^2}, \\ c_{23} &= \frac{(a_2s_3 - d_4)p_z - (-a_3 - a_2c_3)(c_1p_x + s_1p_y)}{p_z^2 + (c_1p_x + s_1p_y)^2}. \end{aligned} \tag{4.67}$$

Since the denominators are equal and positive, we solve for the sum of θ_2 and θ_3 as

$$\begin{aligned} \theta_{23} = \text{Atan2}\big[&(-a_3 - a_2c_3)p_z - (c_1p_x + s_1p_y)(d_4 - a_2s_3), \\ &(a_2s_3 - d_4)p_z + (a_3 + a_2c_3)(c_1p_x + s_1p_y)\big] \end{aligned} \tag{4.68}$$

Equation (4.68) computes four values of θ_{23} according to the four possible combinations of solutions for θ_1 and θ_3. Then, four possible solutions for θ_2 are computed as

$$\theta_2 = \theta_{23} - \theta_3, \tag{4.69}$$

where the appropriate solution for θ_3 is used when forming the difference.

Now the entire left side of (4.65) is known. Equating the $(1,3)$ elements from both sides of (4.65), as well as the $(3,3)$ elements, we get

$$\begin{aligned} r_{13}c_1c_{23} + r_{23}s_1c_{23} - r_{33}s_{23} &= -c_4s_5, \\ -r_{13}s_1 + r_{23}c_1 &= s_4s_5. \end{aligned} \tag{4.70}$$

As long as $s_5 \neq 0$, we can solve for θ_4 as

$$\theta_4 = \text{Atan2}\left(-r_{13}s_1 + r_{23}c_1, -r_{13}c_1c_{23} - r_{23}s_1c_{23} + r_{33}s_{23}\right). \tag{4.71}$$

When $\theta_5 = 0$ the manipulator is in a singular configuration in which joint axes 4 and 6 line up and cause the same motion of the last link of the robot. In this case, all that matters (and all that can be solved for) is the sum or difference of θ_4 and θ_6. This situation is detected by checking whether both arguments of the Atan2 in (4.71) are near zero. If so, θ_4 is chosen arbitrarily,* and when θ_6 is computed later, it will be computed accordingly.

If we consider (4.49) again, we can now rewrite it so that all the left-hand side is a function of only knowns and θ_4 by rewriting it as:

$$\left[{}^0_4T(\theta_4)\right]^{-1} \; {}^0_6T = {}^4_5T(\theta_5)\, {}^5_6T(\theta_6), \tag{4.72}$$

where $\left[{}^0_4T(\theta 4)\right]^{-1}$ is given by

$$\begin{bmatrix} c_1c_{23}c_4 + s_1s_4 & s_1c_{23}c_4 - c_1s_4 & -s_{23}c_4 & -a_2c_3c_4 + d_3s_4 - a_3c_4 \\ -c_1c_{23}s_4 + s_1c_4 & -s_1c_{23}s_4 - c_1c_4 & s_{23}s_4 & a_2c_3s_4 + d_3c_4 + a_3s_4 \\ -c_1s_{23} & -s_1s_{23} & -c_{23} & a_2s_3 - d_4 \\ 0 & 0 & 0 & 1 \end{bmatrix}, \tag{4.73}$$

and ${}^4_6T$ is given by equation (3.10) developed in Chapter 3. Equating the $(1,3)$ elements from both sides of (4.72), as well as the $(3,3)$ elements, we get

$$\begin{aligned} r_{13}(c_1c_{23}c_4 + s_1s_4) + r_{23}(s_1c_{23}c_4 - c_1s_4) - r_{33}(s_{23}c_4) &= -s_5, \\ r_{13}(-c_1s_{23}) + r_{23}(-s_1s_{23}) + r_{33}(-c_{23}) &= c_5. \end{aligned} \tag{4.74}$$

So we can solve for θ_5 as:

$$\theta_5 = \text{Atan2}\left(s_5, c_5\right), \tag{4.75}$$

where s_5 and c_5 are given by (4.74) above.

Applying the same method one more time, we compute $\left({}^0_5T\right)^{-1}$ and write (4.49) in the form

$$\left({}^0_5T\right)^{-1} \; {}^0_6T = {}^5_6T(\theta_6). \tag{4.76}$$

Equating the $(3,1)$ elements from both sides of (4.72), as well as the $(1,1)$ elements as we have done before, we get

$$\theta_6 = \text{Atan2}\left(s_6, c_6\right), \tag{4.77}$$

* It is usually chosen to be equal to the present value of joint 4.

where

$$s_6 = -r_{11}(c_1c_{23}s_4 - s_1c_4) - r_{21}(s_1c_{23}s_4 + c_1c_4) + r_{31}(s_{23}s_4),$$

$$c_6 = r_{11}\left[(c_1c_{23}c_4 + s_1s_4)c_5 - c_1s_{23}s_5\right] + r_{21}\left[(s_1c_{23}c_4 - c_1s_4)c_5 - s_1s_{23}s_5\right]$$
$$- r_{31}(s_{23}c_4c_5 + c_{23}s_5).$$

Because of the plus-or-minus signs appearing in (4.59) and (4.63), these equations compute four solutions. Additionally, there are four more solutions obtained by "flipping" the wrist of the manipulator. For each of the four solutions computed above, we obtain the flipped solution by:

$$\begin{aligned} \theta_4' &= \theta_4 + 180°, \\ \theta_5' &= -\theta_5, \\ \theta_6' &= \theta_6 + 180°. \end{aligned} \tag{4.78}$$

After all eight solutions have been computed some or all of them may have to be discarded because of joint limit violations. Of the remaining valid solutions, usually the one closest to the present manipulator configuration is chosen.

The Yasukawa Motoman L-3

As a second example we will solve the kinematic equations of the Yasukawa Motoman L-3 which were developed in Chapter 3. This solution will be partially algebraic and partially geometric. The Motoman L-3 has three features which make the inverse kinematic problem quite different than that of the PUMA. First, because the manipulator has only five joints, it is not able to position and orient its end-effector in order to attain *general* goal frames. Second, because of the four-bar type of linkages and chain drive scheme, motion at one actuator moves two or more joints. Third, the actuator position limits are not constants, but depend on the position of the other actuators, and so determining whether a computed set of actuator values is in range or not is not trivial.

If we consider the nature of the subspace of the Motoman manipulator (and the same applies to many manipulators with five degrees of freedom), we quickly realize that this subspace can be described by giving one constraint on the attainable orientation: The pointing direction of the tool, that is, the $\hat{Z}_T$ axis, must lie in the "plane of the arm." This plane is the vertical plane which contains the axis of joint 1, and the point where axes 4 and 5 intersect. The nearest orientation to a general orientation is the one obtained by rotating the tool's pointing direction so that it lies in the plane using a minimum amount of rotation.

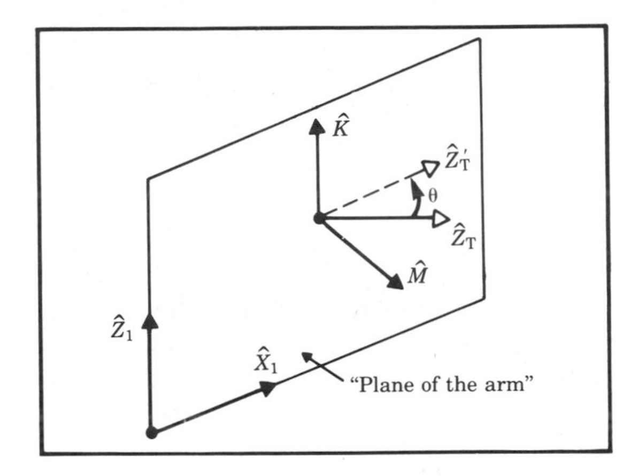

FIGURE 4.9 Rotating a goal frame into the Motoman's subspace.

Without developing an explicit expression for this subspace, we will determine a method of projecting a general goal frame into it. Note that this entire discussion is for the case that the wrist frame and tool frame differ only by a translation along $\hat{Z}_W$.

In Fig. 4.9 we indicate the plane of the arm by its normal, $\hat{M}$, and the desired pointing direction of the tool by $\hat{Z}_T$. This pointing direction must be rotated by angle θ about some vector $\hat{K}$ in order to cause the new pointing direction, $\hat{Z}'_T$, to lie in the plane. It is clear that the $\hat{K}$ which minimizes θ lies in the plane and is orthogonal to both $\hat{Z}_T$ and $\hat{Z}'_T$.

For any given goal frame, $\hat{M}$ is defined as

$$\hat{M} = \frac{1}{\sqrt{p_x^2 + p_y^2}} \begin{bmatrix} -p_y \\ p_x \\ 0 \end{bmatrix}, \tag{4.79}$$

where p_x and p_y are the X and Y coordinates of the desired tool position. Then K is given by

$$K = \hat{M} \times \hat{Z}_T. \tag{4.80}$$

The new $\hat{Z}'_T$ is

$$\hat{Z}'_T = \hat{K} \times \hat{M}. \tag{4.81}$$

The amount of rotation, θ, is given by

$$\cos\theta = \hat{Z}_T \cdot \hat{Z}'_T. \tag{4.82}$$

Using Rodriques' formula (see Exercise 2.20) we have

$$\hat{Y}'_T = c\theta\hat{Y}_T + s\theta(\hat{K} \times \hat{Y}_T) + (1 - c\theta)(\hat{K} \cdot \hat{Y}_T)\hat{K}. \tag{4.83}$$

Finally, we compute the remaining unknown column of the new rotation matrix of the tool as:

$$\hat{X}'_T = \hat{Y}'_T \times \hat{Z}'_T. \tag{4.84}$$

Equations (4.79) through (4.84) describe a method of projecting a given general goal orientation into the subspace of the Motoman robot.

Assuming that the given wrist frame, ${}^B_W T$, lies in the manipulator's subspace, we solve the kinematic equations as follows. In deriving the kinematic equations for the Motoman L-3, we formed the product of link transformations:

$${}^0_5T = {}^0_1T\, {}^1_2T\, {}^2_3T\, {}^3_4T\, {}^4_5T. \tag{4.85}$$

If we let

$${}^0_5T = \begin{bmatrix} r_{11} & r_{12} & r_{13} & p_x \\ r_{21} & r_{22} & r_{23} & p_y \\ r_{31} & r_{32} & r_{33} & p_z \\ 0 & 0 & 0 & 1 \end{bmatrix} \tag{4.86}$$

and premultiply both sides by ${}^0_1T^{-1}$, we have

$${}^0_1T^{-1}\, {}^0_5T = {}^1_2T\, {}^2_3T\, {}^3_4T\, {}^4_5T, \tag{4.87}$$

where the left hand side is

$$\begin{bmatrix} c_1r_{11} + s_1r_{21} & c_1r_{12} + s_1r_{22} & c_1r_{13} + s_1r_{23} & c_1p_x + s_1p_y \\ -r_{31} & -r_{32} & -r_{33} & -p_z \\ -s_1r_{11} + c_1r_{21} & -s_1r_{12} + c_1r_{22} & -s_1r_{13} + c_1r_{23} & -s_1p_x + c_1p_y \\ 0 & 0 & 0 & 1 \end{bmatrix} \tag{4.88}$$

and the right hand side is

$$\begin{bmatrix} * & * & s_{234} & * \\ * & * & -c_{234} & * \\ s_5 & c_5 & 0 & 0 \\ 0 & 0 & 0 & 1 \end{bmatrix}, \tag{4.89}$$

where several of the elements have not been shown. Equating the (3, 4) elements, we get

$$-s_1p_x + c_1p_y = 0, \tag{4.90}$$

which gives us*

$$\theta_1 = \text{Atan2}\,(p_y, p_x). \tag{4.91}$$

* For this manipulator, a second solution would violate joint limits and so is not calculated.

Equating the (3, 1) and (3, 2) elements we get

$$s_5 = -s_1 r_{11} + c_1 r_{21},$$
$$c_5 = -s_1 r_{12} + c_1 r_{22}, \tag{4.92}$$

from which we calculate θ_5 as

$$\theta_5 = \text{Atan2}\,(r_{21}c_1 - r_{11}s_1, r_{22}c_1 - r_{12}s_1). \tag{4.93}$$

Equating the (2, 3) and (1, 3) elements we get

$$c_{234} = r_{33},$$
$$s_{234} = c_1 r_{13} + s_1 r_{23}, \tag{4.94}$$

which leads to

$$\theta_{234} = \text{Atan2}\,(r_{13}c_1 + r_{23}s_1, r_{33}). \tag{4.95}$$

To solve for the individual angles θ_2, θ_3, and θ_4, we will take a geometric approach. Figure 4.10 shows the plane of the arm with point A at joint axis 2, point B at joint axis 3, and point C at joint axis 4.

From the law of cosines applied to triangle ABC we have

$$\cos\theta_3 = \frac{p_x^2 + p_y^2 + p_z^2 - l_2^2 - l_3^2}{2l_2l_3}. \tag{4.96}$$

Then we have*

$$\theta_3 = \text{Atan2}\left(\sqrt{1 - \cos^2\theta_3}, \cos\theta_3\right). \tag{4.97}$$

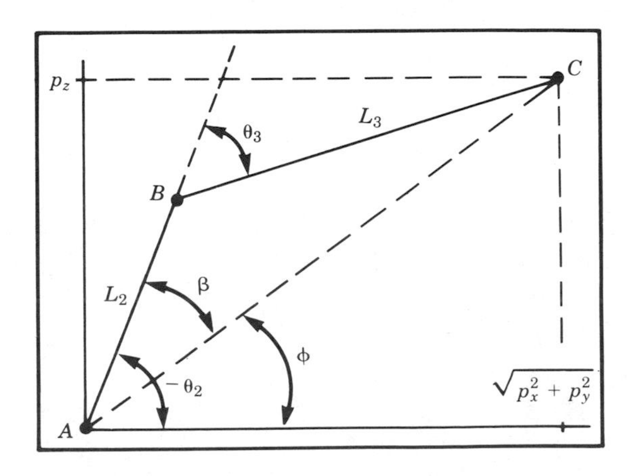

FIGURE 4.10 The plane of the Motoman manipulator.

* For this manipulator, a second solution would violate joint limits and so is not calculated.

From Fig. 4.10 we see that $\theta_2 = -\phi - \beta$ or

$$\theta_2 = -\text{Atan2}\left(p_z, \sqrt{p_x^2 + p_y^2}\right) - \text{Atan2}\left(l_3 \sin\theta_3, l_2 + l_3 \cos\theta_3\right). \quad \textbf{(4.98)}$$

Finally we have

$$\theta_4 = \theta_{234} - \theta_2 - \theta_3. \quad \textbf{(4.99)}$$

Having solved for joint angles we must perform the further computation to obtain the actuator values. Refering to Section 3.7, we solve equation (3.15) for A_i

$$
\begin{aligned}
A_1 &= \frac{1}{k_1}(\theta_1 - \lambda_1), \\
A_2 &= \frac{1}{k_2}\left(\sqrt{-2\alpha_2\beta_2 \cos\left(\theta_2 - \Omega_2 - tan^{-1}(\frac{\phi_2}{\gamma_2}) + 270^\circ\right) + \alpha_2^2 + \beta_2^2} - \lambda_2\right), \\
A_3 &= \frac{1}{k_3}\left(\sqrt{-2\alpha_3\beta_3 \cos\left(\theta_2 + \theta_3 - tan^{-1}(\frac{\phi_3}{\gamma_3}) + 90^\circ\right) + \alpha_3^2 + \beta_3^2} - \lambda_3\right), \\
A_4 &= \frac{1}{k_4}(180^\circ + \lambda_4 - \theta_2 - \theta_3 - \theta_4), \\
A_5 &= \frac{1}{k_5}(\lambda_5 - \theta_5).
\end{aligned} \quad \textbf{(4.100)}
$$

Since the actuators have limited ranges of motion, we must check that our computed solution is in range. This "in range" check is complicated by the fact that due to the mechanical arrangement, actuators interact and affect each other's allowed range of motion. For the Motoman robot, actuators 2 and 3 interact in such a way that the following relationship must always be obeyed:

$$A_2 - 10{,}000 > A_3 > A_2 + 3000. \quad \textbf{(4.101)}$$

That is, the limits of actuator 2 are a function of the position of actuator 3. Similarly:

$$32{,}000 - A_4 < A_5 < 55{,}000. \quad \textbf{(4.102)}$$

Since one revolution of joint 5 corresponds to 25,600 actuator counts, when $A_4 > 2600$ there are two possible solutions for A_5. This is the only situation in which the Yasukawa Motoman L-3 has more than one solution.

4.7 The standard frames

The ability to solve for joint angles is really the central element in many robot control systems. Again, consider the paradigm indicated in Fig. 4.11, which shows the standard frames.

The way these frames are used in a general robot system is as follows:

1. The user specifies to the system where he wishes the station frame to be located. This might be at the corner of a work surface as in Fig. 4.12 or even affixed to a moving conveyor belt. The station frame, $\{S\}$, is defined relative to the base frame, $\{B\}$.
2. The user specifies the description of the tool being used by the robot by giving the $\{T\}$ frame specification. Each tool the robot picks up may have a different $\{T\}$ frame associated with it. Note that the same tool grasped in different ways requires different $\{T\}$ frame definitions. $\{T\}$ is specified relative to $\{W\}$, that is, ${}^W_T T$.
3. The user specifies the goal point for a robot motion by giving the description of the goal frame, $\{G\}$, relative to the station frame. Often the definitions of $\{T\}$ and $\{S\}$ remain fixed for several motions of the robot. In this case, once they are defined the user simply gives a series of $\{G\}$ specifications.

   In many systems, the tool frame definition (${}^W_T T$) is constant (for example, it is defined with its origin at the center of the finger-tips).

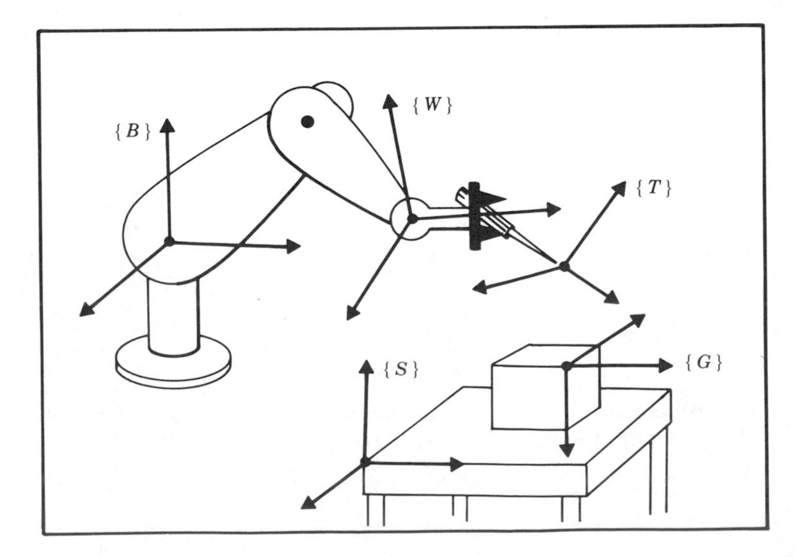

FIGURE 4.11 Location of the "standard" frames.

Also, the station frame may be fixed, or easily taught by the user with the robot itself. In such systems, the user need not be aware of the five standard frames—he or she simply thinks in terms of moving the tool to locations (goals) with respect to the work area specified by station frame.

4. The robot system calculates a series of joint angles to move the joints through in order that the tool frame moves from its initial location in a smooth manner until $\{T\} = \{G\}$ at the end of motion.

4.8 SOLVE-ing a manipulator

The `SOLVE` function implements Cartesian transformations and calls the inverse kinematics function. Thus, the inverse kinematics are generalized so that arbitrary tool frame and station frame definitions may be used with our basic inverse kinematics, which solves for the wrist frame relative to the base frame.

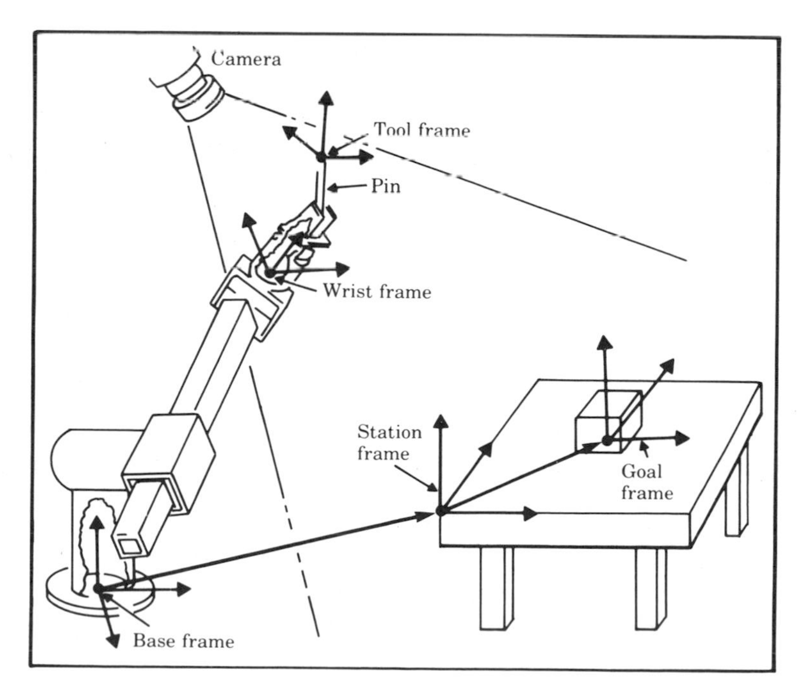

FIGURE 4.12 Example workstation.

Given the goal frame specification, ${}^S_T T$, SOLVE uses the tool and station definitions to calculate the location of $\{W\}$ relative to $\{B\}$, ${}^B_W T$:

$$ {}^B_W T = {}^B_S T \, {}^S_T T \, {}^W_T T^{-1}. \tag{4.103} $$

Then the inverse kinematics take ${}^B_W T$ as an input and calculate θ_1 through θ_n.

4.9 Repeatability and accuracy

Many industrial robots today move to goal points which have been taught. A **taught point** is one that the manipulator is moved to physically, and then the joint position sensors are read, and the joint angles stored. When the robot is commanded to return to that point in space, each joint is moved to the stored value. In simple "teach and playback" manipulators such as these, the inverse kinematic problem never arises because goal points are never specified in Cartesian coordinates. When a manufacturer specifies how precisely a manipulator can return to a taught point, he is specifying the **repeatability** of the manipulator.

Anytime a goal position and orientation are specified in Cartesian terms, the inverse kinematics of the device must be computed in order to solve for the required joint angles. Systems which allow goals to be described in Cartesian terms are capable of moving the manipulator to points which were never taught, points in its workspace to which it has perhaps never gone before. We will call such points **computed points**. Such a capability is necessary for many manipulation tasks. For example, if a computer vision system is used to locate a part which the robot must grasp, the robot must be able to move to the Cartesian coordinates supplied by the vision sensor. The precision with which a computed point can be attained is called the **accuracy** of the manipulator.

The accuracy of a manipulator is lower bounded by the repeatability. Clearly accuracy is affected by the precision of parameters appearing in the kinematic equations of the robot. Errors in knowledge of the Denavit-Hartenberg parameters will cause the inverse kinematic equations to calculate joint angle values which are in error. Hence, while the repeatability of most industrial manipulators is quite good, the accuracy is usually much worse and varies quite a bit from manipulator to manipulator. Calibration techniques can be devised which allow the accuracy of a manipulator to be improved through estimation of that particular manipulator's kinematic parameters [7].

4.10 Computational considerations

In many path control schemes which we will consider in Chapter 7, it is necessary to calculate the inverse kinematics of a manipulator at fairly high rates, for example 30 Hz or faster. Therefore, computational efficiency is an issue. These speed requirements rule out the use of numerical solution techniques which are iterative in nature, and for this reason, we have not considered them.

Most of the general comments of Section 3.10, made for forward kinematics, also hold for the problem of inverse kinematics. For the inverse kinematic case, a table-lookup Atan2 routine is often used to attain higher speeds.

Structure of the computation of multiple solutions is also important. It is generally fairly efficient to generate all of them in parallel, rather than pursuing one after another serially. Of course, in some applications, when all solutions are not required, substantial time is saved by computing only one.

When a geometric approach is used to develop an inverse kinematic solution, it is sometimes possible to calculate multiple solutions by simple operations on the various angles solved for in obtaining the first solution. That is, the first solution is moderately expensive computationally, but the other solutions are found very quickly by summing and differencing angles, subtracting π, and so on.

References

[1] B. Roth, J. Rastegar, and V. Scheinman, "On the Design of Computer Controlled Manipulators," On the Theory and Practice of Robots and Manipulators, Vol. 1, First CISM-IFToMM Symposium, September 1973, pp. 93–113.

[2] B. Roth, "Performance Evaluation of Manipulators from a Kinematic Viewpoint," Peformance Evaluation of Manipulators, National Bureau of Standards, special publication, 1975.

[3] D. Pieper and B. Roth, "The Kinematics of Manipulators Under Computer Control," Proc. of the Second International Congress on Theory of Machines and Mechanisms, Vol. 2, pp 159–169, Zakopane, Poland, 1969.

[4] D. Pieper, "The Kinematics of Manipulators Under Computer Control," Ph.D. Thesis, Stanford University, 1968.

[5] R.P. Paul, B. Shimano, and G. Mayer, "Kinematic Control Equations for Simple Manipulators," *IEEE Trans. on Systems, Man, and Cybernetics*, vol. SMC-11, No. 6, 1981.

[6] L. Tsai and A. Morgan, "Solving the Kinematics of the Most General Six- and Five-degree-of-freedom Manipulators by Continuation Methods," Paper 84-DET-20, ASME Mechanisms Conference, Boston, Massachusetts, October 7–10 1984.

[7] S. Hayati, "Robot Arm Geometric Link Parameter Estimation," Proceedings of the 22nd IEEE Conf. on Decision and Control, December 1983.

Exercises

4.1 [15] Sketch the finger-tip workspace of the 3-link manipulator of Chapter 3, Exercise 3 for the case $l_1 = 15.0$, $l_2 = 10.0$, and $l_3 = 3.0$.

4.2 [26] Derive the inverse kinematics of the 3-link manipulator of Chapter 3, Exercise 3.

4.3 [12] Sketch the finger-tip workspace of the 3-DOF manipulator of Chapter 3, Example 3.4.

4.4 [24] Derive the inverse kinematics of the 3-DOF manipulator of Chapter 3, Example 3.4.

4.5 [38] Write a PASCAL subroutine that computes all possible solutions for the PUMA 560 manipulator which lie within the following joint limits:

$$-170.0 < \theta_1 < 170.0,$$

$$-225.0 < \theta_2 < 45.0,$$

$$-250.0 < \theta_3 < 75.0,$$

$$-135.0 < \theta_4 < 135.0,$$

$$-100.0 < \theta_5 < 100.0,$$

$$-180.0 < \theta_6 < 180.0.$$

Use the equations derived in Section 4.6 with the numerical values (in inches):

$$a_2 = 17.0,$$

$$a_3 = 0.8,$$

$$d_3 = 4.9,$$

$$d_4 = 17.0.$$

4.6 [15] Describe a simple algorithm for choosing the nearest solution from a set of possible solutions.

4.7 [10] Make a list of factors which might affect the repeatability of a manipulator. Make a second list of additional factors which affect the accuracy of a manipulator.

4.8 [12] Given a desired position and orientation of the hand of a 3-link planar rotary jointed manipulator, there are two possible solutions. If we add one more rotational joint (in such a way that the arm is still planar), how many solutions are there?

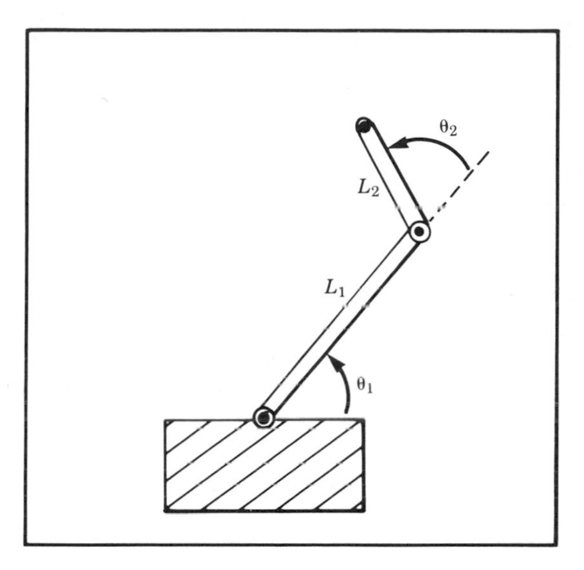

FIGURE 4.13 Two-link planar manipulator.

4.9 [26] Figure 4.13 shows a 2-link planar arm with rotary joints. For this arm, the second link is half as long as the first, that is: $l_1 = 2l_2$. The joint range limits in degrees are

$$0 < \theta_1 < 180,$$

$$-90 < \theta_2 < 180.$$

Sketch the approximate reachable workspace (an area) of the tip of link 2.

4.10 [23] Give an expression for the subspace of the manipulator of Chapter 3, Example 3.4.

Programming Exercise (Part 4)

1. Write a subroutine to calculate the inverse kinematics for the 3-link manipulator (see Section 4.4). The routine should pass arguments as shown below:

```
Procedure INVKIN(VAR wrelb: frame; VAR current,near,far: vec3;
VAR sol:boolean);
```

where "wrelb", an input, is the wrist frame specified relative to the base frame; "current", an input, is the current position of the robot (given as a vector of joint angles); "near" is the nearest solution; "far" is the second solution; and "sol" is a flag which indicates whether solutions were found or not. (sol = FALSE if no solutions were found). The link lengths (meters) are

$$l_1 = l_2 = 0.5.$$

The joint ranges of motion are

$$-170^\circ <= \theta_i <= 170^\circ.$$

Test your routine by calling it back-to-back with `KIN` to test whether they are indeed inverses of one another.

2. A tool is attached to link 3 of the manipulator. This tool is described by ${}^W_T T$, the tool frame relative to the wrist frame. Also, a user has described his work area, the station frame relative to the base of the robot, as ${}^B_S T$. Write the subroutine

```
Procedure SOLVE(VAR trels: frame; VAR current,near,far: vec3;
VAR sol:boolean);
```

where "trels" is the $\{T\}$ frame specified relative to the $\{S\}$ frame. Other parameters are exactly as in the `INVKIN` subroutine. The definitions of $\{T\}$ and $\{S\}$ should be globally defined variables or constants. `SOLVE` should use calls to `TMULT`, `TINVERT`, and `INVKIN`.

3. Write a main program which accepts a goal frame specified in terms of x, y, and ϕ. This goal specification is $\{T\}$ relative to $\{S\}$, which is the way the user wants to specify goals.

The robot is using the same tool in the same working area as in Programming Exercise (Part 2), so $\{T\}$ and $\{S\}$ are defined as:

$$ {}^W_T T = [x \quad y \quad \theta] = [0.1 \quad 0.2 \quad 30.0], $$
$$ {}^B_S T = [x \quad y \quad \theta] = [-0.1 \quad 0.3 \quad 0.0]. $$

Calculate the joint angles for each of the three goal frames given below. Assume that the robot will start with all angles equal to 0.0 and move to these three goals in sequence. The program should find the nearest solution with respect to the previous goal point.

$$ [x_1 \quad y_1 \quad \phi_1] = [0.0 \quad 0.0 \quad -90.0], $$
$$ [x_2 \quad y_2 \quad \phi_2] = [0.6 \quad -0.3 \quad 45.0], $$
$$ [x_3 \quad y_3 \quad \phi_3] = [-0.4 \quad 0.3 \quad 120.0], $$
$$ [x_4 \quad y_4 \quad \phi_4] = [0.8 \quad 1.4 \quad 30.0]. $$

You should call `SOLVE` and `WHERE` back to back to make sure they are truly inverse functions.

5

JACOBIANS: VELOCITIES AND STATIC FORCES

5.1 Introduction

In this chapter we expand our consideration of robot manipulators beyond static positioning problems. We introduce the notions of linear and angular velocity of a rigid body and use these concepts to analyze the motion of a manipulator. We also will consider forces acting on a rigid body, and then show how these ideas can be used to study the application of static forces with manipulators.

It turns out that the study of both velocities and static forces leads to a matrix entity called the **Jacobian*** of the manipulator, which will be introduced in this chapter.

---

* Mathematicians call it the "Jacobian matrix," but roboticists usually shorten it to simply "Jacobian."

The field of kinematics of mechanisms is not treated in great depth here. For the most part, the presentation is restricted to only those ideas which are fundamental to the particular problem of robotics. The interested reader is urged to study further from any of several texts on mechanics [1], [2], [3].

5.2 Notation for time-varying position and orientation

Before investigating the description of the motion of a rigid body, we briefly discuss some basics: the differentiation of vectors, representation of angular velocity, and notation.

Differentiation of position vectors

As a basis for our consideration of velocities (and in Chapter 6, accelerations), we introduce the following notation for the derivative of a vector:

$$ {}^{B}V_{Q} = \frac{d}{dt}\, {}^{B}Q = \lim_{\Delta t \to 0} \frac{{}^{B}Q(t+\Delta t) - {}^{B}Q(t)}{\Delta t}. \tag{5.1} $$

The velocity of a position vector can be thought of as the linear velocity of the point in space repesented by the position vector. From (5.1) we see that we are calculating the derivative of Q relative to frame $\{B\}$. For example, if Q is not changing in time relative to $\{B\}$, then the velocity calculated is zero—even if there is some other frame from which Q is varying. Thus it is important to indicate the frame in which the vector is differentiated.

As with any vector, a velocity vector may be described in terms of any frame, and this frame of reference is noted with a leading superscript. Hence, the velocity vector calculated by (5.1) when expressed in terms of frame $\{A\}$ would be written:

$$ {}^{A}\left({}^{B}V_{Q}\right) = \frac{{}^{A}d}{dt}\, {}^{B}Q. \tag{5.2} $$

So we see that in the general case, a velocity vector is associated with a point in space, but the numerical values describing the velocity of that point depend on two frames: one with respect to which the differentiation was done, and one in which the resulting velocity vector is expressed.

In (5.1) the calculated velocity is written in terms of the frame of differentiation, so the result could be indicated with a leading B

superscript, but for simplicity, when both superscripts are the same, we needn't indicate the outer one; that is, we write

$$ {}^{B}\left({}^{B}V_{Q}\right) = {}^{B}V_{Q}. \quad \mathbf{(5.3)} $$

Finally, we can always remove the outer, leading superscript by explicitly including the rotation matrix which accomplishes the change in reference frame (see Section 2.10); that is, we write

$$ {}^{A}\left({}^{B}V_{Q}\right) = {}^{A}_{B}R\,{}^{B}V_{Q}. \quad \mathbf{(5.4)} $$

We will usually write expressions in the form of the right-hand side of (5.4) so that the symbols representing velocities always mean the velocity in the frame of differentiation, and do not have outer, leading superscripts.

Rather than considering a general point's velocity relative to an arbitrary frame, we will very often consider the velocity of the *origin of a frame* relative to a some understood universe reference frame. For this special case we define a shorthand notation:

$$ v_C = {}^{U}V_{CORG}, \quad \mathbf{(5.5)} $$

where the point in question is the origin of frame $\{C\}$ and the reference frame is $\{U\}$. For example, we can use the notation v_C to refer to the velocity of the origin of frame $\{C\}$, and ${}^{A}v_C$ is the velocity of the origin of frame $\{C\}$ expressed in terms of frame $\{A\}$ (though differentiation was done relative to $\{U\}$).

EXAMPLE 5.1

Figure 5.1 shows a fixed universe frame, $\{U\}$, a frame attached to a train traveling at 100 mph, $\{T\}$, and a frame attached to a car traveling at 30 mph, $\{C\}$. Both vehicles are heading in the $\hat{X}$ direction of $\{U\}$. The rotation matrices, ${}^{U}_{T}R$ and ${}^{U}_{C}R$, are known and constant.

What is $\frac{{}^{U}d}{dt}\,{}^{U}P_{CORG}$?

$$ \frac{{}^{U}d}{dt}\,{}^{U}P_{CORG} = {}^{U}V_{CORG} = v_C = 30\hat{X}. $$

What is ${}^{C}\left({}^{U}V_{TORG}\right)$?

$$ {}^{C}\left({}^{U}V_{TORG}\right) = {}^{C}v_T = {}^{C}_{U}R\,v_T = {}^{C}_{U}R(100\hat{X}) = {}^{U}_{C}R^{-1}\,100\hat{X}. $$

What is ${}^{C}\left({}^{T}V_{CORG}\right)$?

$$ {}^{C}\left({}^{T}V_{CORG}\right) = {}^{C}_{T}R\,{}^{T}V_{CORG} = -{}^{U}_{C}R^{-1}\,{}^{U}_{T}R\,70\hat{X}. \quad \blacksquare $$

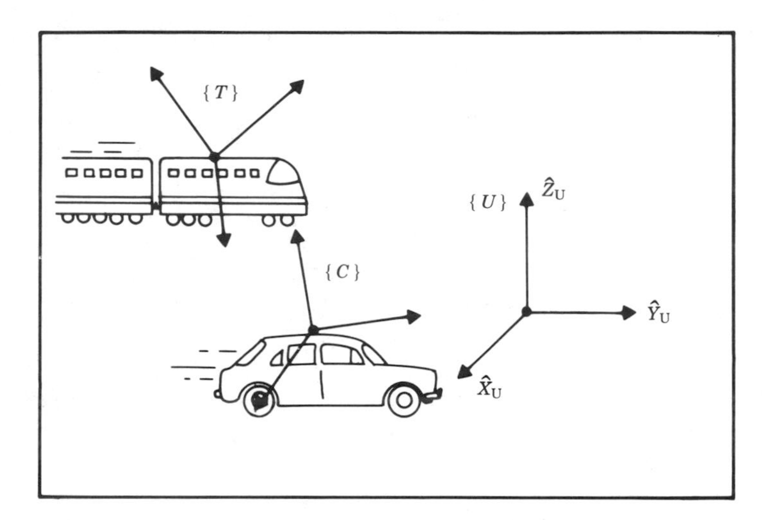

FIGURE 5.1 Example of some frames in linear motion.

The angular velocity vector

We now introduce an **angular velocity vector** using the symbol Ω. Whereas linear velocity describes an attribute of a point, angular velocity describes an attribute of a body. Since we always attach a frame to the bodies we consider, we can also think of angular velocity as describing rotational motion of a frame.

In Fig. 5.2, ${}^{A}\Omega_{B}$ describes the rotation of frame $\{B\}$ relative to $\{A\}$. Physically, at any instant, the direction of ${}^{A}\Omega_{B}$ indicates the

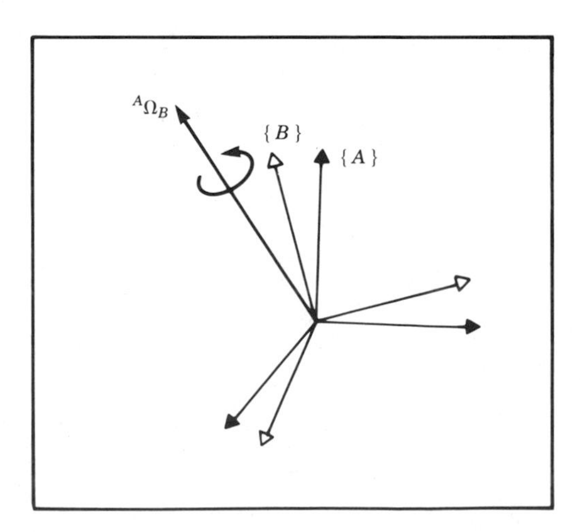

FIGURE 5.2 Frame $\{B\}$ is rotating with angular velocity ${}^{A}\Omega_{B}$ relative to frame $\{A\}$.

instantaneous axis of rotation of $\{B\}$ relative to $\{A\}$, and the magnitude of ${}^{A}\Omega_{B}$ indicates the speed of rotation. Again, like any vector, an angular velocity vector may be expressed in any coordinate system, and so another leading superscript may be added, for example ${}^{C}\left({}^{A}\Omega_{B}\right)$ is the angular velocity of frame $\{B\}$ relative to $\{A\}$ expressed in terms of frame $\{C\}$.

Again, we introduce a simplified notation for an important special case. This is simply the case in which there is an understood reference frame, so that it need not be mentioned in the notation

$$\omega_{C} = {}^{U}\Omega_{C}, \tag{5.6}$$

where ω_C is the angular velocity of frame $\{C\}$ relative to some understood reference frame, $\{U\}$. For example, ${}^{A}\omega_{C}$ is the angular velocity of frame $\{C\}$ expressed in terms of $\{A\}$ (though the angular velocity is with respect to $\{U\}$).

5.3 Linear and rotational velocity of rigid bodies

In this section we investigate the description of motion of a rigid body, at least as far as velocity is concerned. These ideas extend the notions of translations and orientations described in Chapter 2 to the time varying case. In Chapter 6 we will further extend our study to considerations of acceleration.

As in Chapter 2, we attach a coordinate system to any body which we wish to describe. Then, motion of rigid bodies can be equivalently studied as the motion of frames relative to one another.

Linear velocity

Consider a frame $\{B\}$ attached to a rigid body. We wish to describe the motion of ${}^{B}Q$ relative to frame $\{A\}$, as in Fig. 5.3. We may consider $\{A\}$ to be fixed.

Frame $\{B\}$ is located relative to $\{A\}$ as described by a position vector, ${}^{A}P_{BORG}$, and a rotation matrix, ${}^{A}_{B}R$. For the moment we will assume that the orientation, ${}^{A}_{B}R$, is not changing with time. That is, the motion of point Q relative to $\{A\}$ is due to ${}^{A}P_{BORG}$ and/or ${}^{B}Q$ changing in time.

Solving for the linear velocity of point Q in terms of $\{A\}$ is quite simple. Just express both components of the velocity in terms of $\{A\}$ and sum:

$$ {}^{A}V_{Q} = {}^{A}V_{BORG} + {}^{A}_{B}R\,{}^{B}V_{Q}. \tag{5.7}$$

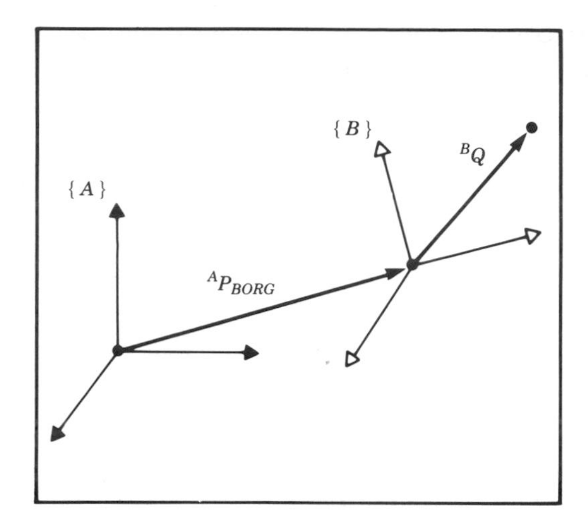

FIGURE 5.3 Frame $\{B\}$ is translating with velocity $^AV_{BORG}$ relative to frame $\{A\}$.

Equation (5.7) is only for the case in which relative orientation of $\{B\}$ and $\{A\}$ remains constant.

Rotational velocity

Now let us consider two frames with coincident origins and with zero linear velocity so that the origins will remain coincident for all time. One or both could be attached to rigid bodies, but for clarity the rigid bodies are not shown in Fig. 5.4.

The orientation of frame $\{B\}$ with respect to frame $\{A\}$ is changing in time. As indicated in Fig. 5.4, rotational velocity of $\{B\}$ relative to $\{A\}$ is described by a vector called $^A\Omega_B$. We also have indicated a vector ^BQ which locates a point which is fixed in $\{B\}$. Now we consider the all important question: How does a vector change with time as viewed from $\{A\}$ when it is fixed in $\{B\}$ and the systems are rotating?

Let us consider that the vector Q is constant as viewed from frame $\{B\}$, that is:

$$^BV_Q = 0. \tag{5.8}$$

Even though it is constant relative to $\{B\}$, it is clear that point Q will have a velocity as seen from $\{A\}$ due to the rotational velocity $^A\Omega_B$. To solve for the velocity of point Q we will use an intuitive approach. In Fig. 5.5 we show two instants of time as vector Q rotates around $^A\Omega_B$. This is what an observer in $\{A\}$ would observe.

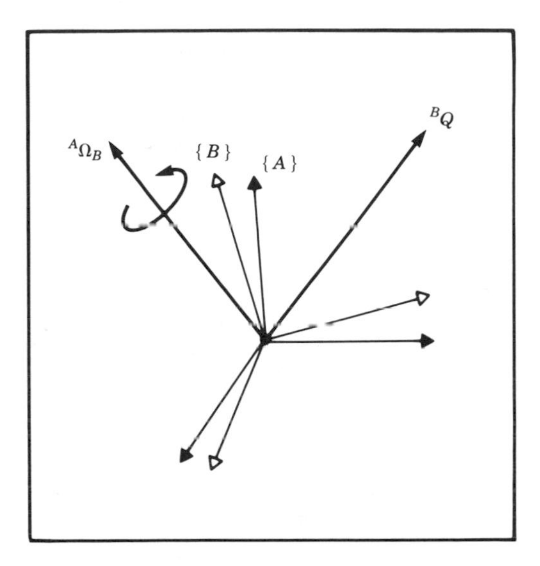

FIGURE 5.4 Vector ^BQ, fixed in frame $\{B\}$, is rotating with respect to frame $\{A\}$ with angular velocity $^A\Omega_B$.

By examining Fig. 5.5 we can determine both the direction and the magnitude of the change in the vector as viewed from $\{A\}$. First, it is clear that the differential change in ^AQ must be perpendicular to both $^A\Omega_B$ and ^AQ. Second, from Fig. 5.5 we see that the magnitude of the differential change is

$$|\Delta Q| = \left(|^AQ| \sin\theta\right) \left(|^A\Omega_B| \Delta t\right). \tag{5.9}$$

These conditions on magnitude and direction immediately suggest the vector cross product. Indeed, our conclusions about direction and magnitude are satisfied by the computational form

$$^AV_Q = {}^A\Omega_B \times {}^AQ. \tag{5.10}$$

In the general case, the vector Q may also be changing with respect to frame $\{B\}$, so adding this component we have

$$^AV_Q = {}^A\left(^BV_Q\right) + {}^A\Omega_B \times {}^AQ. \tag{5.11}$$

Using a rotation matrix to remove the dual-superscript, and since the description of ^AQ at any instant is $^A_BR\ ^BQ$, we have

$$^AV_Q = {}^A_BR\ ^BV_Q + {}^A\Omega_B \times {}^A_BR\ ^BQ. \tag{5.12}$$

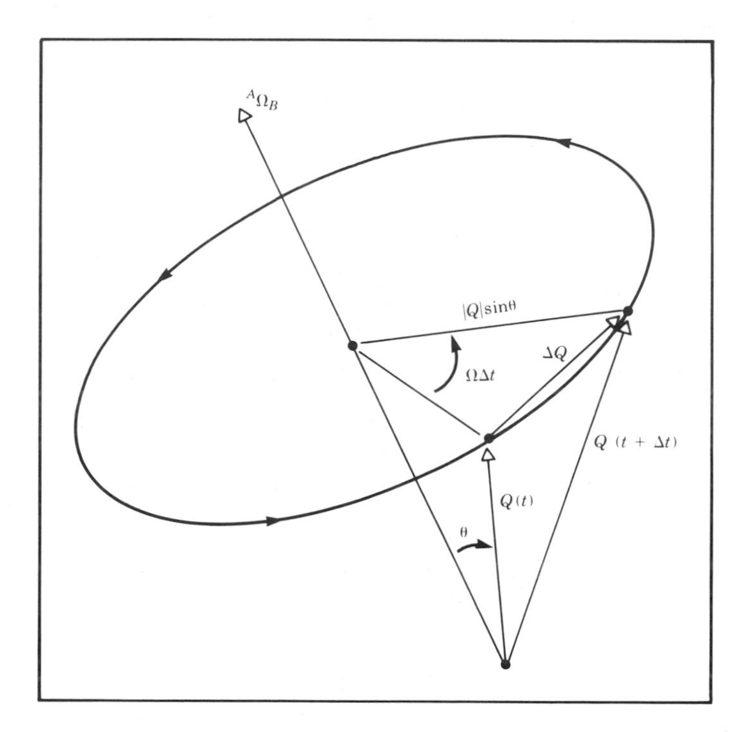

FIGURE 5.5 The velocity of a point due to an angular velocity.

Simultaneous linear and rotational velocity

We can very simply expand (5.12) to the case where origins are not coincident by adding on the linear velocity of the origin to (5.12) to derive the general formula for velocity of a vector fixed in frame $\{B\}$ as seen from frame $\{A\}$:

$$^AV_Q = \ ^AV_{BORG} + {}^A_BR\ ^BV_Q + {}^A\Omega_B \times {}^A_BR\ ^BQ \tag{5.13}$$

Equation (5.13) is the final result for the derivative of a vector in a moving frame as seen from a stationary frame.

5.4 Motion of the links of a robot

In considering the motion of robot links we will always use link frame $\{0\}$ as our reference frame. Hence, v_i is the linear velocity of the origin of link frame $\{i\}$, and ω_i is the angular velocity of link frame $\{i\}$.

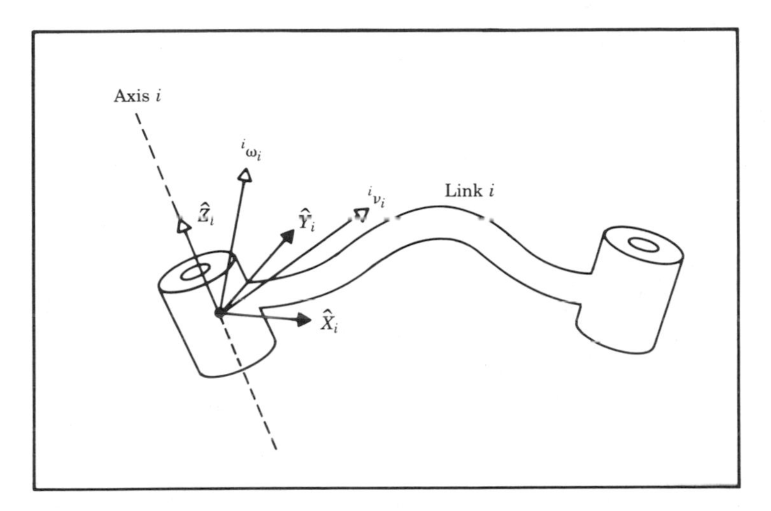

FIGURE 5.6 The velocity of link i is given by vectors v_i and ω_i which may be written in any frame, even frame $\{i\}$.

At any instant, each link of a robot in motion has some linear and angular velocity. Figure 5.6 indicates these vectors for link i. In this case, it is indicated that they are written in frame $\{i\}$.

5.5 Velocity "propagation" from link to link

We now consider the problem of calculating the linear and angular velocities of the links of a robot. A manipulator is a chain of bodies, each one capable of motion relative to its neighbors. Because of this structure we can compute the velocities of each link in order starting from the base. The velocity of link $i+1$ will be that of link i, plus whatever new rotational component was added by joint $i+1$.

As indicated in Fig. 5.6, let us now think of each link of the mechanism as a rigid body with linear and angular velocity vectors describing its motion. Further, we will express these velocities with respect to the link frame itself rather than with respect to the base coordinate system. Figure 5.7 shows links i and $i+1$ along with their velocity vectors defined in the link frames.

Rotational velocities may be added when both ω vectors are written with respect to the same frame. Therefore, the angular velocity of link $i+1$ is the same as that of link i plus a new component caused by rotational velocity at joint $i+1$. This can be written in terms of frame

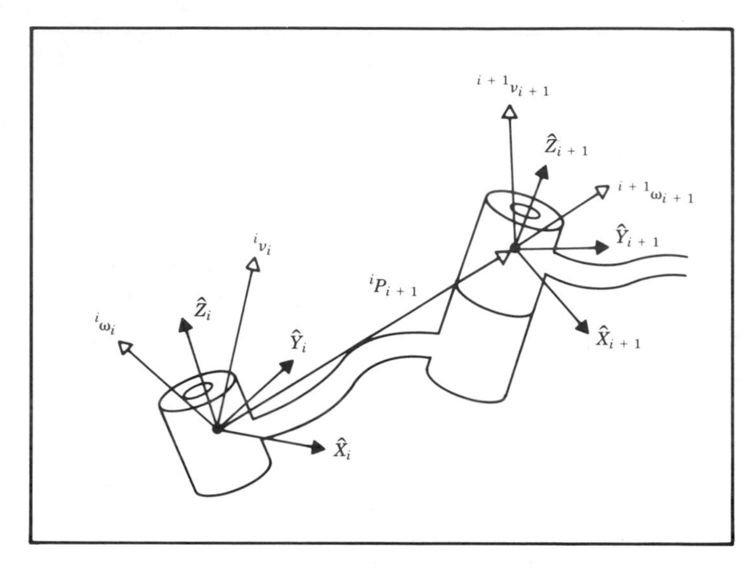

FIGURE 5.7 Velocity vectors of neighboring links.

$\{i\}$ as

$$^{i}\omega_{i+1} = {}^{i}\omega_{i} + {}^{i}_{i+1}R\,\dot{\theta}_{i+1}\,{}^{i+1}\hat{Z}_{i+1}. \tag{5.14}$$

Note that

$$\dot{\theta}_{i+1}\,{}^{i+1}\hat{Z}_{i+1} = {}^{i+1}\begin{bmatrix} 0 \\ 0 \\ \dot{\theta}_{i+1} \end{bmatrix}. \tag{5.15}$$

We have made use of the rotation matrix relating frames $\{i\}$ and $\{i+1\}$ in order to represent the added rotational component due to motion at the joint in frame $\{i\}$. The rotation matrix rotates the axis of rotation of joint $i+1$ into its description in frame $\{i\}$ so that the two components of angular velocity may be added.

By premultiplying both sides of (5.14) by ${}^{i+1}_{i}R$ we can find the description of the angular velocity of link $i+1$ with respect to frame $\{i+1\}$:

$$^{i+1}\omega_{i+1} = {}^{i+1}_{i}R\,{}^{i}\omega_{i} + \dot{\theta}_{i+1}\,{}^{i+1}\hat{Z}_{i+1}. \tag{5.16}$$

The linear velocity of the origin of frame $\{i+1\}$ is the same as that of the origin of frame $\{i\}$ plus a new component caused by rotational velocity of link i. This is exactly the situation described by (5.13), with one term vanishing because ${}^{i}P_{i+1}$ is constant in frame $\{i\}$. Therefore we have

$$^{i}v_{i+1} = {}^{i}v_{i} + {}^{i}\omega_{i} \times {}^{i}P_{i+1}. \tag{5.17}$$

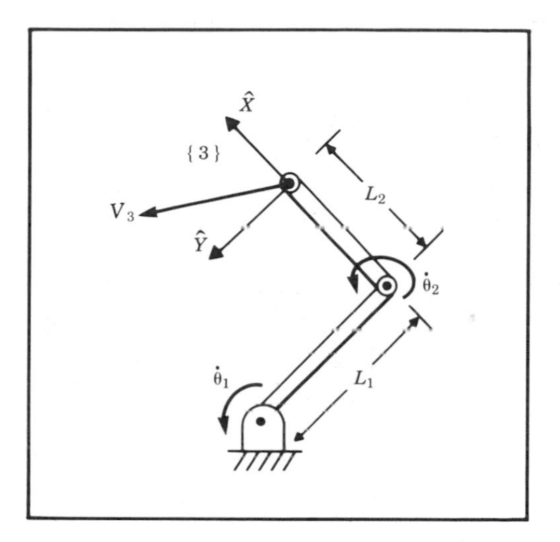

FIGURE 5.8 A two-link manipulator.

Premultiplying both sides by ${}^{i+1}_{i}R$, we compute

$$ {}^{i+1}v_{i+1} = {}^{i+1}_{i}R({}^{i}v_i + {}^{i}\omega_i \times {}^{i}P_{i+1}). \tag{5.18} $$

Equations (5.16) and (5.18) are perhaps the most important results of this chapter. The equivalent relationships for the case that joint $i+1$ is prismatic are:

$$ \begin{aligned} {}^{i+1}\omega_{i+1} &= {}^{i+1}_{i}R\,{}^{i}\omega_i, \\ {}^{i+1}v_{i+1} &= {}^{i+1}_{i}R({}^{i}v_i + {}^{i}\omega_i \times {}^{i}P_{i+1}) + \dot{d}_{i+1}\,{}^{i+1}\hat{Z}_{i+1}. \end{aligned} \tag{5.19} $$

Applying these equations succesively from link to link, we can compute ${}^{N}\omega_N$ and ${}^{N}v_N$, the rotational and linear velocities of the last link. Note that the resulting velocities are expressed in terms of frame $\{N\}$. This turns out to be useful, as we will see later. If the velocities are desired in terms of the base coordinate system, they can be rotated into base coordinates by multiplication with ${}^{0}_{N}R$.

EXAMPLE 5.2

A two-link manipulator with rotational joints is shown in Fig. 5.8. Calculate the velocity of the tip of the arm as a function of joint rates. Give the answer in two forms—in terms of frame $\{3\}$ and also in terms of frame $\{0\}$.

Frame $\{3\}$ has been attached at the end of the manipulator as shown in Fig. 5.9, and we wish to find the velocity of the origin of this frame

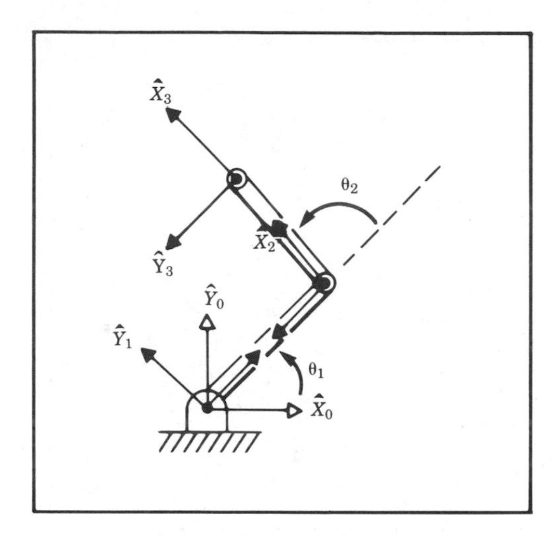

FIGURE 5.9 Frame assignments for the two-link manipulator.

expressed in frame {3}. As a second part of the problem, we will express these velocities in frame {0} as well. We will start by attaching frames to the links as we have done before (see Fig. 5.9)

We will use Eqs. (5.16) and (5.18) to compute the velocity of the origin of each frame starting from the base frame {0}, which has zero velocity. Since (5.16) and (5.18) will make use of the link transformations, we compute them as:

$$
{}^{0}_{1}T = \begin{bmatrix} c_1 & -s_1 & 0 & 0 \\ s_1 & c_1 & 0 & 0 \\ 0 & 0 & 1 & 0 \\ 0 & 0 & 0 & 1 \end{bmatrix},
$$

$$
{}^{1}_{2}T = \begin{bmatrix} c_2 & -s_2 & 0 & l_1 \\ s_2 & c_2 & 0 & 0 \\ 0 & 0 & 1 & 0 \\ 0 & 0 & 0 & 1 \end{bmatrix}, \tag{5.20}
$$

$$
{}^{2}_{3}T = \begin{bmatrix} 1 & 0 & 0 & l_2 \\ 0 & 1 & 0 & 0 \\ 0 & 0 & 1 & 0 \\ 0 & 0 & 0 & 1 \end{bmatrix}.
$$

Note that these correspond to the manipulator of Example 3.3 with joint 3 permanently fixed at zero degrees. The final transformation between frames {2} and {3} need not be cast as a standard link transformation (though it may be helpful to do so). Then using (5.16) and (5.18) sequentially from link to link, we calculate:

$$ {}^{1}\omega_1 = \begin{bmatrix} 0 \\ 0 \\ \dot{\theta}_1 \end{bmatrix}, \tag{5.21} $$

$$ {}^{1}v_1 = \begin{bmatrix} 0 \\ 0 \\ 0 \end{bmatrix}, \tag{5.22} $$

$$ {}^{2}\omega_2 = \begin{bmatrix} 0 \\ 0 \\ \dot{\theta}_1 + \dot{\theta}_2 \end{bmatrix}, \tag{5.23} $$

$$ {}^{2}v_2 = \begin{bmatrix} c_2 & s_2 & 0 \\ -s_2 & c_2 & 0 \\ 0 & 0 & 1 \end{bmatrix} \begin{bmatrix} 0 \\ l_1\dot{\theta}_1 \\ 0 \end{bmatrix} = \begin{bmatrix} l_1 s_2 \dot{\theta}_1 \\ l_1 c_2 \dot{\theta}_1 \\ 0 \end{bmatrix}, \tag{5.24} $$

$$ {}^{3}\omega_3 = {}^{2}\omega_2, \tag{5.25} $$

$$ {}^{3}v_3 = \begin{bmatrix} l_1 s_2 \dot{\theta}_1 \\ l_1 c_2 \dot{\theta}_1 + l_2(\dot{\theta}_1 + \dot{\theta}_2) \\ 0 \end{bmatrix}. \tag{5.26} $$

Equation (5.26) is our answer. Also the rotational velocity of frame $\{3\}$ is found in Eq. (5.25).

To find these velocities with respect to the nonmoving base frame, we rotate them with the rotation matrix ${}^{0}_{3}R$, which is

$$ {}^{0}_{3}R = {}^{0}_{1}R \;\; {}^{1}_{2}R \;\; {}^{2}_{3}R = \begin{bmatrix} c_{12} & -s_{12} & 0 \\ s_{12} & c_{12} & 0 \\ 0 & 0 & 1 \end{bmatrix}. \tag{5.27} $$

This rotation yields

$$ {}^{0}v_3 = \begin{bmatrix} -l_1 s_1 \dot{\theta}_1 - l_2 s_{12}(\dot{\theta}_1 + \dot{\theta}_2) \\ l_1 c_1 \dot{\theta}_1 + l_2 c_{12}(\dot{\theta}_1 + \dot{\theta}_2) \\ 0 \end{bmatrix}. \quad \blacksquare \tag{5.28} $$

It is important to point out the two distinct uses for Eqs. (5.16) and (5.18). First, they may be used as a means of deriving analytical expressions as in Example 5.2 above. Here, we manipulate the symbolic equations until we arrive at a form such as (5.26) which will be evaluated with a computer in some application. Second, they may be used directly to compute (5.16) and (5.18) as they are written. They can easily be written as a subroutine which is then applied iteratively to compute link velocities. As such they could be used for any manipulator without the need of deriving the equations for a particular manipulator. However, the computation then yields a numeric result with the structure of the equations hidden. We are often interested in the structure of an analytic result such as (5.26). Also, if we bother to do the work (that is, (5.21) through (5.28)), we generally will find that there are fewer computations left for the computer to perform in the final application.

5.6 Jacobians

The Jacobian is a multidimensional form of the derivative. Suppose, for example, that we have six functions, each of which is a function of six independent variables:

$$\begin{aligned} y_1 &= f_1(x_1, x_2, x_3, x_4, x_5, x_6), \\ y_2 &= f_2(x_1, x_2, x_3, x_4, x_5, x_6), \\ &\vdots \\ y_6 &= f_6(x_1, x_2, x_3, x_4, x_5, x_6). \end{aligned} \tag{5.29}$$

We could also use vector notation to write these equations as:

$$Y = F(X). \tag{5.30}$$

Now, if we wish to calculate the differentials of y_i as a function of differentials of x_j, we simply use the chain rule to calculate, and we get

$$\begin{aligned} \delta y_1 &= \frac{\partial f_1}{\partial x_1}\delta x_1 + \frac{\partial f_1}{\partial x_2}\delta x_2 + \quad \cdots \quad + \frac{\partial f_1}{\partial x_6}\delta x_6, \\ \delta y_2 &= \frac{\partial f_2}{\partial x_1}\delta x_1 + \frac{\partial f_2}{\partial x_2}\delta x_2 + \quad \cdots \quad + \frac{\partial f_2}{\partial x_6}\delta x_6, \\ &\vdots \\ \delta y_6 &= \frac{\partial f_6}{\partial x_1}\delta x_1 + \frac{\partial f_6}{\partial x_2}\delta x_2 + \quad \cdots \quad + \frac{\partial f_6}{\partial x_6}\delta x_6, \end{aligned} \tag{5.31}$$

which again might be written more simply using vector notation as:

$$\delta Y = \frac{\partial F}{\partial X}\,\delta X. \tag{5.32}$$

The 6×6 matrix of partial derivatives in (5.32) is what we call the Jacobian, J. Note that if the functions $f_1(X)$ through $f_6(X)$ are nonlinear, then the partial derivatives are a function of x_i. So we use the notation

$$\delta Y = J(X)\ \delta X. \tag{5.33}$$

By dividing both sides by the differential time element, we can think of the Jacobian as mapping velocities in X to those in Y:

$$\dot{Y} = J(X)\ \dot{X}. \tag{5.34}$$

At any particular instant, X has a certain value, and $J(X)$ is a linear transformation. At each new time instant, X has changed and therefore so has the linear transformation. Jacobians are time varying linear transformations.

In the field of robotics, we generally speak of Jacobians which relate joint velocities to Cartesian velocities of the tip of the arm. For example:

$${}^{0}\mathcal{V} = {}^{0}J(\Theta)\ \dot{\Theta}, \tag{5.35}$$

where Θ is the vector of joint angles of the manipulator, and $\mathcal{V}$ is a vector of Cartesian velocities. In (5.35) we have added a leading superscript to our Jacobian notation to indicate in which frame the resulting Cartesian velocity is expressed. Sometimes this superscript is omitted when the frame is obvious or when it is unimportant to the development. Note that for any given configuration of the manipulator, joint rates are related to velocity of the tip in a linear fashion. This is only an instantaneous relationship, since in the next instant the Jacobian has changed slightly. For the general case of a six jointed robot, the Jacobian is 6×6, $\dot{\Theta}$ is 6×1, and ${}^{0}\mathcal{V}$ is 6×1. This 6×1 Cartesian velocity vector is the 3×1 linear velocity vector and the 3×1 rotational velocity vector stacked together:

$${}^{0}\mathcal{V} = \begin{bmatrix} {}^{0}v \\ {}^{0}\omega \end{bmatrix}. \tag{5.36}$$

Jacobians of any dimension (including nonsquare) may be defined. The number of rows equals the number of degrees of freedom in the Cartesian space being considered. The number of columns in a Jacobian is equal to the number of joints of the manipulator. In dealing with a planar arm, for example, there is no reason for the Jacobian to have more than three rows, although for redundant planar manipulators, there could be arbitrarily many columns (one for each joint).

In the case of a two-link arm, we can write a 2×2 Jacobian which relates joint rates to end-effector velocity. From the result of Example 5.2

we can easily determine the Jacobian of our two-link arm. The Jacobian written in frame $\{3\}$ is seen (from (5.26)) to be

$$ {}^{3}J(\Theta) = \begin{bmatrix} l_1 s_2 & 0 \\ l_1 c_2 + l_2 & l_2 \end{bmatrix}, \tag{5.37} $$

and the Jacobian written in frame $\{0\}$ is (from (5.28))

$$ {}^{0}J(\Theta) = \begin{bmatrix} -l_1 s_1 - l_2 s_{12} & -l_2 s_{12} \\ l_1 c_1 + l_2 c_{12} & l_2 c_{12} \end{bmatrix}. \tag{5.38} $$

Note that in both cases, we have chosen to write a square matrix which relates joint rates to end-effector velocity. We could also consider the 3×2 Jacobian which would include the angular velocity of the end-effector.

Considering Eqs. (5.29) through (5.33) which define the Jacobian, we see that the Jacobian might also be found by directly differentiating the kinematic equations of the mechanism. However, while this is straightforward for linear velocity, there is no 3×1 orientation vector whose derivative is ω. Hence, we have introduced a method to derive the Jacobian using succesive application of (5.16) and (5.18). There are several other methods which may be used (see, for example, [4]), one of which will be introduced shortly in Section 5.8. One reason for deriving Jacobians using the method presented is that it helps prepare us for material in Chapter 6, in which we will find that similar techniques apply to calculating the dynamic equations of motion of a manipulator.

5.7 Singularities

Given that we have a linear transformation relating joint velocity to Cartesian velocity, a reasonable question to ask is: Is this matrix invertible? That is, is it nonsingular? If the matrix is nonsingular then we can invert it to calculate joint rates given Cartesian velocities:

$$ \dot{\Theta} = J^{-1}(\Theta)\, \mathcal{V}. \tag{5.39} $$

This is an important relationship. For example, say we wish the hand of the robot to move with a certain velocity vector in Cartesian space. Using (5.39) we could calculate the necessary joint rates at each instant along the path. The real question of invertibility is: Is the Jacobian invertible for all values of Θ? If not, where is it not invertible?

Most manipulators have values of Θ where the Jacobian becomes singular. Such locations are called **singularities of the mechanism** or **singularities** for short. All manipulators have singularities at the boundary of their workspace, and most have loci of singularities inside their workspace. An in-depth study of the classification of singularities is beyond the scope of this book—for more information see [5]. For our purposes, and without giving rigorous definitions, we will class singularities into two categories:

1. **Workspace boundary singularities** are those which occur when the manipulator is fully streched out or folded back on itself such that the end-effector is near or at the boundary of the workspace.
2. **Workspace interior singularities** are those which occur away from the workspace boundary and generally are caused by two or more joint axes lining up.

When a manipulator is in a singular configuration, it has lost one or more degrees of freedom as viewed from Cartesian space. This means that there is some direction (or subspace) in Cartesian space along which it is impossible to move the hand of the robot no matter which joint rates are selected. It is obvious that this happens at the workspace boundary of robots.

EXAMPLE 5.3

Where are the singularities of the simple 2-link arm of Example 5.2? What is the physical explanation of the singularities? Are they workspace boundary singularities or are they workspace interior singularities?

To find the singular points of a mechanism we must examine the determinant of its Jacobian. Where the determinant is equal to zero, the Jacobian has lost full rank, and is singular.

$$DET[J(\Theta)] = \begin{vmatrix} l_1 s_2 & 0 \\ l_1 c_2 + l_2 & l_2 \end{vmatrix} = l_1 l_2 s_2 = 0. \tag{5.40}$$

Clearly, a singularity of the mechanism exists when θ_2 is 0 or 180 degrees. Physically, when $\theta_2 = 0$ the arm is stretched straight out. In this configuration, motion of the end-effector is possible only along one Cartesian direction (the one perpendicular to the arm). Therefore, the mechanism has lost one freedom. Likewise when $\theta_2 = 180$ the arm is folded completely back on itself, and motion of the hand is again only possible in one Cartesian direction instead of two. We will class these singularities as workspace boundary singularities because they exist at the edge of the manipulator's workspace. Note that the Jacobian written with respect to frame $\{0\}$, or any other frame, would have yielded the same result. ■

The danger in applying (5.39) in a robot control system is that at a singular point, the inverse Jacobian blows up! This results in joint rates approaching infinity as the singularity is approached.

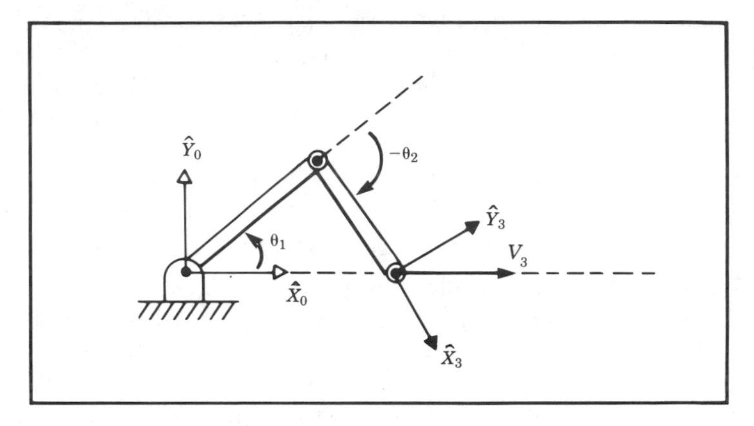

FIGURE 5.10 A two-link manipulator moving its tip at a constant linear velocity.

EXAMPLE 5.4

Consider our two-link robot from Example 5.2 moving its end-effector along the $\hat{X}$ axis at 1.0 m/s as in Fig. 5.10. Show that joint rates are reasonable when far from a singularity, but as a singularity is approached at $\theta_2 = 0$, joint rates tend to infinity.

We start by calculating the inverse of the Jacobian written in $\{0\}$:

$$ {}^0J^{-1}(\Theta) = \frac{1}{l_1 l_2 s_2} \begin{bmatrix} l_2 c_{12} & l_2 s_{12} \\ -l_1 c_1 - l_2 c_{12} & -l_1 s_1 - l_2 s_{12} \end{bmatrix}. \tag{5.41} $$

Then using Eq. (5.41) for a velocity of 1 m/s in the $\hat{X}$ direction we can calculate joint rates as a function of manipulator configuration:

$$ \begin{aligned} \dot{\theta}_1 &= \frac{c_{12}}{l_1 s_2}, \\ \dot{\theta}_2 &= -\frac{c_1}{l_2 s_2} - \frac{c_{12}}{l_1 s_2}. \end{aligned} \tag{5.42} $$

Clearly, as the arm stretches out toward $\theta_2 = 0$ both joint rates go to infinity. ■

EXAMPLE 5.5

For the PUMA 560 manipulator, give two examples of singularities which can occur.

There is singularity when θ_3 is near $-90.0°$. Calculation of the exact value of θ_3 is left as an exercise (see Exercise 5.14). In this situation, links 2 and 3 are "stretched out" just like the singular location of the 2-link manipulator in Example 5.3. This is classed as a workspace boundary singularity.

Whenever $\theta_5 = 0.0°$ the manipulator is in a singular configuration. In this configuration joint axes 4 and 6 line up—both of their actions would result in the same end-effector motion, so it is as if a degree of freedom has been lost. Because this can occur interior to the workspace envelope, we will class it as a workspace interior singularity. ■

5.8 Static forces in manipulators

The chain-like nature of a manipulator leads us quite naturally to consider how forces and moments "propagate" from one link to the next. Typically the robot is pushing on something in the environment with the chain's free end (the end-effector), or is perhaps supporting a load at the hand. We wish to solve for the joint torques which must be acting to keep the system in static equilibrium.

In considering static forces in a manipulator we first lock all the joints so that the manipulator becomes a structure. We then consider each link in this structure and write a force-moment balance relationship in terms of the link frames. Finally, we compute what static torque must be acting about the joint axis in order for the manipulator to be in static equilibrium. In this way, we solve for the set of joint torques needed to support a static load acting at the end-effector.

We define special symbols for the force and torque exerted by a neighbor link:

$$f_i = \text{force exerted on link } i \text{ by link } i-1,$$
$$n_i = \text{torque exerted on link } i \text{ by link } i-1.$$

We will use our usual convention for assigning frames to links. Figure 5.11 shows the static forces and moments acting on link i. Summing the forces and setting them equal to zero we have:

$$^{i}f_i - {}^{i}f_{i+1} = 0. \qquad \textbf{(5.43)}$$

Summing torques about the origin of frame $\{i\}$ we have:

$$^{i}n_i - {}^{i}n_{i+1} - {}^{i}P_{i+1} \times {}^{i}f_{i+1} = 0. \qquad \textbf{(5.44)}$$

If we start with a description of the force and moment applied by the hand, we can calculate the force and moment applied by each link working from the last link down to the base, link 0. To do this, we formulate the force-moment expressions (5.43) and (5.44) such that they

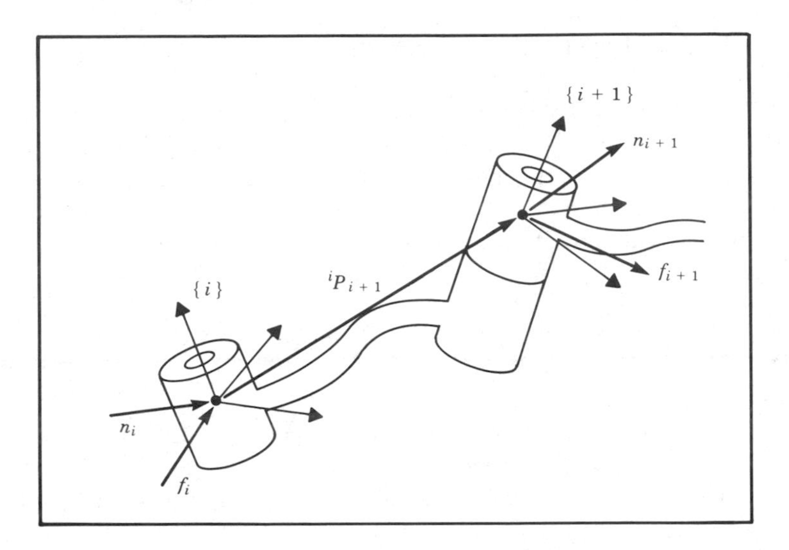

FIGURE 5.11 Static force-moment balance for a single link.

specify iterations from higher numbered links to lower numbered links. The result may be written:

$$ {}^{i}f_i = {}^{i}f_{i+1}, \tag{5.45} $$

$$ {}^{i}n_i = {}^{i}n_{i+1} + {}^{i}P_{i+1} \times {}^{i}f_{i+1}. \tag{5.46} $$

In order to write these equations in terms of only forces and moments defined within their own link frames, we transform with the rotation matrix describing frame $\{i+1\}$ relative to frame $\{i\}$. This leads to our most important result for static force "propagation" from link to link:

$$ {}^{i}f_i = {}^{i}_{i+1}R \; {}^{i+1}f_{i+1}, \tag{5.47} $$

$$ {}^{i}n_i = {}^{i}_{i+1}R \; {}^{i+1}n_{i+1} + {}^{i}P_{i+1} \times {}^{i}f_i. \tag{5.48} $$

Finally, the important question arises: What torques are needed at the joints in order to balance the reaction forces and moments acting on the links? All components of the force and moment vectors are resisted by the structure of the mechanism itself, except for the torque about the joint axis. Therefore, to find the joint torque required to maintain the static equilibrium, the dot product of the joint axis vector with the moment vector acting on the link is computed:

$$ \tau_i = {}^{i}n_i^T \; {}^{i}\hat{Z}_i. \tag{5.49} $$

In the case that joint i is prismatic, we compute the joint actuator force as:

$$ \tau_i = {}^{i}f_i^T \; {}^{i}\hat{Z}_i. \tag{5.50} $$

Note that we are using the symbol τ even for a linear joint force.

As a matter of convention we generally define the positive direction of joint torque as the direction which would tend to move the joint in the direction of increasing joint angle.

Equations (5.47) through (5.50) give us a means to compute the joint torques needed to apply any force or moment with the end-effector of a manipulator in the static case.

EXAMPLE 5.6

The two-link manipulator of Example 5.2 is applying a force vector ${}^3F$ with its end-effector (consider this force to be acting at the origin of $\{3\}$). Find the required joint torques as a function of configuration and of the applied force. See Fig. 5.12.

We apply Eqs. (5.47) through (5.49) starting from the last link and going toward the base of the robot:

$${}^2f_2 = \begin{bmatrix} f_x \\ f_y \\ 0 \end{bmatrix}, \tag{5.51}$$

$${}^2n_2 = l_2\hat{X}_2 \times \begin{bmatrix} f_x \\ f_y \\ 0 \end{bmatrix} = \begin{bmatrix} 0 \\ 0 \\ l_2 f_y \end{bmatrix}, \tag{5.52}$$

$${}^1f_1 = \begin{bmatrix} c_2 & -s_2 & 0 \\ s_2 & c_2 & 0 \\ 0 & 0 & 1 \end{bmatrix} \begin{bmatrix} f_x \\ f_y \\ 0 \end{bmatrix} = \begin{bmatrix} c_2 f_x - s_2 f_y \\ s_2 f_x + c_2 f_y \\ 0 \end{bmatrix}, \tag{5.53}$$

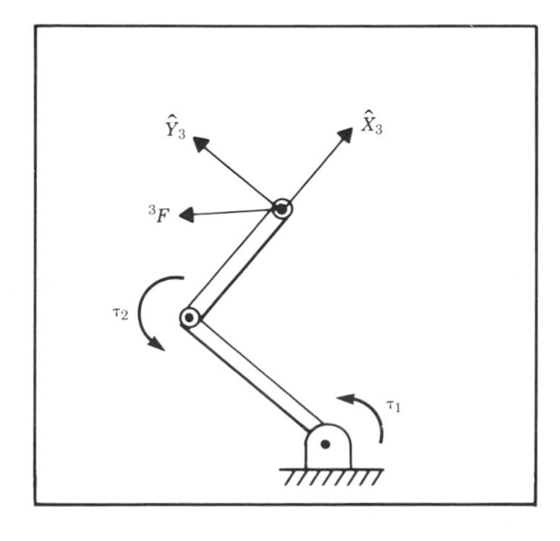

FIGURE 5.12 A two-link manipulator applying a force at its tip.

$$
{}^{1}n_{1} = \begin{bmatrix} 0 \\ 0 \\ l_2 f_y \end{bmatrix} + l_1 \hat{X}_1 \times {}^{1}f_1 = \begin{bmatrix} 0 \\ 0 \\ l_1 s_2 f_x + l_1 c_2 f_y + l_2 f_y \end{bmatrix}. \tag{5.54}
$$

Therefore we have

$$
\tau_1 = l_1 s_2 f_x + (l_2 + l_1 c_2) f_y, \tag{5.55}
$$

$$
\tau_2 = l_2 f_y. \tag{5.56}
$$

This relationship may be written as a matrix operator as:

$$
\tau = \begin{bmatrix} l_1 s_2 & l_2 + l_1 c_2 \\ 0 & l_2 \end{bmatrix} \begin{bmatrix} f_x \\ f_y \end{bmatrix}. \quad \blacksquare \tag{5.57}
$$

It is not a coincidence that this matrix is the transpose of the Jacobian that we found in (5.37)!

5.9 Jacobians in the force domain

We have found joint torques that will exactly balance forces at the hand in the static situation. When forces act on a mechanism, work (in the technical sense) is done if the mechanism moves through a displacement. Work is defined as a force acting through a distance and is a scalar with units of energy. The principle of **virtual work** allows us to make certain statements about the static case by allowing the amount of this displacement to go to an infinitesimal. Since work has units of energy it must be the same measured in any set of generalized coordinates. Specifically, we can equate the work done in Cartesian terms with the work done in joint space terms. In the multi-dimensional case, work is the dot product of a vector force or torque and a vector displacement. Thus we have

$$
\mathcal{F} \cdot \delta \mathcal{X} = \tau \cdot \delta \Theta, \tag{5.58}
$$

where $\mathcal{F}$ is a 6×1 Cartesian force-moment vector acting at the end-effector, $\delta \mathcal{X}$ is a 6×1 infinitesimal Cartesian displacement of the end-effector, τ is a 6×1 vector of torques at the joints, and $\delta\Theta$ is a 6×1 vector of infinitesimal joint displacements. Expression (5.58) can also be written

$$
\mathcal{F}^T \delta \mathcal{X} = \tau^T \delta \Theta. \tag{5.59}
$$

The definition of the Jacobian is

$$
\delta \mathcal{X} = J \delta \Theta, \tag{5.60}
$$

and so we may write

$$\mathcal{F}^T J \delta\Theta = \tau^T \delta\Theta, \tag{5.61}$$

which must hold for all $\delta\Theta$, and so we have

$$\mathcal{F}^T J = \tau^T. \tag{5.62}$$

Transposing both sides yields the result

$$\tau - J^T \ \mathcal{F}. \tag{5.63}$$

Equation (5.63) verifies in general what we saw in the particular case of the two-link manipulator in Example 5.6: The Jacobian transpose maps Cartesian forces acting at the hand into equivalent joint torques. When the Jacobian is written with respect to frame $\{0\}$, then force vectors written in $\{0\}$ may be transformed, as made clear by the notation

$$\tau = {}^0J^T \ {}^0\mathcal{F}. \tag{5.64}$$

When the Jacobian loses full rank, there are certain directions in which the end-effector cannot exert static forces as desired. That is, in (5.64) if the Jacobian is singular, $\mathcal{F}$ could be increased or decreased in certain directions (those defining the null-space of the Jacobian [6]) with no effect on the value calculated for τ. This also means that near singular configurations, mechanical advantage tends toward infinity such that with small joint torques large forces could be generated at the end-effector.* Thus, singularities manifest themselves in the force domain as well as in the position domain.

Note that (5.64) is a very interesting relationship in that it allows us to convert a Cartesian quantity into a joint space quantity without calculating any inverse kinematic functions. We will make use of this when we consider the problem of control in later chapters.

5.10 Cartesian transformation of velocities and static forces

We may wish to think in terms of 6×1 representations of general velocity of a body:

$$\mathcal{V} = \begin{bmatrix} v \\ \omega \end{bmatrix}. \tag{5.65}$$

* Consider a two-link planar manipulator nearly outstreched with the end-effector in contact with a reaction surface. In this configuration arbitrarily large forces could be exerted with "small" joint torques.

Likewise we may consider 6×1 representations of general force vectors:

$$\mathcal{F} = \begin{bmatrix} F \\ N \end{bmatrix}, \tag{5.66}$$

where F is a 3×1 force vector and N is a 3×1 moment vector. It is then natural to think of 6×6 transformations which map these quantities from one frame to another. This is exactly what we have already done in considering the propagation of velocities and forces from link to link. Here we write (5.16) and (5.18) in matrix operator form to transform general velocity vectors in frame $\{A\}$ to their description in frame $\{B\}$. Since the two frames involved here are rigidly connected, $\dot{\theta}_{i+1}$ appearing in (5.16) is set to zero in deriving the following:

$$\begin{bmatrix} {}^{B}v_B \\ {}^{B}\omega_B \end{bmatrix} = \begin{bmatrix} {}^{B}_{A}R & -{}^{B}_{A}R\,{}^{A}P_{BORG}\times \\ 0 & {}^{B}_{A}R \end{bmatrix} \begin{bmatrix} {}^{A}v_A \\ {}^{A}\omega_A \end{bmatrix}, \tag{5.67}$$

where the cross product is understood to be the matrix operator

$$P\times = \begin{bmatrix} 0 & -p_z & p_y \\ p_z & 0 & -p_x \\ -p_y & p_x & 0 \end{bmatrix}. \tag{5.68}$$

Now, (5.67) relates velocities in one frame to those in another, so the 6×6 operator is a Jacobian. In this case, it is a Jacobian which relates two Cartesian frames, so we use the following notation to compactly express (5.67):

$${}^{B}\mathcal{V}_B = {}^{B}_{A}J\;{}^{A}\mathcal{V}_A. \tag{5.69}$$

We may invert (5.67) in order to compute the description of velocity in terms of $\{A\}$ when given the quantities in $\{B\}$:

$$\begin{bmatrix} {}^{A}v_A \\ {}^{A}\omega_A \end{bmatrix} = \begin{bmatrix} {}^{A}_{B}R & {}^{A}P_{BORG}\times {}^{A}_{B}R \\ 0 & {}^{A}_{B}R \end{bmatrix} \begin{bmatrix} {}^{B}v_B \\ {}^{B}\omega_B \end{bmatrix}, \tag{5.70}$$

or

$${}^{A}\mathcal{V}_A = {}^{A}_{B}J\;{}^{B}\mathcal{V}_B. \tag{5.71}$$

Note that these mappings of velocities from frame to frame depend on ${}^{A}_{B}T$ (or its inverse) and so must be interpreted as instantaneous results, unless the relationship between the two frames is static. Similarly, from (5.47) and (5.48) we write the 6×6 matrix which transforms general force vectors written in terms of $\{B\}$ into their description in frame $\{A\}$:

$$\begin{bmatrix} {}^{A}F_A \\ {}^{A}N_A \end{bmatrix} = \begin{bmatrix} {}^{A}_{B}R & 0 \\ {}^{A}P_{BORG}\times {}^{A}_{B}R & {}^{A}_{B}R \end{bmatrix} \begin{bmatrix} {}^{B}F_B \\ {}^{B}N_B \end{bmatrix}, \tag{5.72}$$

which may be written compactly as

$${}^{A}\mathcal{F}_A = {}^{A}_{B}J^{T}\;{}^{B}\mathcal{F}_B. \tag{5.73}$$

We see that the matrix in (5.72) is indeed the transpose of the one in (5.70).

EXAMPLE 5.7

Figure 5.13 shows an end-effector holding a tool. Located at the point where the end-effector attaches to the manipulator, there is a force-sensing wrist. This is a device which can measure the forces and torques which are applied to it.

Consider the output of this sensor to be a 6×1 vector, ${}^{S}\mathcal{F}$, composed of three forces and three torques expressed in the sensor frame, $\{S\}$. Our real interest is to know the forces and torques applied at the tip of the tool, ${}^{T}\mathcal{F}$. Find the 6 × 6 transformation which transforms the force-moment vector from $\{S\}$ to the tool frame, $\{T\}$. The transform relating $\{T\}$ to $\{S\}$, ${}^{S}_{T}T$, is known. (Note that $\{S\}$ here is the sensor frame, not the station frame as usual.)

This is simply an application of (5.73). First, from ${}^{S}_{T}T$ we calculate the inverse, ${}^{T}_{S}T$, which is composed of ${}^{T}_{S}R$, and ${}^{T}P_{SORG}$. Then we apply (5.73) to obtain:

$$ {}^{T}\mathcal{F}_{T} = {}^{T}_{S}J^{T}\; {}^{S}\mathcal{F}_{S}, \tag{5.74} $$

where

$$ {}^{T}_{S}J^{T} = \begin{bmatrix} {}^{T}_{S}R & 0 \\ {}^{T}P_{SORG} \times {}^{T}_{S}R & {}^{T}_{S}R \end{bmatrix}. \quad \blacksquare \tag{5.75} $$

References

[1] K. Hunt, "Kinematic Geometry of Mechanisms," Oxford University Press, 1978.

[2] Symon, "Mechanics," Third Edition, Addison-Wesley, 1971.

[3] I. Shames, "Engineering Mechanics," Second Edition, Prentice-Hall, 1967.

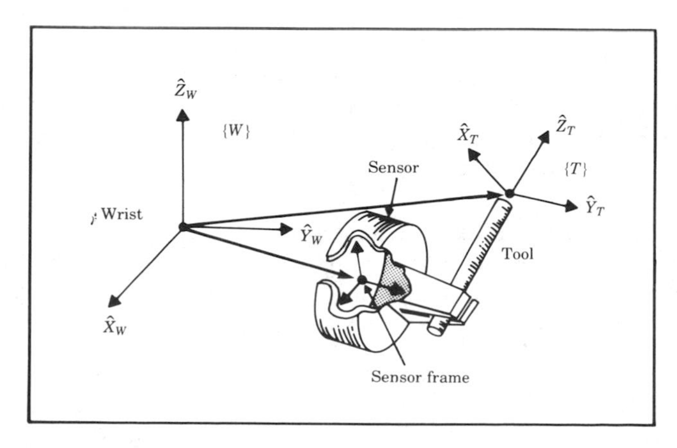

FIGURE 5.13 Frames of interest with a force sensor.

[4] D. Orin, and W. Schrader, "Efficient Jacobian Determination for Robot Manipulators," in "Robotics Research: The First International Symposium," M. Brady, and R. P. Paul, editors, MIT Press, 1984.
[5] B. Gorla, and M. Renaud, "Robots Manipulateurs," Cepadues - Editions, Toulouse, 1984.
[6] B. Noble, "Applied Linear Algebra," Prentice-Hall, 1969.
[7] J.K. Salisbury, and J. Craig, "Articulated Hands: Kinematic and Force Control Issues," *International Journal of Robotics Research,* Vol. 1, No. 1, Spring 1982.

Exercises

5.1 [10] Repeat Example 5.3 using the Jacobian written in frame $\{0\}$. Are the results the same as those of Example 5.3?

5.2 [25] Find the Jacobian of the manipulator with three degrees of freedom from Exercise 3 of Chapter 3. Write it in terms of a frame $\{4\}$ located at the tip of the hand with the same orientation as frame $\{3\}$.

5.3 [35] Find the Jacobian of the manipulator with three degrees of freedom from Exercise 3 of Chapter 3. Write it in terms of a frame $\{4\}$ located at the tip of the hand with the same orientation as frame $\{3\}$. Derive the Jacobian in three different ways: velocity propagation from base to tip, static force propagation from tip to base, and by direct differentiation of the kinematic equations.

5.4 [8] Prove that singularities in the force domain exist at the same configurations as singularities in the position domain.

5.5 [39] Calculate the Jacobian of the PUMA 560 in frame $\{6\}$.

5.6 [47] Is it true that any mechanism with three revolute joints and nonzero link lengths must have a locus of singular points interior to it's workspace?

5.7 [7] Sketch a figure of a mechanism with three degrees of freedom whose linear velocity Jacobian is the 3×3 identity matrix over all configurations of the manipulator. Describe the kinematics in a sentence or two.

5.8 [18] General mechanisms may have certain configurations, called "isotropic points" where the columns of the Jacobian become orthogonal and of equal magnitude [7]. For the two-link manipulator of Example 5.2 determine if any isotropic points exist. Hint: Is there a requirement on l_1 and l_2?

5.9 [50] Find the conditions necessary for isotropic points to exist in a general manipulator with six degrees of freedom (see Exercise 5.8).

5.10 [7] For the two-link manipulator of Example 5.2 give the transformation which would map joint torques into a 2×1 force vector, ${}^3F$, at the hand.

5.11 [14] Given:

$$ {}^{A}_{B}T = \begin{bmatrix} 0.866 & -0.500 & 0.000 & 10.0 \\ 0.500 & 0.866 & 0.000 & 0.0 \\ 0.000 & 0.000 & 1.000 & 5.0 \\ 0 & 0 & 0 & 1 \end{bmatrix}. $$

If the velocity vector at the origin of $\{A\}$ is

$$ {}^{A}\mathcal{V} = \begin{bmatrix} 0.0 \\ 2.0 \\ -3.0 \\ 1.414 \\ 1.414 \\ 0.0 \end{bmatrix}, $$

find the 6×1 velocity vector with reference point the origin of $\{B\}$.

5.12 [15] For the three-link manipulator of Exercise 3.3, give a set of joint angles for which the manipulator is at a workspace boundary singularity, and another set of angles for which the manipulator is at a workspace interior singularity.

5.13 [9] A certain two-link manipulator has the following Jacobian:

$$ {}^{0}J(\Theta) = \begin{bmatrix} -l_1 s_1 - l_2 s_{12} & -l_2 s_{12} \\ l_1 c_1 + l_2 c_{12} & l_2 c_{12} \end{bmatrix}. $$

Ignoring gravity, what are the joint torques required in order that the manipulator apply a static force vector ${}^{0}F = 10\hat{X}_0$.

5.14 [18] If the link parameter a_3 of the PUMA 560 were zero, a workspace boundary singularity would occur when $\theta_3 = -90.0°$. Give an expression for the value of θ_3 where the singularity occurs and show that if a_3 were zero, the result would be $\theta_3 = -90.0°$. Hint: In this configuration a straight line passes through joint axes 2, 3, and the point where axes 4, 5, and 6 intersect.

5.15 [24] Give the 3×3 Jacobian which calculates linear velocity of the tool tip from the three joint rates for the manipulator of Example 3.4 in Chapter 3. Give the Jacobian in frame $\{0\}$.

Programming Exercise (Part 5)

1. Two frames, $\{A\}$ and $\{B\}$ are not moving relative to one another; that is, ${}^{A}_{B}T$ is constant. In the planar case, we define the velocity of frame $\{A\}$ as

$$ {}^{A}\mathcal{V}_A = \begin{bmatrix} {}^{A}\dot{x}_A \\ {}^{A}\dot{y}_A \\ {}^{A}\dot{\theta}_A \end{bmatrix}. $$

Write a routine which, given ${}^{A}_{B}T$ and ${}^{A}\mathcal{V}_A$, computes ${}^{B}\mathcal{V}_B$. Hint: This is the planar analog of (5.67). Use a procedure heading something like:

```
Procedure Veltrans(VAR brela: frame; VAR vrela, vrelb: vec3);
```

where "vrela" is the velocity relative to frame $\{A\}$, that is, ${}^A\mathcal{V}_A$; "vrelb" is the output of the routine, its the velocity relative to frame $\{B\}$, that is, ${}^B\mathcal{V}_B$.

2. Determine the 3×3 Jacobian of the 3-link planar manipulator (from Example 3.3). In order to derive the Jacobian you should use velocity propagation analysis (as in Example 5.2) or static force analysis (as in Example 5.6). Hand in your work showing how you derived the Jacobian.

Write a routine to compute the Jacobian in frame $\{3\}$, that is, ${}^3J(\Theta)$, as a function of the joint angles. Note that frame $\{3\}$ is the standard link frame with origin on the axis of joint 3. Use a procedure heading something like:

```
Procedure Jacobian(VAR theta: vec3; Var Jac: mat33);
```

The manipulator data are: $l_1 = l_2 = 0.5$ meters.

3. A tool frame and a station frame are defined by the user for a certain task as below (units are meters and degrees):

$$ {}^W_T T = [x \quad y \quad \theta] = [0.1 \quad 0.2 \quad 30.0], $$

$$ {}^B_S T = [x \quad y \quad \theta] = [0.0 \quad 0.0 \quad 0.0]. $$

At a certain particular instant, the tool tip is at the position

$$ {}^S_T T = [x \; y \; \theta] = [0.6 \; -0.3 \; 45.0]. $$

At the same instant, the joint rates (in deg/sec) are measured to be:

$$ \dot{\Theta} = \left[\dot{\theta}_1 \; \dot{\theta}_2 \; \dot{\theta}_3\right] = [20.0 \; -10.0 \; 12.0]. $$

Calculate the linear and angular velocity of the tool tip relative to its own frame, that is, ${}^T\mathcal{V}_T$. If there is more than one possible answer calculate all possible answers.

6

MANIPULATOR DYNAMICS

6.1 Introduction

Our study of manipulators so far has focused on kinematic considerations only. We have studied static positions, static forces, and velocities; but we have never considered *the forces required to cause motion.* In this chapter we consider the equations of motion for a manipulator—the way in which motion of the manipulator arises from torques applied by the actuators, or from external forces applied to the manipulator.

Dynamics of mechanisms is a field in which many books have been written. Indeed, one can spend years studying the field. Obviously, we cannot cover the material in the completeness it deserves. However, certain formulations of the dynamics problem seem particularly well suited to application to manipulators. In particular, methods which make use of the serial chain nature of manipulators are natural candidates for our study.

There are two problems related to the dynamics of a manipulator that we wish to solve. In the first problem we are given a trajectory point,

Θ, $\dot{\Theta}$, and $\ddot{\Theta}$, and we wish to find the required vector of joint torques, τ. This formulation of dynamics is useful for the problem of controlling the manipulator (Chapter 8). The second problem is to calculate how the mechanism will move under application of a set of joint torques. That is, given a torque vector, τ, calculate the resulting motion of the manipulator, Θ, $\dot{\Theta}$, and $\ddot{\Theta}$. This is useful for simulating the manipulator.

6.2 Acceleration of a rigid body

We now extend our analysis of rigid body motion to the case of accelerations. At any instant, the linear and angular velocity vectors have derivatives which are called the linear and angular accelerations, respectively. That is,

$$^{B}\dot{V}_Q = \frac{d}{dt}\,{}^{B}V_Q = \lim_{\Delta t \to 0} \frac{{}^{B}V_Q(t+\Delta t) - {}^{B}V_Q(t)}{\Delta t}, \tag{6.1}$$

and

$$^{A}\dot{\Omega}_B = \frac{d}{dt}\,{}^{A}\Omega_B = \lim_{\Delta t \to 0} \frac{{}^{A}\Omega_B(t+\Delta t) - {}^{A}\Omega_B(t)}{\Delta t}. \tag{6.2}$$

As with velocities, when the reference frame of the differentiation is understood to be some universal reference frame, $\{U\}$, we will use the notation

$$\dot{v}_A = {}^{U}\dot{V}_{AORG} \tag{6.3}$$

and

$$\dot{\omega}_A = {}^{U}\dot{\Omega}_A. \tag{6.4}$$

Linear acceleration

We start by restating an important result from Chapter 5, (5.12), which describes the velocity of a vector ^{B}Q as seen from frame $\{A\}$ when the origins are coincident:

$$^{A}V_Q = {}^{A}_{B}R\,{}^{B}V_Q + {}^{A}\Omega_B \times {}^{A}_{B}R\,{}^{B}Q. \tag{6.5}$$

The left-hand side of this equation describes how ^{A}Q is changing in time. So, since origins are coincident, we could rewrite (6.5) as:

$$\frac{d}{dt}\left({}^{A}_{B}R\,{}^{B}Q\right) = {}^{A}_{B}R\,{}^{B}V_Q + {}^{A}\Omega_B \times {}^{A}_{B}R\,{}^{B}Q. \tag{6.6}$$

This form of the equation will be particularly useful when deriving the corresponding acceleration equation.

By differentiating (6.5), we can derive expressions for the acceleration of ${}^{B}Q$ as viewed from $\{A\}$ when origins of $\{A\}$ and $\{B\}$ coincide:

$$ {}^{A}\dot{V}_{Q} = \frac{d}{dt}\left({}^{A}_{B}R\,{}^{B}V_{Q}\right) + {}^{A}\dot{\Omega}_{B} \times {}^{A}_{B}R\,{}^{B}Q + {}^{A}\Omega_{B} \times \frac{d}{dt}\left({}^{A}_{B}R\,{}^{B}Q\right) \qquad (6.7) $$

Now we apply (6.6) twice—once to the first term, and once to the last term. The right side of equation (6.7) becomes:

$$ \begin{aligned} &{}^{A}_{B}R\,{}^{B}\dot{V}_{Q} + {}^{A}\Omega_{B} \times {}^{A}_{B}R\,{}^{B}V_{Q} + {}^{A}\dot{\Omega}_{B} \times {}^{A}_{B}R\,{}^{B}Q \\ &\quad + {}^{A}\Omega_{B} \times \left({}^{A}_{B}R\,{}^{B}V_{Q} + {}^{A}\Omega_{B} \times {}^{A}_{B}R\,{}^{B}Q\right). \end{aligned} \qquad (6.8) $$

Combining two terms, we get

$$ {}^{A}_{B}R\,{}^{B}\dot{V}_{Q} + 2\,{}^{A}\Omega_{B} \times {}^{A}_{B}R\,{}^{B}V_{Q} + {}^{A}\dot{\Omega}_{B} \times {}^{A}_{B}R\,{}^{B}Q + {}^{A}\Omega_{B} \times \left({}^{A}\Omega_{B} \times {}^{A}_{B}R\,{}^{B}Q\right). \qquad (6.9) $$

Finally, to generalize to the case in which the origins are not coincident, we add one term which gives the linear acceleration of the origin of $\{B\}$, resulting in the final general formula:

$$ \begin{aligned} &{}^{A}\dot{V}_{BORG} + {}^{A}_{B}R\,{}^{B}\dot{V}_{Q} + 2\,{}^{A}\Omega_{B} \times {}^{A}_{B}R\,{}^{B}V_{Q} + {}^{A}\dot{\Omega}_{B} \times {}^{A}_{B}R\,{}^{B}Q \\ &\quad + {}^{A}\Omega_{B} \times \left({}^{A}\Omega_{B} \times {}^{A}_{B}R\,{}^{B}Q\right). \end{aligned} \qquad (6.10) $$

A particular case that is worth pointing out is when ${}^{B}Q$ is constant, or

$$ {}^{B}V_{Q} = {}^{B}\dot{V}_{Q} = 0. \qquad (6.11) $$

In this case, (6.10) simplifies to

$$ {}^{A}\dot{V}_{Q} = {}^{A}\dot{V}_{BORG} + {}^{A}\Omega_{B} \times \left({}^{A}\Omega_{B} \times {}^{A}_{B}R\,{}^{B}Q\right) + {}^{A}\dot{\Omega}_{B} \times {}^{A}_{B}R\,{}^{B}Q. \qquad (6.12) $$

We will use this result in calculating the linear acceleration of the links of a manipulator with rotational joints. When a prismatic joint is present, the more general form of (6.10) will be used.

Angular acceleration

Consider the case of $\{B\}$ rotating relative to $\{A\}$ with ${}^{A}\Omega_{B}$, and $\{C\}$ rotating relative to $\{B\}$ with ${}^{B}\Omega_{C}$. To calculate ${}^{A}\Omega_{C}$ we sum the vectors in frame $\{A\}$:

$$ {}^{A}\Omega_{C} = {}^{A}\Omega_{B} + {}^{A}_{B}R\,{}^{B}\Omega_{C}. \qquad (6.13) $$

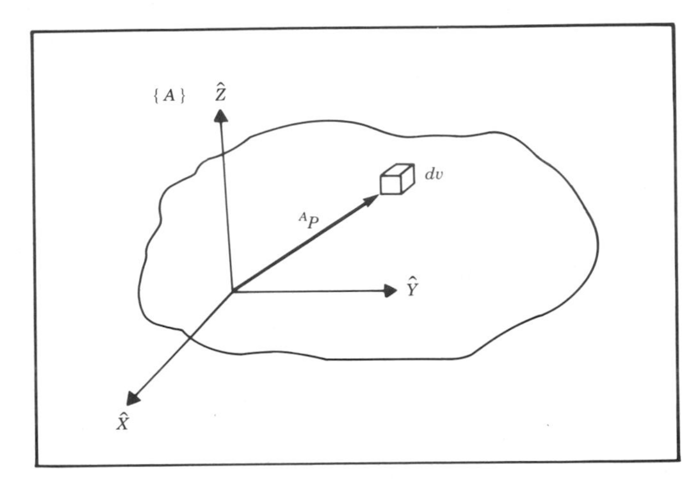

FIGURE 6.1 The inertia tensor of an object describes the object's mass distribution. Here a vector ^{A}P locates the differential volume element, dv.

By differentiating, we obtain

$$^{A}\dot{\Omega}_{C} = {}^{A}\dot{\Omega}_{B} + \frac{d}{dt}\left({}^{A}_{B}R\, {}^{B}\Omega_{C}\right). \tag{6.14}$$

Now, applying (6.6) to the last term of (6.14), we obtain

$$^{A}\dot{\Omega}_{C} = {}^{A}\dot{\Omega}_{B} + {}^{A}_{B}R\, {}^{B}\dot{\Omega}_{C} + {}^{A}\Omega_{B} \times {}^{A}_{B}R\, {}^{B}\Omega_{C}. \tag{6.15}$$

We will use this result to calculate the angular acceleration of the links of a manipulator.

6.3 Mass distribution

In systems with a single degree of freedom, we often talk about the mass of a rigid body. In the case of rotational motion about a single axis, the notion of the *moment of inertia* is a familiar one. For a rigid body which is free to move in three dimensions there are an infinite number of possible rotation axes. In the case of rotation about an arbitrary axis, we need a complete way of characterising the mass distribution of a rigid body. Here we introduce the **inertia tensor** which for our purposes can be thought of as a generalization of the scalar moment of inertia of an object.

We shall now define a set of quantities which give information about the distribution of mass of a rigid body relative to a reference frame.

Fig. 6.1 shows a rigid body with an attached frame. While inertia tensors may be defined relative to any frame, we will always consider the case of an inertia tensor defined for a frame attached to the rigid body. Where important we will indicate, with a leading superscript, the frame of reference of a given inertia tensor. The inertia tensor relative to frame $\{A\}$ is expressed in the matrix form as the 3×3 matrix:

$$^{A}I = \begin{bmatrix} I_{xx} & -I_{xy} & -I_{xz} \\ -I_{xy} & I_{yy} & -I_{yz} \\ -I_{xz} & -I_{yz} & I_{zz} \end{bmatrix}, \tag{6.16}$$

where the scalar elements are given by:

$$\begin{aligned} I_{xx} &= \iiint_V (y^2 + z^2)\rho dv, \\ I_{yy} &= \iiint_V (x^2 + z^2)\rho dv, \\ I_{zz} &= \iiint_V (x^2 + y^2)\rho dv, \\ I_{xy} &= \iiint_V xy\ \rho dv, \\ I_{xz} &= \iiint_V xz\ \rho dv, \\ I_{yz} &= \iiint_V yz\ \rho dv, \end{aligned} \tag{6.17}$$

where the rigid body is composed of differential volume elements, dv, containing material of density ρ. Each volume element is located with a vector, $^{A}P = [x\ y\ z]^T$, as shown in Fig. 6.1.

The elements I_{xx}, I_{yy}, and I_{zz} are called the **mass moments of inertia**. Note that in each case we are integrating the mass elements, ρdv, times the square of the perpendicular distance from the corresponding axis. The elements with mixed indices are called the **mass products of inertia**. This set of six independent quantities will, for a given body, depend on the position and orientation of the frame in which they are defined. If we are free to choose the orientation of the reference frame, it is possible to cause the products of inertia to be zero. The axes of the reference frame when so aligned are called the **principal axes** and the corresponding mass moments are the **principal moments** of inertia.

EXAMPLE 6.1

Find the inertia tensor for the rectangular body of uniform density ρ with respect to the coordinate system shown in Fig. 6.2.

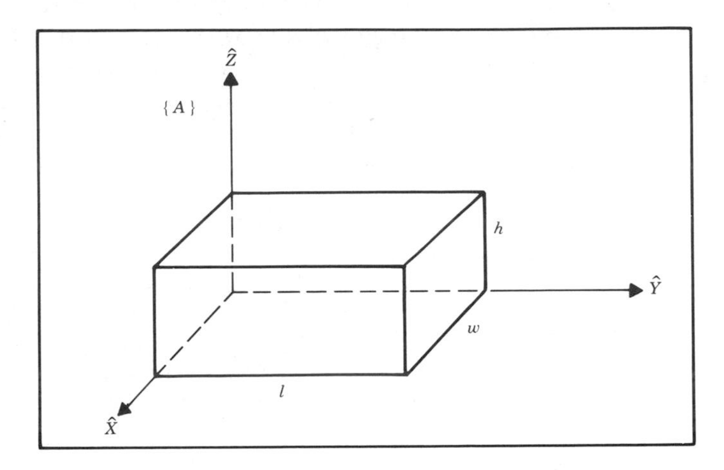

FIGURE 6.2 A body of uniform density.

First, we compute I_{xx}. Using volume element $dv = dx\ dy\ dz$, we get

$$\begin{aligned} I_{xx} &= \int_0^h \int_0^l \int_0^w (y^2 + z^2)\rho dx dy dz \\ &= \int_0^h \int_0^l (y^2 + z^2) w \rho dy dz \\ &= \int_0^h \left(\frac{l^3}{3} + z^2 l \right) w \rho dz \\ &= \left(\frac{hl^3 w}{3} + \frac{h^3 lw}{3} \right) \rho \\ &= \frac{m}{3} \left(l^2 + h^2 \right), \end{aligned} \tag{6.18}$$

where m is the total mass of the body. Permuting the terms, we can get I_{yy} and I_{zz} by inspection:

$$I_{yy} = \frac{m}{3} \left(w^2 + h^2 \right) \tag{6.19}$$

and

$$I_{zz} = \frac{m}{3} \left(l^2 + w^2 \right). \tag{6.20}$$

We next compute I_{xy}:

$$
\begin{aligned}
I_{xy} &= \int_0^h \int_0^l \int_0^w xy\rho dx dy dz \\
&= \int_0^h \int_0^l \frac{w^2}{2} y\rho dy dz \\
&= \int_0^h \frac{w^2 l^2}{4} \rho dz \\
&= \frac{m}{4} wl.
\end{aligned}
\tag{6.21}
$$

Permuting the terms, we get

$$I_{xz} = \frac{m}{4} hw \tag{6.22}$$

and

$$I_{yz} = \frac{m}{4} hl. \tag{6.23}$$

Hence the inertia tensor for this object is

$${}^{A}I = \begin{bmatrix} \frac{m}{3}\left(l^2+h^2\right) & -\frac{m}{4}wl & -\frac{m}{4}hw \\ -\frac{m}{4}wl & \frac{m}{3}\left(w^2+h^2\right) & -\frac{m}{4}hl \\ -\frac{m}{4}hw & -\frac{m}{4}hl & \frac{m}{3}\left(l^2+w^2\right) \end{bmatrix}. \quad \blacksquare \tag{6.24}$$

As we said, the inertia tensor is a function of the location and orientation of the reference frame. A well known result, the **parallel axis theorem**, is one way of computing how the inertia tensor changes under *translations* of the reference coordinate system. The parallel axis theorem relates the inertia tensor in a frame with origin at the center of mass to the inertia tensor with respect to another reference frame. Where $\{C\}$ is located at the center of mass of the body, and $\{A\}$ is an arbitrarily translated frame, the theorem is stated in two equations [1]:

$$
\begin{aligned}
{}^{A}I_{zz} &= {}^{C}I_{zz} + m(x_c^2 + y_c^2), \\
{}^{A}I_{xy} &= {}^{C}I_{xy} + m x_c y_c,
\end{aligned}
\tag{6.25}
$$

where x_c, y_c, and z_c locate the center of mass relative to $\{A\}$. The remaining moments and products of inertia are computed from permutations of x, y, and z in (6.25).

EXAMPLE 6.2

Find the inertia tensor for the same solid body described for Example 6.1 when it is described in a coordinate system with origin at the body's center of mass.

We can apply the parallel axis theorem, (6.25), where

$$\begin{bmatrix} x_c \\ y_c \\ z_c \end{bmatrix} = \frac{1}{2} \begin{bmatrix} w \\ l \\ h \end{bmatrix}$$

Then we find

$$\begin{aligned} {}^{C}I_{zz} &= \frac{m}{12}(w^2 + l^2), \\ {}^{C}I_{xy} &= 0. \end{aligned} \tag{6.26}$$

The other elements are found by symmetry. The resulting inertia tensor written in the frame at the center of mass is:

$${}^{C}I = \begin{bmatrix} \frac{m}{12}(h^2 + l^2) & 0 & 0 \\ 0 & \frac{m}{12}(w^2 + h^2) & 0 \\ 0 & 0 & \frac{m}{12}(l^2 + w^2) \end{bmatrix}. \tag{6.27}$$

Since the result is diagonal, frame $\{C\}$ must represent the principal axes of this body. ■

Some additional facts about inertia tensors are as follows:

1. If two axes of the reference frame form a plane of symmetry for the mass distribution of the body, the products of inertia having as an index the coordinate which is normal to the plane of symmetry will be zero.
2. Moments of inertia must always be positive. Products of inertia may have either sign.
3. The sum of the three moments of inertia are invariant under orientation changes in the reference frame.
4. The eigenvalues of an inertia tensor are the principal moments for the body. The associated eigenvectors are the principal axes.

Most manipulators have links whose geometry and composition is somewhat complex so that the application of (6.17) is difficult in practice. A pragmatic option is actually to measure rather than to calculate the moments of inertia of each link using a measuring device (eg., an *inertia pendulum*).

6.4 Newton's equation, Euler's equation

We will consider each link of a manipulator as a rigid body. If we know the location of the center of mass and the inertia tensor of the link, then its mass distribution is completely characterised. In order to move the links, we must accelerate and decelerate them. The forces required for such motion are a function of the acceleration desired and of the mass distribution of the links. Newton's equation along with its rotational analog, Euler's equation, describe how forces, inertias, and accelerations relate.

Newton's equation

Figure 6.3 shows a rigid body whose center of mass is accelerating with acceleration $\dot{v}_C$. In such a situation, the force, F, acting at the center of mass which causes this acceleration is given by Newton's equation,

$$F = m\dot{v}_C, \tag{6.28}$$

where m is the total mass of the body.

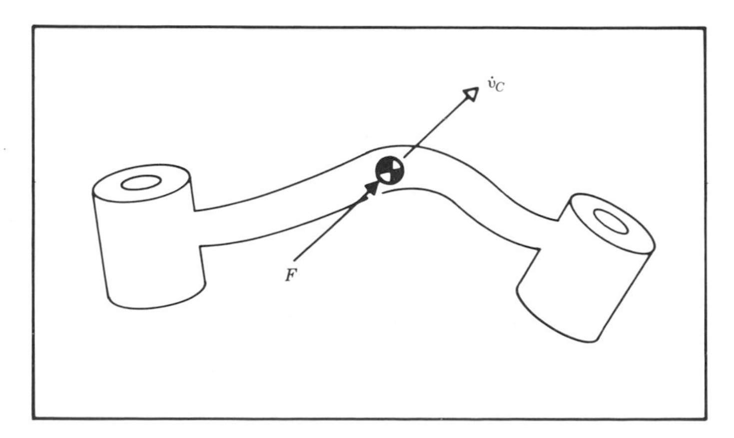

FIGURE 6.3 A force F acting at the center of mass of a body causes the body to accelerate at $\dot{v}_C$.

Euler's equation

Figure 6.4 shows a rigid body rotating with angular velocity, ω, and with angular acceleration, $\dot{\omega}$. In such a situation, the moment N, which must be acting on the body to cause this motion, is given by Euler's equation

$$N = {}^{C}I\dot{\omega} + \omega \times {}^{C}I\omega, \tag{6.29}$$

where ${}^{C}I$ is the inertia tensor of the body written in a frame, $\{C\}$, whose origin is located at the center of mass.

6.5 Iterative Newton-Euler dynamic formulation

We now consider the problem of computing the torques that correspond to a given trajectory of a manipulator. We assume we know the position, velocity, and acceleration of the joints, $(\Theta, \dot{\Theta}, \ddot{\Theta})$. With this knowledge, and with knowledge of the kinematics and mass distribution information of the robot, we can calculate the joint torques required to cause this motion.

Outward iterations to compute velocities and accelerations

In order to compute inertial forces acting on the links it is necessary to compute the rotational velocity and linear and rotational acceleration of the center of mass of each link of the manipulator at any given instant. These computations will be done in an iterative nature starting with link 1 and moving succesively, link by link, *outward* to link n.

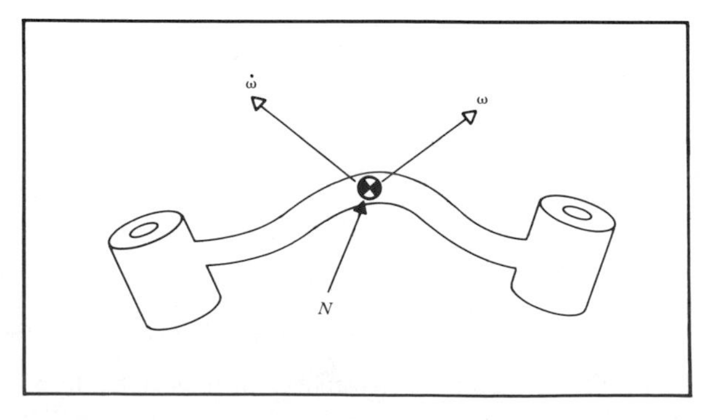

FIGURE 6.4 A moment N is acting on a body, and the body is rotating with velocity ω and accelerating at $\dot{\omega}$.

The "propagation" of rotational velocity from link to link was discussed in Chapter 5, and is given (for joint $i+1$ rotational) by

$$^{i+1}\omega_{i+1} = {}^{i+1}_{i}R\,{}^{i}\omega_i + \dot{\theta}_{i+1}\,{}^{i+1}\hat{Z}_{i+1}. \qquad \textbf{(6.30)}$$

From (6.15) we obtain the equation for transforming angular acceleration from one link to the next,

$$^{i+1}\dot{\omega}_{i+1} = {}^{i+1}_{i}R\,{}^{i}\dot{\omega}_i + {}^{i+1}_{i}R\,{}^{i}\omega_i \times \dot{\theta}_{i+1}\,{}^{i+1}\hat{Z}_{i+1} + \ddot{\theta}_{i+1}\,{}^{i+1}\hat{Z}_{i+1}. \qquad \textbf{(6.31)}$$

When joint $i+1$ is prismatic, this simplifies to

$$^{i+1}\dot{\omega}_{i+1} = {}^{i+1}_{i}R\,{}^{i}\dot{\omega}_i. \qquad \textbf{(6.32)}$$

The linear acceleration of each link frame origin is obtained by application of (6.12):

$$^{i+1}\dot{v}_{i+1} = {}^{i+1}_{i}R\left[{}^{i}\dot{\omega}_i \times {}^{i}P_{i+1} + {}^{i}\omega_i \times \left({}^{i}\omega_i \times {}^{i}P_{i+1}\right) + {}^{i}\dot{v}_i\right], \qquad \textbf{(6.33)}$$

which, for prismatic joint $i+1$, becomes (from 6.10):

$$\begin{aligned} ^{i+1}\dot{v}_{i+1} = {}& {}^{i+1}_{i}R\left({}^{i}\dot{\omega}_i \times {}^{i}P_{i+1} + {}^{i}\omega_i \times \left({}^{i}\omega_i \times {}^{i}P_{i+1}\right) + {}^{i}\dot{v}_i\right) \\ & + 2\,{}^{i+1}\omega_{i+1} \times \dot{d}_{i+1}\,{}^{i+1}\hat{Z}_{i+1} + \ddot{d}_{i+1}\,{}^{i+1}\hat{Z}_{i+1}. \end{aligned} \qquad \textbf{(6.34)}$$

We also will need the linear acceleration of the center of mass of each link, which also can be found by applying (6.12):

$$^{i}\dot{v}_{C_i} = {}^{i}\dot{\omega}_i \times {}^{i}P_{C_i} + {}^{i}\omega_i \times \left({}^{i}\omega_i \times {}^{i}P_{C_i}\right) + {}^{i}\dot{v}_i, \qquad \textbf{(6.35)}$$

where we imagine a frame, $\{C_i\}$, attached to each link with its origin located at the center of mass of the link, and with the same orientation as the link frame, $\{i\}$. Equation (6.35) doesn't involve joint motion at all, and so is valid for joint $i+1$ revolute or prismatic.

Note that the application of the equations to link 1 are particularly simple since $^{0}\omega_0 = {}^{0}\dot{\omega}_0 = 0$.

The force and torque acting on a link

Having computed the linear and angular accelerations of the mass center of each link, we can apply the Newton-Euler equations (Section 6.4) to compute the inertial force and torque acting at the center of mass of each link. Thus we have

$$\begin{aligned} F_i &= m\dot{v}_{C_i}, \\ N_i &= {}^{C_i}I\dot{\omega}_i + \omega_i \times {}^{C_i}I\omega_i, \end{aligned} \qquad \textbf{(6.36)}$$

where $\{C_i\}$ has it's origin at the center of mass of the link, and has the same orientation as the link frame, $\{i\}$.

Inward iterations to compute forces and torques

Having computed the forces and torques acting on each link, it now remains to calculate the joint torques which will result in these net forces and torques being applied to each link.

We can do this by writing a force balance and moment balance equation based on a free body diagram of a typical link (see Fig. 6.5). Each link has forces and torques exerted on it by its neighbors, and in addition experiences an inertial force and torque. In Chapter 5 we defined special symbols for the force and torque exerted by a neighbor link, which we repeat here:

$$f_i = \text{force exerted on link } i \text{ by link } i-1,$$
$$n_i = \text{torque exerted on link } i \text{ by link } i-1.$$

By summing forces acting on link i we arrive at a force balance relationship,

$$^{i}F_i = {}^{i}f_i - {}^{i}_{i+1}R\, {}^{i+1}f_{i+1}. \tag{6.37}$$

By summing torques about the center of mass and setting them equal to zero we arrive at the torque balance equation:

$$^{i}N_i = {}^{i}n_i - {}^{i}n_{i+1} + \left(-{}^{i}P_{C_i}\right) \times {}^{i}f_i - \left({}^{i}P_{i+1} - {}^{i}P_{C_i}\right) \times {}^{i}f_{i+1}. \tag{6.38}$$

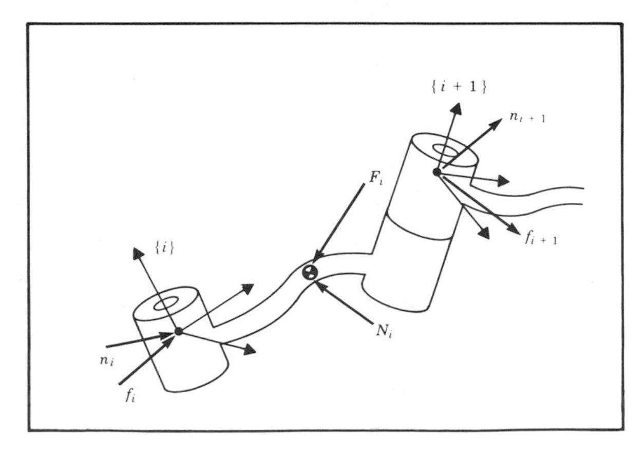

FIGURE 6.5 The force balance, including inertial forces, for a single manipulator link.

Using the result from the force balance relation (6.37) and adding a few rotation matrices, we can write (6.38) as:

$$ {}^{i}N_i = {}^{i}n_i - {}^{i}_{i+1}R\, {}^{i+1}n_{i+1} - {}^{i}P_{C_i} \times {}^{i}F_i - {}^{i}P_{i+1} \times {}^{i}_{i+1}R\, {}^{i+1}f_{i+1}. \quad \textbf{(6.39)} $$

Finally, we can rearrange the force and torque equations so that they appear as iterative relationships from higher numbered neighbor to lower-numbered neighbor.

$$ {}^{i}f_i = {}^{i}_{i+1}R\, {}^{i+1}f_{i+1} + {}^{i}F_i, \quad \textbf{(6.40)} $$

$$ {}^{i}n_i = {}^{i}N_i + {}^{i}_{i+1}R\, {}^{i+1}n_{i+1} + {}^{i}P_{C_i} \times {}^{i}F_i + {}^{i}P_{i+1} \times {}^{i}_{i+1}R\, {}^{i+1}f_{i+1}. \quad \textbf{(6.41)} $$

These equations are evaluated link by link starting from link n and working inward toward the base of the robot. These *inward force iterations* are analogous to the static force iterations introduced in Chapter 5, except that inertial forces and torques are now considered at each link.

As in the static case, the required joint torques are found by taking the $\hat{Z}$ component of the torque applied by one link on it's neighbor:

$$ \tau_i = {}^{i}n_i^T\, {}^{i}\hat{Z}_i. \quad \textbf{(6.42)} $$

For joint i prismatic, we use

$$ \tau_i = {}^{i}f_i^T\, {}^{i}\hat{Z}_i, \quad \textbf{(6.43)} $$

where we have used the symbol τ for a linear actuator force.

Note that for a robot moving in free space, ${}^{N+1}f_{N+1}$ and ${}^{N+1}n_{N+1}$ are set equal to zero, and so the first application of the equations for link n is very simple. If the robot is contacting the environment, the forces and torques due to this contact may be included in the force balance by having nonzero ${}^{N+1}f_{N+1}$ and ${}^{N+1}n_{N+1}$.

The iterative Newton-Euler dynamics algorithm

The complete algorithm for computing joint torques from the motion of the joints is composed of two parts. First, link velocities and accelerations are iteratively computed from link 1 out to link n and the Newton-Euler equations are applied to each link. Second, forces and torques of interaction and joint actuator torques are computed recursively from link n back to link 1. The equations are summarized below for the case of all joints rotational.

Outward iterations: $i : 0 \to 5$

$$ {}^{i+1}\omega_{i+1} = {}^{i+1}_{i}R\,{}^{i}\omega_i + \dot{\theta}_{i+1}\,{}^{i+1}\hat{Z}_{i+1}, \tag{6.44} $$

$$ {}^{i+1}\dot{\omega}_{i+1} = {}^{i+1}_{i}R\,{}^{i}\dot{\omega}_i + {}^{i+1}_{i}R\,{}^{i}\omega_i \times \dot{\theta}_{i+1}\,{}^{i+1}\hat{Z}_{i+1} + \ddot{\theta}_{i+1}\,{}^{i+1}\hat{Z}_{i+1}, \tag{6.45} $$

$$ {}^{i+1}\dot{v}_{i+1} = {}^{i+1}_{i}R\left({}^{i}\dot{\omega}_i \times {}^{i}P_{i+1} + {}^{i}\omega_i \times \left({}^{i}\omega_i \times {}^{i}P_{i+1}\right) + {}^{i}\dot{v}_i\right), \tag{6.46} $$

$$ {}^{i+1}\dot{v}_{C_{i+1}} = {}^{i+1}\dot{\omega}_{i+1} \times {}^{i+1}P_{C_{i+1}} + {}^{i+1}\omega_{i+1} \times \left({}^{i+1}\omega_{i+1} \times {}^{i+1}P_{C_{i+1}}\right) + {}^{i+1}\dot{v}_{i+1}, \tag{6.47} $$

$$ {}^{i+1}F_{i+1} = m_{i+1}\,{}^{i+1}\dot{v}_{C_{i+1}}, \tag{6.48} $$

$$ {}^{i+1}N_{i+1} = {}^{C_{i+1}}I_{i+1}\,{}^{i+1}\dot{\omega}_{i+1} + {}^{i+1}\omega_{i+1} \times {}^{C_{i+1}}I_{i+1}\,{}^{i+1}\omega_{i+1}. \tag{6.49} $$

Inward iterations: $i : 6 \to 1$

$$ {}^{i}f_i = {}^{i}_{i+1}R\,{}^{i+1}f_{i+1} + {}^{i}F_i, \tag{6.50} $$

$$ {}^{i}n_i = {}^{i}N_i + {}^{i}_{i+1}R\,{}^{i+1}n_{i+1} + {}^{i}P_{C_i} \times {}^{i}F_i + {}^{i}P_{i+1} \times {}^{i}_{i+1}R\,{}^{i+1}f_{i+1}, \tag{6.51} $$

$$ \tau_i = {}^{i}n_i^T\,{}^{i}\hat{Z}_i. \tag{6.52} $$

Inclusion of gravity forces in the dynamics algorithm

The effect of gravity loading on the links can be included quite simply by setting ${}^{0}\dot{v}_0 = G$, where G is the gravity vector. This is equivalent to saying that the base of the robot is accelerating upward with 1 G acceleration. This fictitious upward acceleration causes exactly the same effect on the links as gravity would. So, with no extra computational expense, the gravity effect is calculated.

6.6 Iterative vs. closed form

Equations (6.44) through (6.52) give a computational scheme whereby given the joint positions, velocities, and accelerations, we can compute the required joint torques. As with our development of equations to compute the Jacobian in Chapter 5, these relations can be used in two ways: as a numerical computational algorithm, or as an algorithm used analytically to develop symbolic equations.

Use of the equations as a numerical computational algorithm is attractive because the equations apply to any robot. Once the inertia tensors, link masses, P_{C_i} vectors, and ${}^{i+1}_{i}R$ matrices are specified for a particular manipulator, the equations may be applied directly to compute the joint torques corresponding to any motion.

However, we often are interested in obtaining better insight to the structure of the equations. For example, what is the form of the gravity terms? How does the magnitude of the gravity effects compare with the magnitude of the inertial effects? To investigate these and other questions, we often wish to write **closed form** dynamic equations. These closed form equations can be derived by applying the recursive Newton-Euler equations symbolically to Θ, $\dot{\Theta}$, and $\ddot{\Theta}$. This is analogous to what we did in Chapter 5 to derive the symbolic form of the Jacobian.

6.7 An example of closed form dynamic equations

Here we compute the closed form dynamic equations for the 2-link planar manipulator shown in Fig. 6.6. For simplicity, we assume that the mass distribution is extremely simple: All mass exists as a point mass at the distal end of each link. These masses are m_1 and m_2.

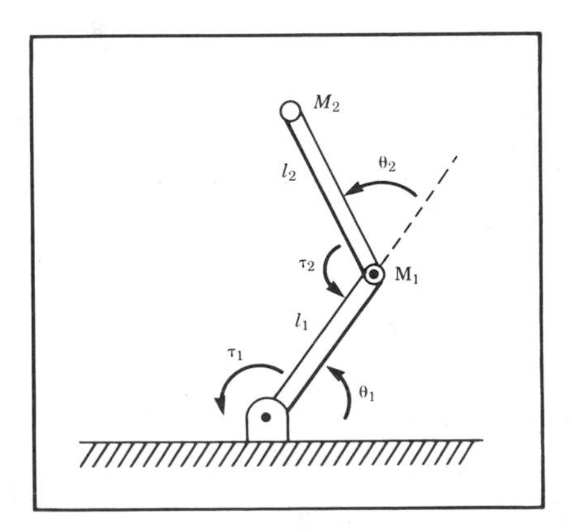

FIGURE 6.6 Two-link with point masses at distal end of links.

First we determine the value of the various quantities which will appear in the recursive Newton-Euler equations. The vectors which locate the center of mass for each link are

$$^{1}P_{C_1} = l_1 \hat{X}_1,$$

$$^{2}P_{C_2} = l_2 \hat{X}_2.$$

Because of the point mass assumption, the inertia tensor written at the center of mass for each link is the zero matrix:

$$^{C_1}I_1 = 0,$$

$$^{C_2}I_2 = 0.$$

There are no forces acting on the end-effector, and so we have

$$f_3 = 0,$$

$$n_3 = 0.$$

The base of the robot is not rotating, and hence we have

$$\omega_0 = 0,$$

$$\dot{\omega}_0 = 0.$$

To include gravity forces we will use

$$^{0}\dot{v}_0 = g\hat{Y}_0.$$

The rotation between successive link frames is given by

$$^{i}_{i+1}R = \begin{bmatrix} c_{i+1} & -s_{i+1} & 0.0 \\ s_{i+1} & c_{i+1} & 0.0 \\ 0.0 & 0.0 & 1.0 \end{bmatrix},$$

$$^{i+1}_{i}R = \begin{bmatrix} c_{i+1} & s_{i+1} & 0.0 \\ -s_{i+1} & c_{i+1} & 0.0 \\ 0.0 & 0.0 & 1.0 \end{bmatrix}.$$

We now apply equations (6.44) through (6.52).

The outward iterations for link 1 are as follows:

$$^{1}\omega_1 = \dot{\theta}_1 {}^{1}\hat{Z}_1 = \begin{bmatrix} 0 \\ 0 \\ \dot{\theta}_1 \end{bmatrix},$$

$$^{1}\dot{\omega}_1 = \ddot{\theta}_1 {}^{1}\hat{Z}_1 = \begin{bmatrix} 0 \\ 0 \\ \ddot{\theta}_1 \end{bmatrix},$$

$$^{1}\dot{v}_1 = \begin{bmatrix} c_1 & s_1 & 0 \\ -s_1 & c_1 & 0 \\ 0 & 0 & 1 \end{bmatrix} \begin{bmatrix} 0 \\ g \\ 0 \end{bmatrix} = \begin{bmatrix} gs_1 \\ gc_1 \\ 0 \end{bmatrix},$$

$$
{}^{1}\dot{v}_{C_1} = \begin{bmatrix} 0 \\ l_1\ddot{\theta}_1 \\ 0 \end{bmatrix} + \begin{bmatrix} -l_1\dot{\theta}_1^2 \\ 0 \\ 0 \end{bmatrix} + \begin{bmatrix} gs_1 \\ gc_1 \\ 0 \end{bmatrix} = \begin{bmatrix} -l_1\dot{\theta}_1^2 + gs_1 \\ l_1\ddot{\theta}_1 + gc_1 \\ 0 \end{bmatrix},
$$

$$
{}^{1}F_1 = \begin{bmatrix} -m_1 l_1\dot{\theta}_1^2 + m_1 gs_1 \\ m_1 l_1\ddot{\theta}_1 + m_1 gc_1 \\ 0 \end{bmatrix},
$$

$$
{}^{1}N_1 = \begin{bmatrix} 0 \\ 0 \\ 0 \end{bmatrix}.
$$

(6.53a–f)

The outward iterations for link 2 are as follows:

$$
{}^{2}\omega_2 = \begin{bmatrix} 0 \\ 0 \\ \dot{\theta}_1 + \dot{\theta}_2 \end{bmatrix},
$$

$$
{}^{2}\dot{\omega}_2 = \begin{bmatrix} 0 \\ 0 \\ \ddot{\theta}_1 + \ddot{\theta}_2 \end{bmatrix},
$$

$$
{}^{2}\dot{v}_2 = \begin{bmatrix} c_2 & s_2 & 0 \\ -s_2 & c_2 & 0 \\ 0 & 0 & 1 \end{bmatrix} \begin{bmatrix} -l_1\dot{\theta}_1^2 + gs_1 \\ l_1\ddot{\theta}_1 + gc_1 \\ 0 \end{bmatrix} = \begin{bmatrix} l_1\ddot{\theta}_1 s_2 - l_1\dot{\theta}_1^2 c_2 + gs_{12} \\ l_1\ddot{\theta}_1 c_2 + l_1\dot{\theta}_1^2 s_2 + gc_{12} \\ 0 \end{bmatrix},
$$

$$
{}^{2}\dot{v}_{C_2} = \begin{bmatrix} 0 \\ l_2(\ddot{\theta}_1 + \ddot{\theta}_2) \\ 0 \end{bmatrix} + \begin{bmatrix} -l_2(\dot{\theta}_1 + \dot{\theta}_2)^2 \\ 0 \\ 0 \end{bmatrix} + \begin{bmatrix} l_1\ddot{\theta}_1 s_2 - l_1\dot{\theta}_1^2 c_2 + gs_{12} \\ l_1\ddot{\theta}_1 c_2 + l_1\dot{\theta}_1^2 s_2 + gc_{12} \\ 0 \end{bmatrix},
$$

$$
{}^{2}F_2 = \begin{bmatrix} m_2 l_1\ddot{\theta}_1 s_2 - m_2 l_1\dot{\theta}_1^2 c_2 + m_2 gs_{12} - m_2 l_2(\dot{\theta}_1 + \dot{\theta}_2)^2 \\ m_2 l_1\ddot{\theta}_1 c_2 + m_2 l_1\dot{\theta}_1^2 s_2 + m_2 gc_{12} + m_2 l_2(\ddot{\theta}_1 + \ddot{\theta}_2) \\ 0 \end{bmatrix},
$$

$$
{}^{2}N_2 = \begin{bmatrix} 0 \\ 0 \\ 0 \end{bmatrix}.
$$

(6.54a – f)

Inward iterations for link 2 are as follows:

$$
{}^{2}f_2 = {}^{2}F_2
$$

$$
{}^{2}n_2 = \begin{bmatrix} 0 \\ 0 \\ m_2 l_1 l_2 c_2\ddot{\theta}_1 + m_2 l_1 l_2 s_2\dot{\theta}_1^2 + m_2 l_2 gc_{12} + m_2 l_2^2(\ddot{\theta}_1 + \ddot{\theta}_2) \end{bmatrix}
$$

(6.55a – b)

Inward iterations for link 1 are as follows:

$$
{}^1f_1 = \begin{bmatrix} c_2 & -s_2 & 0 \\ s_2 & c_2 & 0 \\ 0 & 0 & 1 \end{bmatrix} \begin{bmatrix} m_2 l_1 s_2 \ddot{\theta}_1 - m_2 l_1 c_2 \dot{\theta}_1^2 + m_2 g s_{12} - m_2 l_2 (\dot{\theta}_1 + \dot{\theta}_2)^2 \\ m_2 l_1 c_2 \ddot{\theta}_1 + m_2 l_1 s_2 \dot{\theta}_1^2 + m_2 g c_{12} + m_2 l_2 (\ddot{\theta}_1 + \ddot{\theta}_2) \\ 0 \end{bmatrix}
+ \begin{bmatrix} -m_1 l_1 \dot{\theta}_1^2 + m_1 g s_1 \\ m_1 l_1 \ddot{\theta}_1 + m_1 g c_1 \\ 0 \end{bmatrix},
$$

$$
{}^1n_1 = \begin{bmatrix} 0 \\ 0 \\ m_2 l_1 l_2 c_2 \ddot{\theta}_1 + m_2 l_1 l_2 s_2 \dot{\theta}_1^2 + m_2 l_2 g c_{12} + m_2 l_2^2 (\ddot{\theta}_1 + \ddot{\theta}_2) \end{bmatrix}
+ \begin{bmatrix} 0 \\ 0 \\ m_1 l_1^2 \ddot{\theta}_1 + m_1 l_1 g c_1 \end{bmatrix}
+ \begin{bmatrix} 0 \\ 0 \\ m_2 l_1^2 \ddot{\theta}_1 - m_2 l_1 l_2 s_2 (\dot{\theta}_1 + \dot{\theta}_2)^2 + m_2 l_1 g s_2 s_{12} \\ \qquad + m_2 l_1 l_2 c_2 (\ddot{\theta}_1 + \ddot{\theta}_2) + m_2 l_1 g c_2 c_{12} \end{bmatrix}.
$$

(6.56a – b)

Extracting the $\hat{Z}$ components of the ${}^i n_i$, we find the joint torques:

$$
\begin{aligned}
\tau_1 = {} & m_2 l_2^2 (\ddot{\theta}_1 + \ddot{\theta}_2) + m_2 l_1 l_2 c_2 (2\ddot{\theta}_1 + \ddot{\theta}_2) + (m_1 + m_2) l_1^2 \ddot{\theta}_1 - m_2 l_1 l_2 s_2 \dot{\theta}_2^2 \\
& - 2 m_2 l_1 l_2 s_2 \dot{\theta}_1 \dot{\theta}_2 + m_2 l_2 g c_{12} + (m_1 + m_2) l_1 g c_1, \\
\tau_2 = {} & m_2 l_1 l_2 c_2 \ddot{\theta}_1 + m_2 l_1 l_2 s_2 \dot{\theta}_1^2 + m_2 l_2 g c_{12} + m_2 l_2^2 (\ddot{\theta}_1 + \ddot{\theta}_2).
\end{aligned}
$$

(6.57a – b)

Equations (6.57) give expressions for the torque at the actuators as a function of joint position, velocity, and acceleration. Note that these rather complex functions arose from one of the simplest manipulators imaginable. Obviously, the closed form equations for a manipulator with six degrees of freedom are quite complex.

6.8 The general structure of the manipulator dynamic equations

It is often convenient to express the dynamic equations of a manipulator in a single equation which hides the details, but shows some of the structure of the equations.

The state space equation

When the Newton-Euler equations are evaluated symbolically for any manipulator, they yield a dynamic equation which can be written in the form

$$\tau = M(\Theta)\ddot{\Theta} + V(\Theta, \dot{\Theta}) + G(\Theta), \tag{6.58}$$

where $M(\Theta)$ is the $n \times n$ **mass matrix** of the manipulator, $V(\Theta, \dot{\Theta})$ is an $n \times 1$ vector of centrifugal and Coriolis terms, and $G(\Theta)$ is an $n \times 1$ vector of gravity terms. We use the term **state space equation** because the term $V(\Theta, \dot{\Theta})$, appearing in (6.58) has both position and velocity dependence [2].

Each element of $M(\Theta)$ and $G(\Theta)$ is a complex function which depends on Θ, the position of all the joints of the manipulator. Each element of $V(\Theta, \dot{\Theta})$ is a complex function of both Θ and $\dot{\Theta}$.

We may separate the various types of terms appearing in the dynamic equations and form the mass matrix of the manipulator, the centrifugal and Coriolis vector, and the gravity vector.

EXAMPLE 6.3

Give $M(\Theta)$, $V(\Theta, \dot{\Theta})$, and $G(\Theta)$ for the manipulator of Section 6.7.

Equation (6.58) defines the manipulator mass matrix, $M(\Theta)$: it is composed of all those terms which multiply $\ddot{\Theta}$, and is a function of Θ. Therefore we have

$$M(\Theta) = \begin{bmatrix} l_2^2 m_2 + 2l_1 l_2 m_2 c_2 + l_1^2(m_1 + m_2) & l_2^2 m_2 + l_1 l_2 m_2 c_2 \\ l_2^2 m_2 + l_1 l_2 m_2 c_2 & l_2^2 m_2 \end{bmatrix}. \tag{6.59}$$

Any manipulator mass matrix is symmetric and positive definite, and is, therefore, always invertible.

The velocity term, $V(\Theta, \dot{\Theta})$, contains all those terms which have any dependence on joint velocity. Therefore we have

$$V(\Theta, \dot{\Theta}) = \begin{bmatrix} -m_2 l_1 l_2 s_2 \dot{\theta}_2^2 - 2m_2 l_1 l_2 s_2 \dot{\theta}_1 \dot{\theta}_2 \\ m_2 l_1 l_2 s_2 \dot{\theta}_1^2 \end{bmatrix}. \tag{6.60}$$

A term like $-m_2 l_1 l_2 s_2 \dot{\theta}_2^2$ is caused by a **centrifugal force**, and is recognized as such because it depends on the square of a joint velocity. A term such as $-2m_2 l_1 l_2 s_2 \dot{\theta}_1 \dot{\theta}_2$ is caused by a **Coriolis force** and will always contain the product of two different joint velocities.

The gravity term, $G(\Theta)$, contains all those terms in which the gravitational constant, g, appears. Therefore we have

$$G(\Theta) = \begin{bmatrix} m_2 l_2 g c_{12} + (m_1 + m_2) l_1 g c_1 \\ m_2 l_2 g c_{12} \end{bmatrix}. \tag{6.61}$$

Note that the gravity term depends only on Θ, and not on its derivatives. ■

The configuration space equation

By writing the velocity dependent term, $V(\Theta, \dot{\Theta})$, in a different form, we can write the dynamic equations as

$$\tau = M(\Theta)\ddot{\Theta} + B(\Theta)\left[\dot{\Theta}\dot{\Theta}\right] + C(\Theta)\left[\dot{\Theta}^2\right] + G(\Theta), \qquad (6.62)$$

where $B(\Theta)$ is a matrix of dimensions $n \times n(n-1)/2$ of Coriolis coefficients, $\left[\dot{\Theta}\dot{\Theta}\right]$ is an $n(n-1)/2 \times 1$ vector of joint velocity products given by

$$\left[\dot{\Theta}\dot{\Theta}\right] = \left[\dot{\theta}_1\dot{\theta}_2 \ \dot{\theta}_1\dot{\theta}_3 \ \cdots \ \dot{\theta}_{n-1}\dot{\theta}_n\right]^T, \qquad (6.63)$$

$C(\Theta)$ is an $n \times n$ matrix of centrifugal coefficients, and $\left[\dot{\Theta}^2\right]$ is an $n \times 1$ vector given by

$$\left[\dot{\theta}_1^2 \ \dot{\theta}_2^2 \ \cdots \ \dot{\theta}_n^2\right]^T. \qquad (6.64)$$

We will call (6.62) the **configuration space equation** since the matrices are functions only of manipulator position [2].

In this form of the dynamic equations, the complexity of the computation is seen to be in the form of computing various parameters which are a function of only the manipulator position, Θ. This is important in applications (such as computer control of a manipulator) in which the dynamic equations must be updated as the manipulator moves. (Equation (6.62) gives a form in which parameters which are only a function of joint position, and can be updated at a rate related to how fast the manipulator is changing configuration.) We will consider this form again when we discuss the problem of manipulator control in Chapter 8.

EXAMPLE 6.4

Give $B(\Theta)$ and $C(\Theta)$ (from (6.62)) for the manipulator of Section 6.7.

For this simple two-link manipulator, we have

$$\left[\dot{\Theta}\dot{\Theta}\right] = \left[\dot{\theta}_1\dot{\theta}_2\right],$$
$$\left[\dot{\Theta}^2\right] = \begin{bmatrix} \dot{\theta}_1^2 \\ \dot{\theta}_2^2 \end{bmatrix}. \qquad (6.65)$$

So we see that

$$B(\Theta) = \begin{bmatrix} -2m_2l_1l_2s_2 \\ 0 \end{bmatrix} \qquad (6.66)$$

and

$$C(\Theta) = \begin{bmatrix} 0 & -m_2 l_1 l_2 s_2 \\ m_2 l_1 l_2 s_2 & 0 \end{bmatrix}. \quad \blacksquare \tag{6.67}$$

6.9 Formulating manipulator dynamics in Cartesian space

Our dynamic equations have been developed in terms of the position and time derivatives of the manipulator joint angles, or in **joint space**, with the general form:

$$\tau = M(\Theta)\ddot{\Theta} + V(\Theta,\dot{\Theta}) + G(\Theta). \tag{6.68}$$

We developed this equation in joint space because we could use the serial link nature of the mechanism to advantage in deriving the equations. In this section we discuss the formulation of the dynamic equations which relate acceleration of the end-effector expressed in Cartesian space to Cartesian forces and moments acting at the end-effector.

The Cartesian state space equation

As we will see in Chapters 8 and 9, it may be desirable to express the dynamics of a manipulator with respect to Cartesian variables in the general form [3]

$$\mathcal{F} = M_x(\Theta)\ddot{\mathcal{X}} + V_x(\Theta,\dot{\Theta}) + G_x(\Theta), \tag{6.69}$$

where $\mathcal{F}$ is a force-torque vector acting on the end-effector of the robot, and $\mathcal{X}$ is an appropriate Cartesian vector representing position and orientation of the end-effector [4]. Analogous to the joint space quantities, $M_x(\Theta)$ is the **Cartesian mass matrix**, $V_x(\Theta,\dot{\Theta})$ is a vector of velocity terms in Cartesian space, and $G_x(\Theta)$ is a vector of gravity terms in Cartesian space. Note that the forces acting on the end-effector, $\mathcal{F}$, could in fact be applied by the actuators at the joints using the relationship

$$\tau = J^T(\Theta)\ \mathcal{F}, \tag{6.70}$$

where the Jacobian, $J(\Theta)$, is written in the same frame as $\mathcal{F}$ and $\ddot{\mathcal{X}}$, usually the tool frame, $\{T\}$.

Just as a scalar mass, m, has kinetic energy given by $\frac{1}{2}mv^2$, the kinetic energy of a manipulator is given by the quadratic form

$$\frac{1}{2}\dot{\Theta}^T\ M(\Theta)\ \dot{\Theta}. \tag{6.71}$$

Expressing the kinetic energy in Cartesian space we have

$$\frac{1}{2}\dot{\mathcal{X}}^T\, M_x(\Theta)\, \dot{\mathcal{X}}. \tag{6.72}$$

Setting (6.71) and (6.72) equal and using the definition of the Jacobian

$$\dot{\mathcal{X}} = J(\Theta)\, \dot{\Theta}, \tag{6.73}$$

we derive the Cartesian mass matrix as

$$M_x(\Theta) = J^{-T}(\Theta)\, M(\Theta)\, J^{-1}(\Theta). \tag{6.74}$$

The remaining Cartesian space quantities may similarly be expressed in terms of the joint space quantities in combination with the Jacobian matrix. Without proof, we summarize the formulas for computing the Cartesian dynamics below. For the derivation and more details, see [4].

$$\begin{aligned} M_x(\Theta) &= J^{-T}(\Theta)\, M(\Theta)\, J^{-1}(\Theta), \\ V_x(\Theta,\dot{\Theta}) &= J^{-T}(\Theta)\, \left(V(\Theta,\dot{\Theta}) - M(\Theta)\, J^{-1}(\Theta)\, \dot{J}(\Theta)\, \dot{\Theta}\right), \\ G_x(\Theta) &= J^{-T}(\Theta)\, G(\Theta). \end{aligned} \tag{6.75}$$

The Jacobian appearing in equations (6.75) is written in the same frame as $\mathcal{F}$ and $\mathcal{X}$ in (6.69) though the choice of this frame is arbitrary.* Note that when the manipulator approaches a singularity, certain quantities in the Cartesian dynamics become infinite.

EXAMPLE 6.5

Derive the Cartesian space form of the dynamics for the two-link planar arm of Section 6.7. Write the dynamics in terms of a frame attached to the end of the second link.

For this manipulator we have already obtained the dynamics (in Section 6.7), and the Jacobian (equation (5.37)), which we restate here:

$$J(\Theta) = \begin{bmatrix} l_1 s_2 & 0 \\ l_1 c_2 + l_2 & l_2 \end{bmatrix}. \tag{6.76}$$

First compute the inverse Jacobian:

$$J^{-1}(\Theta) = \frac{1}{l_1 l_2 s_2} \begin{bmatrix} l_2 & 0 \\ -l_1 c_2 - l_2 & l_1 s_2 \end{bmatrix} \tag{6.77}$$

* Certain choices may facilitate computation.

and the time derivative of the Jacobian:

$$\dot{J}(\Theta) = \begin{bmatrix} l_1 c_2 \dot{\theta}_2 & 0 \\ -l_1 s_2 \dot{\theta}_2 & 0 \end{bmatrix}. \qquad \textbf{(6.78)}$$

Using (6.75) and the results of Section 6.7 we obtain

$$M_x(\Theta) = \begin{bmatrix} m_2 + \frac{m_1}{s_2^2} & 0 \\ 0 & m_2 \end{bmatrix},$$

$$V_x(\Theta, \dot{\Theta}) = \begin{bmatrix} -(m_2 l_1 c_2 + m_2 l_2)\dot{\theta}_1^2 - m_2 l_2 \dot{\theta}_2^2 - (2 m_2 l_2 + m_2 l_1 c_2 + m_1 l_1 \frac{c_2}{s_2^2})\dot{\theta}_1 \dot{\theta}_2 \\ m_2 l_1 s_2 \dot{\theta}_1^2 + l_1 m_2 s_2 \dot{\theta}_1 \dot{\theta}_2 \end{bmatrix},$$

$$G_x(\Theta) = \begin{bmatrix} m_1 g \frac{c_1}{s_2} + m_2 g s_{12} \\ m_2 g c_{12} \end{bmatrix}. \qquad \textbf{(6.79)}$$

When $s_2 = 0$ the manipulator is in a singular position and some of the dynamic terms go to infinity. For example, when $\theta_2 = 0$ (arm stretched straight out), the effective Cartesian mass of the end-effector becomes infinite in the $\hat{X}_2$ direction of the link 2 tip frame, as expected. In general, at a singular configuration there is a certain direction, the *singular direction* in which motion is impossible, but general motion in the subspace orthogonal to this direction is possible [5]. ■

The Cartesian configuration space torque equation

Combining (6.69) and (6.70) we can write equivalent joint torques with the dynamics expressed in Cartesian space:

$$\tau = J^T(\Theta)\left(M_x(\Theta)\ddot{\mathcal{X}} + V_x(\Theta, \dot{\Theta}) + G_x(\Theta)\right). \qquad \textbf{(6.80)}$$

We will find it usefull to write this equation in the form

$$\tau = J^T(\Theta) M_x(\Theta)\ddot{\mathcal{X}} + B_x(\Theta)\left[\dot{\Theta}\dot{\Theta}\right] + C_x(\Theta)\left[\dot{\Theta}^2\right] + G(\Theta), \qquad \textbf{(6.81)}$$

where $B_x(\Theta)$ is a matrix of dimensions $n \times n(n-1)/2$ of Coriolis coefficients, $\left[\dot{\Theta}\dot{\Theta}\right]$ is an $n(n-1)/2 \times 1$ vector of joint velocity products given by

$$\left[\dot{\Theta}\dot{\Theta}\right] = \left[\dot{\theta}_1\dot{\theta}_2 \; \dot{\theta}_1\dot{\theta}_3 \; \cdots \; \dot{\theta}_{n-1}\dot{\theta}_n\right]^T, \qquad \textbf{(6.82)}$$

$C_x(\Theta)$ is an $n \times n$ matrix of centrifugal coefficients, and $\left[\dot{\Theta}^2\right]$ is an $n \times 1$ vector given by

$$\left[\dot{\theta}_1^2 \; \dot{\theta}_2^2 \; \cdots \; \dot{\theta}_n^2\right]^T. \qquad \textbf{(6.83)}$$

Note that in (6.81), $G(\Theta)$ is the same as in the joint space equation, but in general $B_x(\Theta) \neq B(\Theta)$ and $C_x(\Theta) \neq C(\Theta)$.

EXAMPLE 6.6

Determine $B_x(\Theta)$ and $C_x(\Theta)$ (from (6.81)) for the manipulator of Section 6.7.

If we form the product $J^T(\Theta)V_x(\Theta,\dot{\Theta})$ we find that

$$B_x(\Theta) = \begin{bmatrix} m_1 l_1^2 \frac{c_2}{s_2} - m_2 l_1 l_2 s_2 \\ m_2 l_1 l_2 s_2 \end{bmatrix} \tag{6.84}$$

and

$$C_x(\Theta) = \begin{bmatrix} 0 & -m_2 l_1 l_2 s_2 \\ m_2 l_1 l_2 s_2 & 0 \end{bmatrix}. \quad \blacksquare \tag{6.85}$$

6.10 Inclusion of nonrigid body effects

It is important to realize that the dynamic equations we have derived *do not encompass all the effects acting on a manipulator.* They include just those forces which arise from rigid body mechanics. The most important source of forces that are *not* included is friction. All mechanisms are, of course, affected by frictional forces. In present day manipulators in which significant gearing is typical, the forces due to friction can actually be quite large—perhaps equalling 25% of the torque required to move the manipulator in typical situations.

In order to make dynamic equations reflect the reality of the physical device, it is important to model (at least approximately) these forces of friction. A very simple model for friction is **viscous friction** in which the torque due to friction is proportional to the velocity of joint motion. Thus we have

$$\tau_{friction} = v\dot{\theta}, \tag{6.86}$$

where v is a viscous friction constant. Another possible simple model for friction, **Coulomb friction**, is sometimes used. Coulomb friction is constant except for a sign dependence on the joint velocity:

$$\tau_{friction} = c\ sgn(\dot{\theta}), \tag{6.87}$$

where c is a Coulomb friction constant. The value of c is often taken at one value when $\dot{\theta} = 0$, the static coefficient, and at a lower value, the dynamic coefficient, when $\dot{\theta} \neq 0$. Whether a joint of a particular manipulator exhibits viscous or Coulomb friction is a complicated issue of lubrication and other effects. A reasonable model is to include both, since both effects are likely:

$$\tau_{friction} = c\ sgn(\dot{\theta}) + v\dot{\theta}. \tag{6.88}$$

It turns out that in many manipulator joints, friction also displays a dependence on the joint position. A major cause of this effect might be gears which are not perfectly round—their eccentricity would cause friction to change according to joint position. So a fairly complex friction model would have the form

$$\tau_{friction} = f(\Theta, \dot{\Theta}). \tag{6.89}$$

These friction models are then added to the other dynamic terms derived from the rigid body model, yielding the more complete model

$$\tau = M(\Theta)\ddot{\Theta} + V(\Theta, \dot{\Theta}) + G(\Theta) + F(\Theta, \dot{\Theta}). \tag{6.90}$$

There are other effects which are also neglected in this model. For example, the assumption of rigid body links means that we have failed to include bending effects (which give rise to resonances) in our equations of motion. However, these effects are extremely difficult to model, and are beyond the scope of this book (see [6]).

6.11 Dynamic simulation

To simulate the motion of a manipulator we must make use of a model of the dynamics, such as we have just developed. Given the dynamics written in closed form as in (6.58), the most common way of simulating the motion is to solve for the acceleration (which involves inverting $M(\Theta)$):

$$\ddot{\Theta} = M^{-1}(\Theta)\left[\tau - V(\Theta, \dot{\Theta}) - G(\Theta) - F(\Theta, \dot{\Theta})\right]. \tag{6.91}$$

We may then apply any of several known numerical integration techniques to integrate the equations forward in time.

Given initial conditions on the motion of the manipulator, usually in the form:

$$\begin{aligned} \Theta(0) &= \Theta_0, \\ \dot{\Theta}(0) &= 0, \\ \ddot{\Theta}(0) &= 0, \end{aligned} \tag{6.92}$$

we can numerically integrate (6.91) forward in time by steps of size Δt. There are many methods of performing numerical integration [7]. Here we introduce perhaps the simplest integration scheme, called *Euler integration*, which is accomplished as follows: Starting with $t = 0$, iteratively compute

$$\dot{\Theta}(t + \Delta t) = \dot{\Theta}(t) + \ddot{\Theta}(t)\Delta t,$$
$$\Theta(t + \Delta t) = \Theta(t) + \dot{\Theta}(t)\Delta t + \frac{1}{2}\ddot{\Theta}(t)\Delta t^2, \tag{6.93}$$

where for each iteration, (6.91) is computed to calculate $\ddot{\Theta}$. In this way, the position, velocity, and acceleration of the manipulator caused by a certain input torque function can be computed numerically.

The selection of Δt is an issue that is often discussed. It should be sufficiently small that breaking continuous time into these small increments is a reasonable approximation. It should be sufficiently large that an excessive amount of computer time is not required to compute a simulation.

6.12 Computational considerations

Because the dynamic equations of motion for typical manipulators are so complex, it is important to consider computational issues. In this section we restrict our attention to joint space dynamics. Some issues of computational efficiency of Cartesian dynamics are discussed in [4].

A historical note concerning efficiency

Counting the number of multiplications and additions for the equations (6.44)–(6.52) when taking into consideration the simple first outward computation and simple last inward computation we get

$$126n - 99 \text{ multiplications},$$
$$106n - 92 \text{ additions},$$

where n is the number of links (here, at least two). While still somewhat complex, the formulation is tremendously efficient in comparison with some previously suggested formulations of manipulator dynamics. The first formulation of the dynamics for a manipulator [8] was done using a fairly straightforward Lagrangian approach whose required computations came out to be approximately [9]:

$$32n^4 + 86n^3 + 171n^2 + 53n - 128 \text{ multiplications},$$
$$25n^4 + 66n^3 + 129n^2 + 42n - 96 \text{ additions}.$$

For the typical case of $n = 6$, the iterative Newton-Euler scheme is about 100 times more efficient! The two approaches must of course yield equivalent equations, and numeric calculations would yield exactly the same results, but the structure of the equations is quite different.

This is not to say that a Lagrangian approach cannot be made to produce efficient equations. Rather, this comparison indicates that in formulating a computational scheme for this problem, care must be taken as regards efficiency. The relative efficiency of the method we have presented stems from posing the computations as iterations from link to link, and in the particulars of how the various quantities are represented [10].

Efficiency of closed form vs. iterative form

While the iterative scheme introduced in this chapter is quite efficient as a general means of computing the dynamics of any manipulator, closed form equations derived for a particular manipulator will usually be more efficient. Consider the two-link planar manipulator of Section 6.7. Plugging in $n = 2$ into the formulas given in Section 6.12, we find that our iterative scheme would require 153 multiplications and 120 additions to compute the dynamics of a general two-link. However, our particular two-link arm happens to be quite simple since it is planar and the masses are considered point masses. So if we consider the closed form equations which we worked out in Section 6.7, we see that computation of the dynamics in this form requires about 30 multiplications and 13 additions. This is an extreme case, because the particular manipulator is so simple, but it illustrates the point that symbolic closed form equations are likely to be the most efficient formulation of dynamics. One example of an actual manipulator with six degrees of freedom is given in [11] where closed form equations required only about 50% of the computation of the general Newton-Euler method.

Hence if manipulators are designed to be *simple* in the kinematic and dynamic sense, they will have dynamic equations which are simple. We might define a **kinematically simple** manipulator to be one which has many (or all) link twists equal to 0°, 90°, or −90°, and many link lengths and offsets equal to zero. We might define a **dynamically simple** manipulator as one for which each link inertia tensor is diagonal in frame $\{C_i\}$.

The drawback of formulating closed form equations is simply that it currently requires a fair amount of human effort. However, symbolic manipulation programs which can derive the closed form equations of motion of a device and automatically factor common terms and perform trigonometric substitutions are under development [12].

Memorization schemes

In any computational scheme there can be a trade-off made between computations and memory usage. In the problem of computing the dynamic equation of a manipulator, (6.58), we have implicitly assumed that when a value of τ is needed, it is computed as quickly as possible from $\Theta, \dot{\Theta}$, and $\ddot{\Theta}$ at run time. If we wish, we can trade off this computational burden at the cost of a tremendously large memory by precomputing (6.58) for all possible $\Theta, \dot{\Theta}$ and $\ddot{\Theta}$ values (suitably quantized). Then, when dynamic information is needed, the answer is found by table lookup.

The size of the memory required is large. Assume that each joint angle range is quantized to 10 discrete values; likewise, assume that velocities and accelerations are quantized to 10 ranges each. For a six-jointed manipulator, the number of cells in the $(\Theta, \dot{\Theta}, \ddot{\Theta})$ quantized space is $(10 \times 10 \times 10)^6$. In each of these cells, there are 6 torque values. Assuming each torque value requires one computer word, this memory size is 6×10^{18} words! Also, note that the table needs to be recomputed for a change in the mass of the load, or another dimension can be added to account for all possible loads.

There are many intermediate solutions which trade off memory for computation in various ways. For example, if the matrices appearing in equation (6.62) were precomputed, the table would only have one dimension (in Θ) rather than three. After the functions of Θ are looked up, a modest amount of computation (given by (6.62)) is done. For more details, and other possible parameterizations of this problem, see [2], [4].

References

[1] I. Shames, *Engineering Mechanics,* Second Edition, Prentice-Hall, 1967.

[2] M. Raibert, "Mechanical Arm Control Using a State Space Memory," SME paper ms77-750, 1977.

[3] O. Khatib, "Dynamic Control of Manipulators in Operational Space," Sixth IFTOMM Congress on Theory of Machines and Mechanisms, New Delhi, December 15-20, 1983.

[4] O. Khatib, "Commande Dynamique dans L'Espace Operationnel des Robots Manipulateurs en Presence d'Obstacles," These de Docteur-Ingenieur. Ecole Nationale Superieure de l'Aeronautique et de L'Espace (ENSAE). Toulouse, France.

[5] O. Khatib, "The Operational Space Formulation in Robot Manipulator Control," 15th ISIR, Tokyo, Japan, September 11-13, 1985.

[6] E. Schmitz, "Experiments on the End-Point Position Control of a Very Flexible One-Link Manipulator," Ph.D. Thesis, Department of Aero. and Astro., Stanford University, SUDAAR No. 547, June 1985.

[7] S. Conte, and C. DeBoor, *Elementary Numerical Analysis: An Algorithmic Approach,* Second Edition, McGraw-Hill, 1972.

[8] J. Uicker, "Dynamic Behaviour of Spatial Linkages," ASME Mech. Vol. 5, No. 68, pp. 1-15.

[9] Hollerbach, "A Recursive Lagrangian Formulation of Manipulator Dynamics and a Comparative Study of Dynamics Formulation Complexity," in *Robot Motion*, M. Brady et al., editors, MIT Press, 1983.

[10] W. Silver, "On the Equivalence of Lagrangian and Newton-Euler Dynamics for Manipulators," International Journal of Robotics Research, Vol. 1, No. 2, pp. 60-70.

[11] T. Kanade, et al., "The CMU Direct-Drive Arm II Project," CMU Robotics Institute Research Review, 1984.

[12] W. Schiehlen, "Computer Generation of Equations of Motion" in "Computer Aided Analysis and Optimization of Mechanical System Dynamics," E.J. Haug, editor, Springer-Verlag, 1984.

Exercises

6.1 [12] Find the inertia tensor of a right cylinder of homogeneous density with respect to a frame with origin at the center of mass of the body.

6.2 [32] Determine the dynamic equations for the 2-link manipulator in Section 6.7 when each link is modelled as a rectangular solid of homogeneous density. Each link has dimensions l_i, w_i, and h_i, and total mass m_i.

6.3 [43] Determine the dynamic equations for the 3-link manipulator of Chapter 3, Exercise 3.3. Consider each link to be a rectangular solid of homogeneous density with dimensions l_i, w_i, and h_i, and total mass m_i.

6.4 [13] Write the set of equations which correspond to (6.44) – (6.52) for the case where the mechanism may have sliding joints.

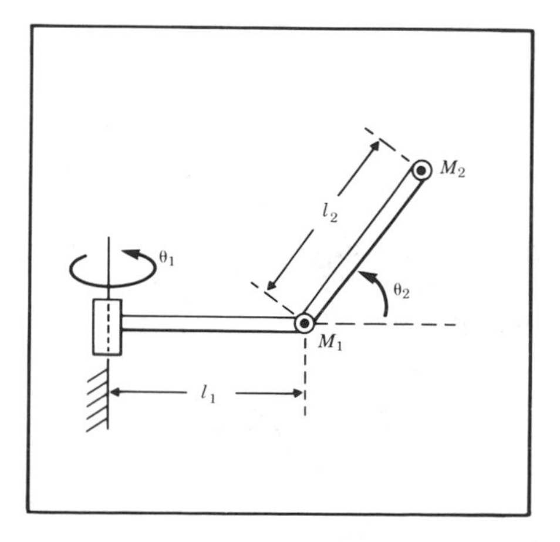

FIGURE 6.7 Two-link with point masses at distal end of links.

6.5 [30] Determine the dynamic equations for the 2-link nonplanar manipulator shown in Fig. 6.7. Assume that all the mass of the links can be considered as a point mass located at the distal (outermost) end of the link. The mass values are m_1 and m_2 and the link lengths are l_1 and l_2. This manipulator is like the first two links of the arm in Exercise 3.3. Also assume that viscous friction is acting at each joint with coefficients v_1 and v_2.

6.6 [32] Derive the Cartesian space form of the dynamics for the two-link planar manpulator of Section 6.7 in terms of the base frame. Hint: See Example 6.5 but use the Jacobian written in the base frame.

6.7 [18] How many memory locations would be required to store the dynamic equations of a general 3-link manipulator in a table? Quantize each joint's position, velocity, and acceleration into 16 ranges. Make any assumptions needed.

6.8 [32] Derive the dynamic equations for the two-link manipulator shown in Fig. 4.6. Link 1 has an inertia tensor given by

$$^{C_1}I = \begin{bmatrix} I_{xx1} & 0 & 0 \\ 0 & I_{yy1} & 0 \\ 0 & 0 & I_{zz1} \end{bmatrix}.$$

Assume that link 2 has all its mass, m_2, located at a point at the end-effector. Assume that gravity is directed downward (opposite $\hat{Z}_1$).

6.9 [37] Derive the dynamic equations for the three-link manipulator with one prismatic joint shown in Fig. 3.8. Link 1 has an inertia tensor given by

$$^{C_1}I = \begin{bmatrix} I_{xx1} & 0 & 0 \\ 0 & I_{yy1} & 0 \\ 0 & 0 & I_{zz1} \end{bmatrix}.$$

Link 2 has point mass m_2 located at the origin of its link frame. Link 3 has an inertia tensor given by

$$^{C_3}I = \begin{bmatrix} I_{xx3} & 0 & 0 \\ 0 & I_{yy3} & 0 \\ 0 & 0 & I_{zz3} \end{bmatrix}.$$

Assume that gravity is directed opposite $\hat{Z}_1$, and viscous friction of magnitude v_i is active at each joint.

6.10 [35] Derive the dynamic equations in Cartesian space for the manipulator of Exercise 6.8. Write the equations in frame $\{2\}$.

6.11 [20] A certain one link manipulator has

$$^{C_1}I = \begin{bmatrix} I_{xx1} & 0 & 0 \\ 0 & I_{yy1} & 0 \\ 0 & 0 & I_{zz1} \end{bmatrix}.$$

Assume that this is just the inertia of the link itself. If the motor armature has a moment of inertia I_m and the gear ratio is 100, what is the total inertia as seen from the motor shaft [1]?

Programming Exercise (Part 6)

1. Derive the dynamic equations of motion for the 3-link manipulator (from Example 3.3). That is, expand Section 6.7 for the 3-link case. The following numerical values describe the manipulator:

$$l_1 = l_2 = 0.5\text{m},$$
$$m_1 = 4.6\text{Kg},$$
$$m_2 = 2.3\text{Kg},$$
$$m_3 = 1.0\text{Kg},$$
$$g = 9.8\text{M/s}^2.$$

For the first two links, we assume that the mass is all concentrated at the distal end of the link. For link 3, we assume that the center of mass is located at the origin of frame {3}, that is, at the proximal end of the link. The inertia tensor for link 3 is

$$^{C_3}I = \begin{bmatrix} 0.05 & 0 & 0 \\ 0 & 0.1 & 0 \\ 0 & 0 & 0.1 \end{bmatrix} \text{KgM}^2.$$

The vectors which locate each center of mass relative to each link frame are

$$^1P_{C_1} = l_1\hat{X}_1,$$
$$^2P_{C_2} = l_2\hat{X}_2,$$
$$^3P_{C_3} = 0.$$

2. Write a simulator for the 3-link manipulator. A simple Euler-integration routine is sufficient for performing the numerical integration (see Section 6.11). To keep your code modular, it may be helpful to define a routine:

```
Procedure UPDATE(VAR tau: vec3; VAR period: real; VAR theta,
thetadot: vec3);
```

Where "tau" is the torque command to the manipulator (always zero for this assignment), "period" is the length of time you wish to advance time (in seconds), and "theta" and "thetadot" are the *state* of the manipulator. Theta and thetadot are updated by "period" seconds each time you call `UPDATE`. Note that "period" would typically be longer than the integration step size, Δt, used in the numerical integration. For example, although the step size for numerical integration might be 0.001 second, you might wish only to print out the manipulator position and velocity each 0.1 seconds.

To test your simulation, set the joint torque commands to zero (for all time) and perform these tests:

a) Set the initial position of the manipulator to

$$[\theta_1 \ \theta_2 \ \theta_3] = [-90 \ 0 \ 0] \, .$$

Simulate for a few seconds. Is the motion of the manipulator what you would expect?

b) Set the initial position of the manipulator to

$$[\theta_1 \ \theta_2 \ \theta_3] = [30 \ 30 \ 10] \, .$$

Simulate for a few seconds. Is the motion of the manipulator what you would expect?

c) Introduce some viscous friction at each joint of the simulated manipulator. That is, add a term to the dynamics of each joint in the form $\tau_f = v\dot{\theta}$, where $v = 5.0$ Newton-meter seconds for each joint. Repeat test (b) above. Is the motion what you would expect?

7

TRAJECTORY GENERATION

7.1 Introduction

In this chapter, we concern ourselves with methods of computing a trajectory in multidimensional space which describes the desired motion of a manipulator. Here, **trajectory** refers to a time history of position, velocity, and acceleration for each degree of freedom.

This problem includes the human interface problem of how we wish to *specify* a trajectory or path through space. In order to make the description of manipulator motion easy for a human user of a robot system, the user shouldn't be required to write down complicated functions of space and time to specify the task. Rather, we must allow the capability of specifying trajectories with simple descriptions of the desired motion, and let the system figure out the details. For example, the user may just specify the desired goal position and orientation of the end-effector, and leave it to the system to decide on the exact shape of the path to get there, the duration, the velocity profile, and other details.

We also are concerned with how trajectories are *represented* in the computer after they have been planned. Finally, there is the problem of actually computing the trajectory from the internal representation, or *generating* the trajectory. Generation occurs at *run-time* and, in the most general case, position, velocity, and acceleration are computed. Since these trajectories are computed on digital computers, the trajectory points are computed at a certain rate, called the **path update rate**. In typical manipulator systems this rate lies between 20 and 200 Hz.

7.2 General considerations in path description and generation

For the most part, we will consider motions of a manipulator as motions of the tool frame, $\{T\}$, relative to the station frame, $\{S\}$. This is the same manner in which an eventual user of the system would think, and designing a path description and generation system in these terms will result in a few important advantages.

When we specify paths as motions of the tool frame relative to the station frame, we decouple the motion description from any particular robot, end- effector, or workpieces. This results in a certain modularity, and would allow the same path description to be used with a different manipulator, or with the same manipulator with a different tool size. Further, we can specify and plan motions relative to a moving workstation (perhaps a conveyor belt) by planning motions relative to the station frame as always, and at run time causing the definition of $\{S\}$ to be changing with time.

As shown in Fig. 7.1, the basic problem is to move the manipulator from an initial position to some desired final position. That is, we wish to move the tool frame from its current value, $\{T_{initial}\}$, to a desired final value, $\{T_{final}\}$. Note that this motion in general involves a change in orientation as well as a change in position of the tool relative to the station.

Sometimes it is necessary to specify the motion in much more detail than simply stating the desired final configuration. One way to include more detail in a path description is to give a sequence of desired **via points** or intermediate points between the initial and final positions. Thus, in completing the motion, the tool frame must pass through a set of intermediate positions and orientations as described by the via points. Each of these via points is actually a frame which specifies both the position and orientation of the tool relative to the station. The name **path points** includes all the via points plus the initial and final points. Remember that although we generally use the term "points"

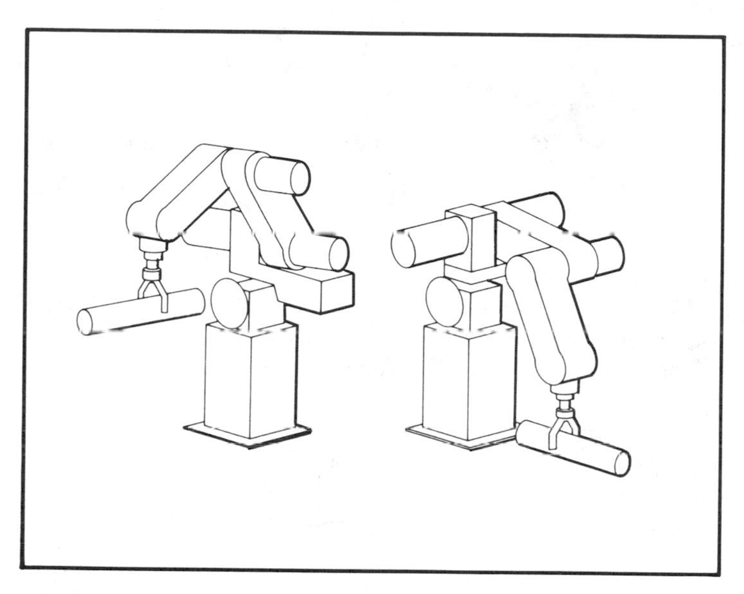

FIGURE 7.1 In executing a trajectory, a manipulator moves from its initial position to a desired goal position in a smooth manner.

these are actually frames which give both position and orientation. Along with these *spatial* constraints on the motion, the user may also wish to specify *temporal* attributes of the motion. For example, the time elapsed between via points might be specified in the description of the path.

Usually, it is desirable for the motion of the manipulator to be *smooth*. For our purposes, we will define a smooth function as one which is continuous and has a continuous first derivative. Sometimes, a continous second derivative is also desirable. Rough, jerky motions tend to cause increased wear on the mechanism, and cause vibrations by exciting resonances in the manipulator. In order to guarantee smooth paths, we must put some sort of constraints on the spatial and temporal qualities of the path *between* the via points.

At this point there are many choices that may be made, and consequently a great variety in the ways that paths might be specified and planned. Any smooth functions of time which pass through the via points could be used to specify the exact path shape. In this chapter we will discuss a couple of frequently used choices for these functions. Other approaches may be found in [1], [2].

7.3 Joint space schemes

In this section we consider methods of path generation in which the path shapes (in space and in time) are described in terms of functions of joint angles.

Each path point is usually specified in terms of a desired position and orientation of the tool frame, $\{T\}$, relative to the station frame, $\{S\}$. Each of these via points is "converted" into a set of desired joint angles by application of the inverse kinematics. Then a smooth function is found for each of the n joints which pass through the via points and end at the goal point. The time required for each segment is the same for each joint so that all joints will reach the via point at the same time, thus resulting in the desired Cartesian position of $\{T\}$ at each via point. Other than specifying the same duration for each joint, the determination of the desired joint angle function for a particular joint does not depend on the functions for the other joints.

Hence, joint space schemes achieve the desired position and orientation at the via points. In between via points the shape of the path, while rather simple in joint space, is complex if described in Cartesian space. Joint space schemes are usually the easiest to compute, and, because we make no continuous correspondence between joint space and Cartesian space, there is essentially no problem with singularities of the mechanism.

Cubic polynomials

Consider the problem of moving the tool from its initial position to a goal position in a certain amount of time. Using the inverse kinematics the set of joint angles that correspond to the goal position and orientation can be calculated. The initial position of the manipulator is also known in the form of a set of joint angles. What is required is a function for each joint whose value at t_0 is the initial position of the joint, and whose value at t_f is the desired goal position of that joint. As shown in Fig. 7.2, there are many smooth functions, $\theta(t)$, which might be used to interpolate the joint value.

In making a single smooth motion, at least four constraints on $\theta(t)$ are evident. Two constraints on the function's value come from the selection of initial and final values:

$$\begin{aligned} \theta(0) &= \theta_0, \\ \theta(t_f) &= \theta_f. \end{aligned} \tag{7.1}$$

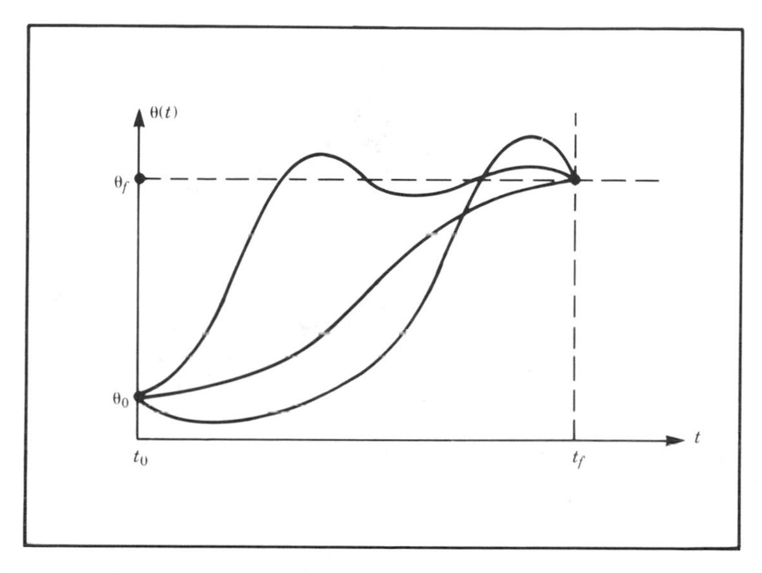

FIGURE 7.2 Several possible path shapes for a single joint.

An additional two constraints are that the function is continuous in velocity, which in this case means the the initial and final velocity are zero:

$$\dot{\theta}(0) = 0,$$
$$\dot{\theta}(t_f) = 0. \tag{7.2}$$

These four constraints can be satisfied by a polynomial of at least third degree. Since a cubic polynomial has four coefficients, it can be made to satisfy the four constraints given by (7.1) and (7.2). These constraints uniquely specify a particular cubic. A cubic has the form

$$\theta(t) = a_0 + a_1 t + a_2 t^2 + a_3 t^3, \tag{7.3}$$

and so the joint velocity and acceleration along this path are clearly

$$\dot{\theta}(t) = a_1 + 2a_2 t + 3a_3 t^2,$$
$$\ddot{\theta}(t) = 2a_2 + 6a_3 t. \tag{7.4}$$

Combining (7.3) and (7.4) with the four desired constraints yields four equations in four unknowns:

$$\theta_0 = a_0,$$
$$\theta_f = a_0 + a_1 t_f + a_2 t_f^2 + a_3 t_f^3,$$
$$0 = a_1, \tag{7.5}$$
$$0 = a_1 + 2a_2 t_f + 3a_3 t_f^2.$$

Solving these equations for the a_i we obtain

$$\begin{aligned} a_0 &= \theta_0, \\ a_1 &= 0, \\ a_2 &= \frac{3}{t_f^2}(\theta_f - \theta_0), \\ a_3 &= -\frac{2}{t_f^3}(\theta_f - \theta_0). \end{aligned} \tag{7.6}$$

Using (7.6) we can calculate the cubic polynomial that connects any initial joint angle position with any desired final position. This solution is for the case when the joint starts and finishes at zero velocity.

EXAMPLE 7.1

A single-link robot with a rotary joint is motionless at $\theta = 15$ degrees. It is desired to move the joint in a smooth manner to $\theta = 75$ degrees in 3 seconds. Find the coefficients of a cubic which accomplishes this motion and brings the manipulator to rest at the goal. Plot the position, velocity, and acceleration of the joint as a function of time.

Plugging into (7.6) we find

$$\begin{aligned} a_0 &= 15.0, \\ a_1 &= 0.0, \\ a_2 &= 20.0, \\ a_3 &= -4.44. \end{aligned} \tag{7.7}$$

Using (7.3) and 7.4) we obtain:

$$\begin{aligned} \theta(t) &= 15.0 + 20.0t^2 - 4.44t^3, \\ \dot{\theta}(t) &= 40.0t - 13.33t^2, \\ \ddot{\theta}(t) &= 40.0 - 26.66t. \end{aligned} \tag{7.8}$$

Figure 7.3 shows the position, velocity, and acceleration functions for this motion sampled at 40 Hz. Note that the velocity profile for any cubic function is a parabola, and the acceleration profile is linear. ■

Cubic polynomials for a path with via points

So far we have considered motions described by a desired duration and a final goal point. In general, we wish to allow paths to be specified which include intermediate via points. If the manipulator is to come to rest at each via point, then we can use the cubic solution of Section 7.3.

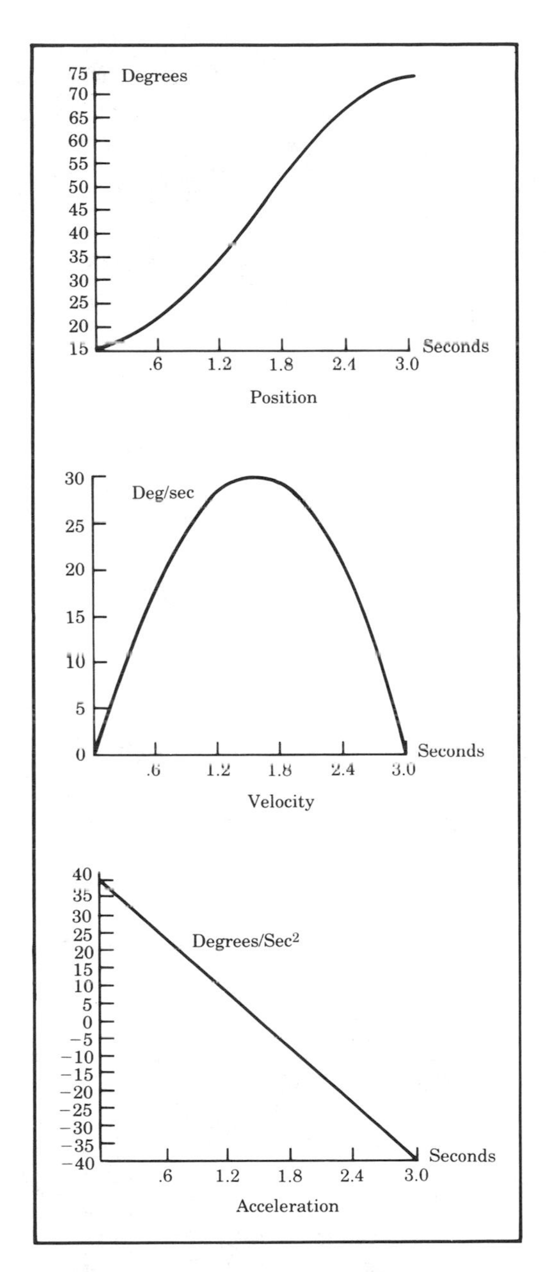

FIGURE 7.3 Position, velocity, and acceleration profiles for a single cubic segment which starts and ends at rest.

Usually, we wish to be able to pass through a via point without stopping, and so we need to generalize the way in which we fit cubics to the path constraints.

As in the case of a single goal point, each via point is usually specified in terms of a desired position and orientation of the tool frame relative to the station frame. Each of these via points is "converted" into a set of desired joint angles by application of the inverse kinematics. We then consider the problem of computing cubics which connect the via point values for each joint together in a smooth way.

If desired velocities of the joints at the via points are known, then we can determine cubic polynomials as before, but now the velocity constraints at each end are not zero, but rather, some known velocity. The constraints of (7.3) become

$$
\begin{aligned}
\dot{\theta}(0) &= \dot{\theta}_0, \\
\dot{\theta}(t_f) &= \dot{\theta}_f.
\end{aligned} \tag{7.9}
$$

The four equations describing this general cubic are

$$
\begin{aligned}
\theta_0 &= a_0, \\
\theta_f &= a_0 + a_1 t_f + a_2 t_f^2 + a_3 t_f^3, \\
\dot{\theta}_0 &= a_1, \\
\dot{\theta}_f &= a_1 + 2a_2 t_f + 3a_3 t_f^2.
\end{aligned} \tag{7.10}
$$

Solving these equations for the a_i we obtain

$$
\begin{aligned}
a_0 &= \theta_0, \\
a_1 &= \dot{\theta}_0, \\
a_2 &= \frac{3}{t_f^2}(\theta_f - \theta_0) - \frac{2}{t_f}\dot{\theta}_0 - \frac{1}{t_f}\dot{\theta}_f, \\
a_3 &= -\frac{2}{t_f^3}(\theta_f - \theta_0) + \frac{1}{t_f^2}(\dot{\theta}_f + \dot{\theta}_0).
\end{aligned} \tag{7.11}
$$

Using (7.11) we can calculate the cubic polynomial that connects any initial and final positions with any initial and final velocities.

If we have the desired joint velocities at each via point, then we simply apply (7.11) to each segment to find the required cubics. There are several ways in which desired velocity at the via points might be specified.

1. The user specifies the desired velocity at each via point in terms of a Cartesian linear and angular velocity of the tool frame at that instant.
2. The system automatically chooses the velocities at the via points by applying a suitable heuristic in either Cartesian space or joint space.
3. The system automatically chooses the velocities at the via points in such a way as to cause the acceleration at the via points to be continuous.

In the first option, Cartesian desired velocities at the via points are "mapped" to desired joint rates using the inverse Jacobian of the manipulator evaluated at the via point. If the manipulator is at a singular point at a particular via point, then the user is not free to assign an arbitrary velocity at this point. While it is a useful capability of a path generation scheme to be able to meet a desired velocity which the user specifies, it would be a burden to require that the user always make these specifications. Therefore, a convenient system should include either option 2 or 3 (or both).

In option 2, the system automatically chooses reasonable intermediate velocities using some kind of heuristic. Consider the path specified by the via points shown for some joint, θ, in Fig. 7.4.

In Fig. 7.4 we have made a reasonable choice of joint velocities at the via points, as indicated with small line segments representing tangents to the curve at each via point. This choice is the result of applying a conceptually and computationally simple heuristic. Imagine the via points connected with straight line segments—if the slope of these lines changes sign at the via point, choose zero velocity, if the slope of these

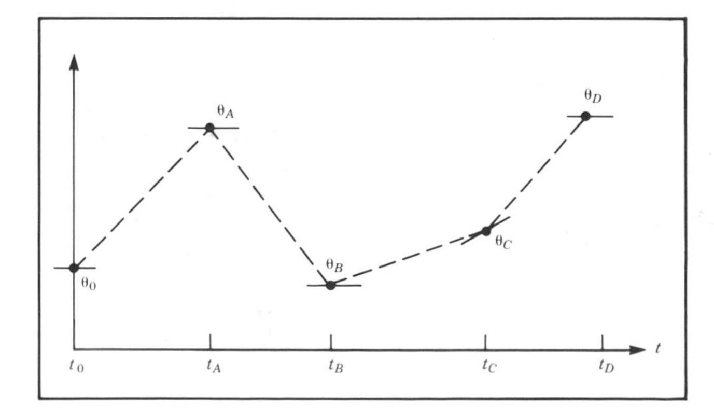

FIGURE 7.4 Via points with desired velocities at the points indicated by tangents.

lines does not change sign, choose the average of the two slopes as the via velocity. In this way, from specification of the desired via points alone, the system can choose the velocities at each point.

In option 3, the system chooses velocities such that acceleration is continuous at the via point. To do this, a new splining solution is needed. In this kind of spline, we replace the (two) velocity constraints at the connection of two cubics with the (two) constraints that a) velocity be continuous and b) acceleration be continuous.

EXAMPLE 7.2

Solve for the coefficients of two cubics which are connected in a two-segment spline with continuous acceleration at the intermediate via point. The initial angle is θ_0, the via point is θ_v, and the goal point is θ_g.

The first cubic is

$$\theta(t) = a_{10} + a_{11}t + a_{12}t^2 + a_{13}t^3, \tag{7.12}$$

and the second is

$$\theta(t) = a_{20} + a_{21}t + a_{22}t^2 + a_{23}t^3. \tag{7.13}$$

Each cubic will be evaluated over an interval starting at $t = 0$ and ending at $t = t_{fi}$, where $i = 1$ or $i = 2$.

The constraints we wish to enforce are

$$\begin{aligned}
\theta_0 &= a_{10}, \\
\theta_v &= a_{10} + a_{11}t_{f1} + a_{12}t_{f1}^2 + a_{13}t_{f1}^3, \\
\theta_v &= a_{20}, \\
\theta_g &= a_{20} + a_{21}t_{f2} + a_{22}t_{f2}^2 + a_{23}t_{f2}^3, \\
0 &= a_{11}, \\
0 &= a_{21} + 2a_{22}t_{f2} + 3a_{23}t_{f2}^2, \\
a_{11} + 2a_{12}t_{f1} + 3a_{13}t_{f1}^2 &= a_{21}, \\
2a_{12} + 6a_{13}t_{f1} &= 2a_{22}.
\end{aligned} \tag{7.14}$$

These constraints specify a linear equation problem of eight equations and eight unknowns. Solving for the case $t_f = t_{f1} = t_{f2}$ we obtain

$$\begin{aligned}
a_{10} &= \theta_0, \\
a_{11} &= 0, \\
a_{12} &= \frac{12\theta_v - 3\theta_g - 9\theta_0}{4t_f^2}, \\
a_{13} &= \frac{-8\theta_v + 3\theta_g + 5\theta_0}{4t_f^3}, \\
a_{20} &= \theta_v, \\
a_{21} &= \frac{3\theta_g - 3\theta_0}{4t_f}, \\
a_{22} &= \frac{-12\theta_v + 6\theta_g + 6\theta_0}{4t_f^2}, \\
a_{23} &= \frac{8\theta_v - 5\theta_g - 3\theta_0}{4t_f^3}. \quad \blacksquare
\end{aligned} \tag{7.15}$$

For the general case of n cubic segments the equations which arise from insisting on continuous acceleration at the via points may be cast in matrix form which is solved to compute the velocities at the via points. The matrix turns out to be tridiagonal and easily solved [4].

Higher order polynomials

Higher order polynomials are sometimes used for path segments. For example, if we wish to be able to specify the position, velocity, *and* acceleration at the beginning and end of a path segment, a quintic polynomial is required:

$$\theta(t) = a_0 + a_1 t + a_2 t^2 + a_3 t^3 + a_4 t^4 + a_5 t^5, \tag{7.16}$$

where the constraints are given as

$$\begin{aligned}
\theta_0 &= a_0, \\
\theta_f &= a_0 + a_1 t_f + a_2 t_f^2 + a_3 t_f^3 + a_4 t_f^4 + a_5 t_f^5, \\
\dot{\theta}_0 &= a_1, \\
\dot{\theta}_f &= a_1 + 2a_2 t_f + 3a_3 t_f^2 + 4a_4 t_f^3 + 5a_5 t_f^4, \\
\ddot{\theta}_0 &= 2a_2, \\
\ddot{\theta}_f &= 2a_2 + 6a_3 t_f + 12a_4 t_f^2 + 20a_5 t_f^3.
\end{aligned} \tag{7.17}$$

These constraints specify a linear set of six equations with six unknowns whose solution is

$$
\begin{aligned}
a_0 &= \theta_0, \\
a_1 &= \dot{\theta}_0, \\
a_2 &= \frac{\ddot{\theta}_0}{2}, \\
a_3 &= \frac{20\theta_f - 20\theta_0 - (8\dot{\theta}_f + 12\dot{\theta}_0)t_f - (3\ddot{\theta}_0 - \ddot{\theta}_f)t_f^2}{2t_f^3}, \\
a_4 &= \frac{30\theta_0 - 30\theta_f + (14\dot{\theta}_f + 16\dot{\theta}_0)t_f + (3\ddot{\theta}_0 - 2\ddot{\theta}_f)t_f^2}{2t_f^4}, \\
a_5 &= \frac{12\theta_f - 12\theta_0 - (6\dot{\theta}_f + 6\dot{\theta}_0)t_f - (\ddot{\theta}_0 - \ddot{\theta}_f)t_f^2}{2t_f^5}.
\end{aligned}
\tag{7.18}
$$

Various algorithms are available for computing smooth functions (polynomial or otherwise) which pass through a given set of data points [3], [4]. Complete coverage is beyond the scope of this book.

Linear function with parabolic blends

Another choice of path shape is linear. That is, we simply linearly interpolate to move from the present joint position to the final position as in Fig. 7.5. Remember that although the motion of each joint in this scheme is linear, the end-effector in general does not move in a straight line in space.

However, straightforward linear interpolation would cause the velocity to be discontinous at the beginning and end of the motion. To create a smooth path with continous position and velocity, we start with the linear function but add a parabolic *blend* region at each path point.

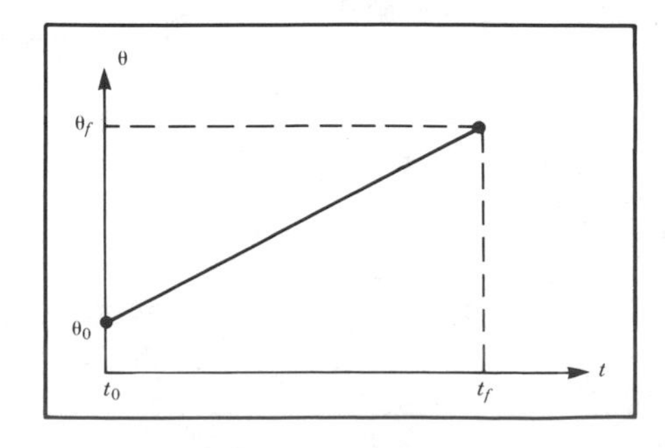

FIGURE 7.5 Linear interpolation requiring infinite acceleration.

During the blend portion of the trajectory, constant acceleration is used to change velocity smoothly. Figure 7.6 shows a simple path constructed in this way. The linear function and the two parabolic functions are "splined" together so that the entire path is continuous in position and velocity.

In order to construct this single segment we will assume that the parabolic blends both have the same duration, and therefore the same constant acceleration (modulo a sign) is used during both blends. As indicated in Fig. 7.7, there are many solutions to the problem—but note that the answer is always symmetric about the halfway point in time, t_h, and about the halfway point in position, θ_h. The velocity at the end of the blend region must equal the velocity of the linear section, and so we have

$$\ddot{\theta} t_b = \frac{\theta_h - \theta_b}{t_h - t_b}, \tag{7.19}$$

where θ_b is the value of θ at the end of the blend region, and $\ddot{\theta}$ is the acceleration acting during the blend region. The value of θ_b is given by

$$\theta_b = \theta_0 + \frac{1}{2}\ddot{\theta} t_b^2. \tag{7.20}$$

Combining (7.19) and (7.20) and $t = 2t_h$, we get

$$\ddot{\theta} t_b^2 - \ddot{\theta} t t_b + (\theta_f - \theta_0) = 0, \tag{7.21}$$

where t is the desired duration of the motion. Given any θ_f, θ_0, and t, we can follow any of the paths given by choice of $\ddot{\theta}$ and t_b which satisfy (7.21). Usually, an acceleration, $\ddot{\theta}$, is chosen and (7.21) is solved for the corresponding t_b. The acceleration chosen must be sufficiently high, or a

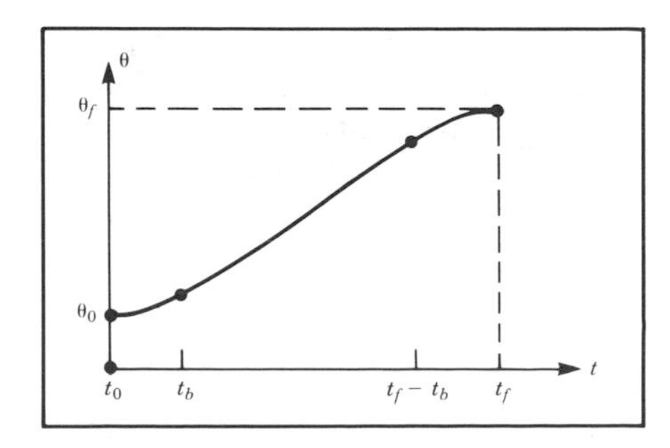

FIGURE 7.6 Linear segment with parabolic blends.

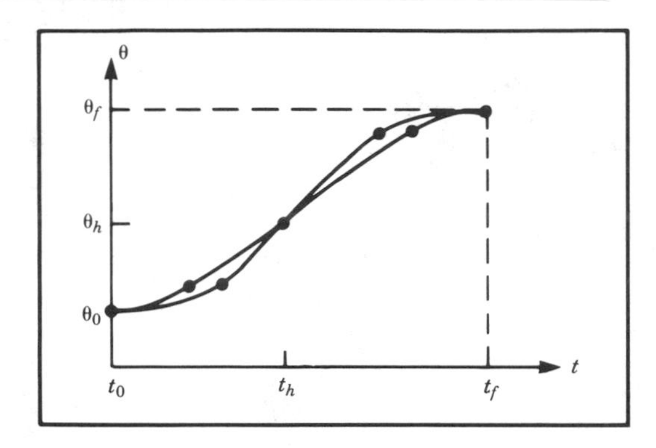

FIGURE 7.7 Linear segment with parabolic blends.

solution will not exist. Solving (7.21) for t_b in terms of the acceleration and other known parameters, we obtain

$$t_b = \frac{t}{2} - \frac{\sqrt{\ddot{\theta}^2 t^2 - 4\ddot{\theta}(\theta_f - \theta_0)}}{2\ddot{\theta}}. \tag{7.22}$$

The constraint on the acceleration used in the blend is

$$\ddot{\theta} \geq \frac{4(\theta_f - \theta_0)}{t^2}. \tag{7.23}$$

When equality occurs in (7.23) the linear portion has shrunk to zero length and the path is composed of two blends which connect with equivalent slope. As the acceleration used becomes larger and larger, the length of the blend region becomes shorter and shorter. In the limit of infinite acceleration we are back to the simple linear interpolation case.

EXAMPLE 7.3

For the same single segment path discussed in Example 7.1, show two examples of a linear path with parabolic blends.

Figure 7.8a shows one possibility where $\ddot{\theta}$ was chosen quite high. In this case we quickly accelerate, then coast at constant velocity, and then decelerate. Figure 7.8b shows a trajectory where acceleration is kept quite low, so that the linear section almost disappears. ■

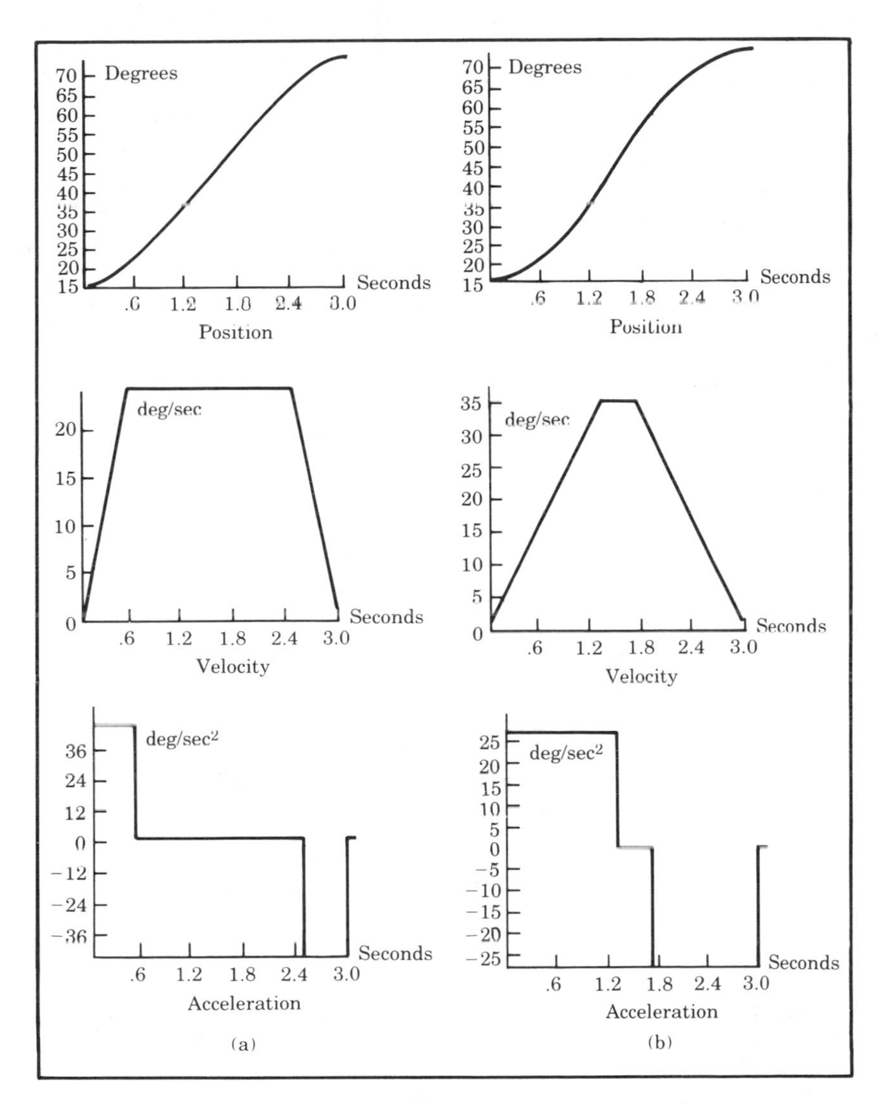

FIGURE 7.8 Position, velocity, and acceleration profiles for linear interpolation with parablic blends. The set of curves on the left are based on a higher acceleration during the blends than those on the right.

Linear function with parabolic blends for a path with via points

We now consider the case of linear paths with parabolic blends for the case in which there are an arbitrary number of via points specified. Figure 7.9 shows a set of joint space via points for some joint θ. Linear functions connect the via points, and parabolic blend regions are added around each via point.

We will use notation as follows. Consider three neighboring path points which we will call points j, k, and l. The duration of the blend region at path point k is t_k. The duration of the linear portion between points j and k is t_{jk}. The overall duration of the segment connecting points j and k is t_{djk}. The velocity during the linear portion is $\dot{\theta}_{jk}$, and the acceleration during the blend at point j is $\ddot{\theta}_j$. See Fig. 7.9 for an example.

As with the single segment case, there are many possible solutions depending on the value of acceleration used at each blend. Given all the path points θ_k, the desired durations t_{djk}, and the magnitude of acceleration to use at each path point $|\ddot{\theta}_k|$, we can compute the blend times t_k. For interior path points this follows simply from the equations

$$\begin{aligned} \dot{\theta}_{jk} &= \frac{\theta_k - \theta_j}{t_{djk}}, \\ \ddot{\theta}_k &= SGN(\dot{\theta}_{kl} - \dot{\theta}_{jk}) \left|\ddot{\theta}_k\right|, \\ t_k &= \frac{\dot{\theta}_{kl} - \dot{\theta}_{jk}}{\ddot{\theta}_k}, \\ t_{jk} &= t_{djk} - \frac{1}{2}t_j - \frac{1}{2}t_k. \end{aligned} \tag{7.24}$$

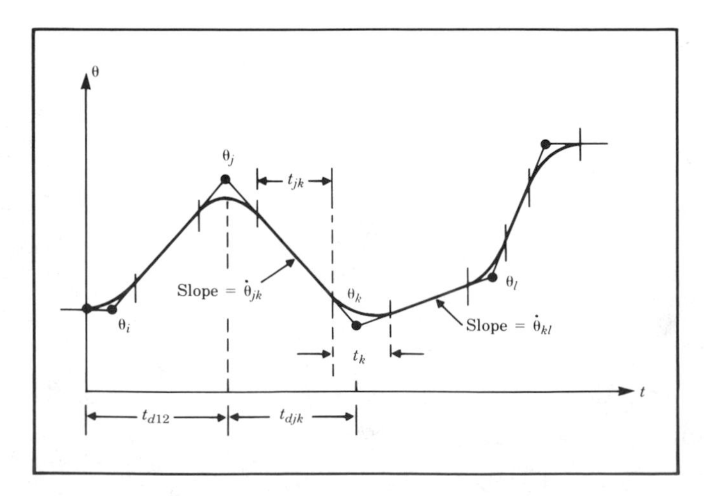

FIGURE 7.9 Multisegment linear path with blends.

The first and last segments must be handled slightly differently since an entire blend region at one end of the segment must be counted in the total segment's time duration.

For the first segment, we solve for t_1 by equating two expressions for the velocity during the linear phase of the segment:

$$\frac{\theta_2 - \theta_1}{t_{d12} - \frac{1}{2}t_1} = \ddot{\theta}_1 t_1. \tag{7.25}$$

This can be solved for t_1, the blend time at the initial point, and then $\dot{\theta}_{12}$ and t_{12} are easily computed:

$$\begin{aligned}
\ddot{\theta}_1 &= SGN(\theta_2 - \theta_1)\left|\ddot{\theta}_1\right|, \\
t_1 &= t_{d12} - \sqrt{t_{d12}^2 - \frac{2(\theta_2 - \theta_1)}{\ddot{\theta}_1}}, \\
\dot{\theta}_{12} &= \frac{\theta_2 - \theta_1}{t_{d12} - \frac{1}{2}t_1}, \\
t_{12} &= t_{d12} - t_1 - \frac{1}{2}t_2.
\end{aligned} \tag{7.26}$$

Likewise, for the last segment (the one connecting points $n-1$ and n) we have

$$\frac{\theta_{n-1} - \theta_n}{t_{d(n-1)n} - \frac{1}{2}t_n} = \ddot{\theta}_n t_n, \tag{7.27}$$

which leads to the solution

$$\begin{aligned}
\ddot{\theta}_n &= SGN(\theta_{n-1} - \theta_n)\left|\ddot{\theta}_n\right|, \\
t_n &= t_{d(n-1)n} - \sqrt{t_{d(n-1)n}^2 + \frac{2(\theta_n - \theta_{n-1})}{\ddot{\theta}_n}}, \\
\dot{\theta}_{(n-1)n} &= \frac{\theta_n - \theta_{n-1}}{t_{d(n-1)n} - \frac{1}{2}t_n}, \\
t_{(n-1)n} &= t_{d(n-1)n} - t_n - \frac{1}{2}t_{n-1}.
\end{aligned} \tag{7.28}$$

Using (7.24) through (7.28) we can solve for the blend times and velocities for a multisegment path. Usually, the user specifies only the via points and the desired duration of the segments. In this case, the system uses default values for acceleration for each joint. Sometimes, to make things even simpler for the user, the system will calculate durations based on default velocities. At all blends, sufficiently large acceleration must be used so that there is sufficient time to get into the linear portion of the segment before the next blend region starts.

EXAMPLE 7.4

The trajectory of a particular joint is specified as follows: Path points in degrees: 10, 35, 25, 10. The duration of these three segments should be: 2, 1, 3 seconds, respectively. The magnitude of the default acceleration to use at all blend points is 50 degrees/second^2. Calculate all segment velocities, blend times, and linear times.

For the first segment we apply (7.26a) to find

$$\ddot{\theta}_1 = 50.0 . \tag{7.29}$$

Applying (7.26b) to calculate the blend time at the initial point, we get

$$t_1 = 2 - \sqrt{4 - \frac{2(35-10)}{50.0}} = 0.27 . \tag{7.30}$$

The velocity, $\dot{\theta}_{12}$, is calculated from (7.26c) as

$$\dot{\theta}_{12} = \frac{35-10}{2-0.5(0.27)} = 13.50 . \tag{7.31}$$

The velocity, $\dot{\theta}_{23}$, is calculated from (7.24a) as

$$\dot{\theta}_{23} = \frac{25-35}{1} = -10.0 . \tag{7.32}$$

Next, we apply (7.24b) to find

$$\ddot{\theta}_2 = -50.0 . \tag{7.33}$$

Then t_2 is calculated from (7.24c), and we get

$$t_2 = \frac{-10.0-13.50}{-50.0} = 0.47 . \tag{7.34}$$

The linear portion length of segment 1 is then calculated from (7.26d) and we get

$$t_{12} = 2 - 0.27 - \frac{1}{2}(0.47) = 1.50 . \tag{7.35}$$

Next, from (7.28a) we have

$$\ddot{\theta}_4 = 50.0 . \tag{7.36}$$

So for the last segment (7.28b) is used to compute t_4, and we have

$$t_4 = 3 - \sqrt{9 + \frac{2(10-25)}{50.0}} = 0.102 . \tag{7.37}$$

The velocity, $\dot{\theta}_{34}$, is calculated from (7.28c) as

$$\dot{\theta}_{34} = \frac{10 - 25}{3 - 0.050} = -5.10\,. \qquad (7.38)$$

Next, (7.24b) is used to obtain

$$\ddot{\theta}_3 = 50.0\,. \qquad (7.39)$$

Then t_3 is calculated from (7.24c), and we get

$$t_3 = \frac{-5.10 - (-10.0)}{50} = 0.098\,. \qquad (7.40)$$

Finally, from (7.24d) we compute

$$t_{23} = 1 - \frac{1}{2}(0.47) - \frac{1}{2}(0.098) = 0.716, \qquad (7.41)$$

$$t_{34} = 3 - \frac{1}{2}(0.098) - 0.102 = 2.849\,. \qquad (7.42)$$

The results of these computations constitute a "plan" for the trajectory. At execution time, these numbers would be used by the **path generator** to compute values of θ, $\dot{\theta}$, and $\ddot{\theta}$ at the path update rate. ■

In these linear–parabolic-blend splines, note that the via points are not actually reached unless the manipulator comes to a stop. Often, when acceleration capability is sufficiently high, the paths will come quite close to the desired via point. If we wish to pass through a point by coming to a stop, the via point is simply repeated in the path specification.

If the user wishes to specify that the manipulator pass *exactly* through a via point without stopping, this can be accomodated using the same formulation as before with the following addition: The system automatically replaces the via point through which we wish the manipulator to pass with two *pseudo via points* on either side of the original (see Fig. 7.10). Then path generation takes place as before. The original via point will now lie in the linear region of the path connecting the two pseudo via points. In addition to requesting that the manipulator pass exactly through a via point, the user can also request that it pass through with a certain velocity. If the user does not specify this velocity the system chooses it based on a suitable heuristic. The term **through point** might be used (rather than via point) to specify a path point *through* which we force the manipulator to pass exactly.

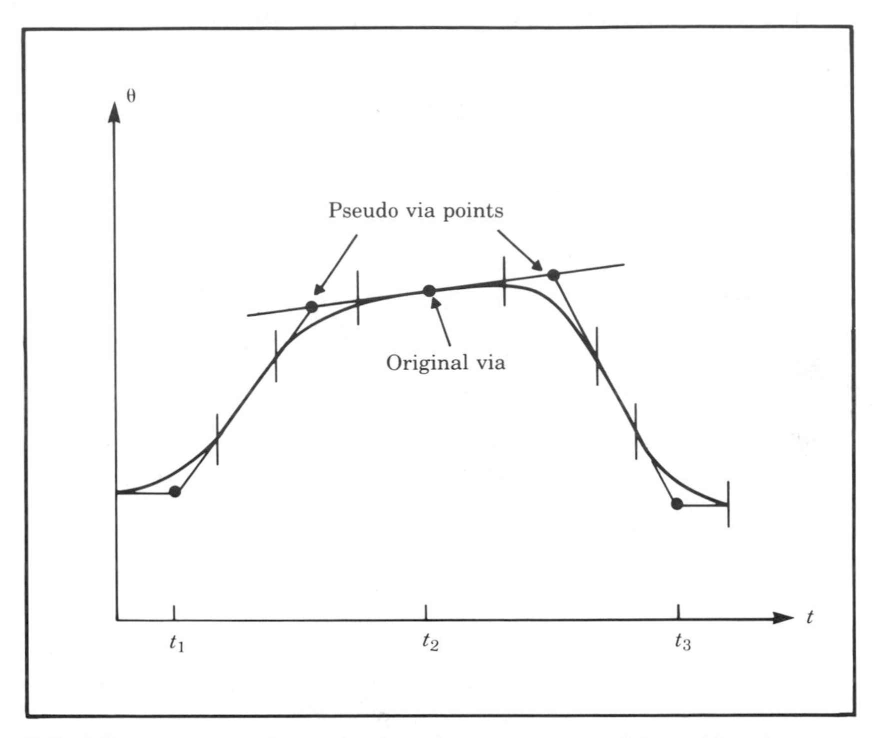

FIGURE 7.10 Use of pseudo via points to create a "through" point.

7.4 Cartesian space schemes

As mentioned in Section 7.3, paths computed in joint space can insure that via and goal points are attained, even when these path points were specified by means of Cartesian frames. However, the spatial shape of the path taken by the end-effector is not a straight line through space, but rather, it is some complicated shape which depends on the particular kinematics of the manipulator being used. In this section we consider methods of path generation in which the path shapes are described in terms of functions which compute Cartesian position and orientation as functions of time. In this way, we can also specify the spatial shape of the path between path points. The most common path shape is a straight line; but circlular, sinusoidal, or other path shapes could be used.

Each path point is usually specified in terms of a desired position and orientation of the tool frame relative to the station frame. In Cartesian based path generation schemes, the functions which are splined together to form a trajectory are functions of time which represent Cartesian variables. These paths can be *planned* directly from the user's definition of path points which are $\{T\}$ specifications relative to $\{S\}$ without first performing inverse kinematics. However, Cartesian schemes are more computationally expensive to execute since at run time, inverse kinematics must be solved at the path update rate. That is, after the path is generated in Cartesian space, as a last step the inverse kinematic calculation is performed to calculate desired joint angles.

Several schemes for generating Cartesian paths have been proposed in literature from the research and industrial robotics community [1], [2]. In the following section we introduce one scheme as an example. In this scheme, we are able to use the same linear/parabolic spliner which we developed for the joint space case.

Cartesian straight line motion

Often we would like to be able easily to specify a spatial path which causes the tip of the tool to move through space in a straight line. Obviously, if we specify many closely separated via points which lie on a straight line, then the tool tip will appear to follow a straight line regardless of the choice of smooth function which interconnects the via points. However, it is much more convenient if the tool follows straight line paths between even widely separated via points. This mode of path specification and execution is called **Cartesian straight line motion**. Defining motions in terms of straight lines is a subset of the more general capability of **Cartesian motion** in which arbitrary functions of Cartesian variables as functions of time could be used to specify a path. In a system which allowed general Cartesian motion, path shapes such as ellipses or sinusoids could be executed.

In planning and generating Cartesian straight line paths, a spline of linear functions with parabolic blends is appropriate. During the linear portion of each segment, since all three components of position change in a linear fashion, the end-effector will move along a linear path in space. However, if we are specifying the orientation as a rotation matrix at each via point, we cannot linearly interpolate its elements as this would not result in a valid rotation matrix at all times. A rotation matrix must be composed of orthonormal columns, and this condition would not be guaranteed if it was constructed by linear interpolation of matrix elements between two valid matrices. Instead, we will use another representation of orientation.

As we studied in Chapter 2, the so-called **angle-axis** representation can be used to specify an orientation with three numbers. If we combine this representation of orientation with the 3×1 Cartesian position representation, we have a 6×1 representation of Cartesian position and orientation. Consider a via point specified relative to the station frame as ${}^S_AT$. That is, the frame $\{A\}$ specifies a via point with position of the end-effector given by ${}^{S}P_{AORG}$, and orientation of the end-effector given by ${}^{S}_{A}R$. This rotation matrix can be converted to the angle-axis representation $ROT({}^{S}\hat{K}_{A}, \theta_{SA})$ or simply ${}^{S}K_{A}$. We will use the symbol $\mathcal{X}$ to represent this 6×1 vector of Cartesian position and orientation. Thus we have

$$
{}^{S}\mathcal{X}_{A} = \begin{bmatrix} {}^{S}P_{AORG} \\ {}^{S}K_{A} \end{bmatrix}, \qquad \textbf{(7.43)}
$$

where ${}^{S}K_{A}$ is formed by scaling the unit vector ${}^{S}\hat{K}_{A}$ by the amount of rotation, θ_{SA}. If every path point is specified in this representation, we then need to describe spline functions which smoothly move these six quantities from path point to path point as functions of time. If linear splines with parabolic blends are used, the path shape between via points will be linear. When via points are passed, the linear and angular velocity of the end-effector are smoothly changed.

One slight complication arises from the fact that the angle-axis representation of orientation is not unique:

$$
({}^{S}\hat{K}_{A}, \ \theta_{SA}) = ({}^{S}\hat{K}_{A}, \theta_{SA} + n360^{\circ}), \qquad \textbf{(7.44)}
$$

where n is any positive or negative integer. In going from a via point $\{A\}$ to a via point $\{B\}$ the total amount of rotation should be minimized. That is, the total amount of rotation should be less than 180°. Assuming that our representation of the orientation of $\{A\}$ is given as ${}^{S}K_{A}$, we must choose the particular ${}^{S}K_{B}$ such that $\left|{}^{S}K_{B} - {}^{S}K_{A}\right|$ is minimized. For example, Fig. 7.11 shows four different possible ${}^{S}K_{B}$'s and their relation to the given ${}^{S}K_{A}$. The difference vectors (broken lines) are compared to determine the ${}^{S}K_{B}$ which will result in minimum rotation—in this case, ${}^{S}K_{B(-1)}$.

Once we select the six values of $\mathcal{X}$ for each via point, we can use the same mathematics we have already developed for generating splines which are composed of linear and parabolic sections. However, we must add one more constraint: The blend times for each degree of freedom must be the same. This will insure that the resultant motion of all the degrees of freedom will be a straight line in space. Since all blend times must be the same, the acceleration used during the blend for each degree of freedom will differ. Hence, we specify a duration of blend, and using (7.24c) we compute the needed acceleration (instead of the other way around). The blend time can be chosen so that a certain upper bound on acceleration is not exceeded.

Many other schemes for representing and interpolating the orientation portion of a Cartesian path may be used. Among these are the use of some of the other 3×1 representations of orientation introduced in Section 2.8. For example, the Intelledex 605T manipulator (Fig. 7.12)

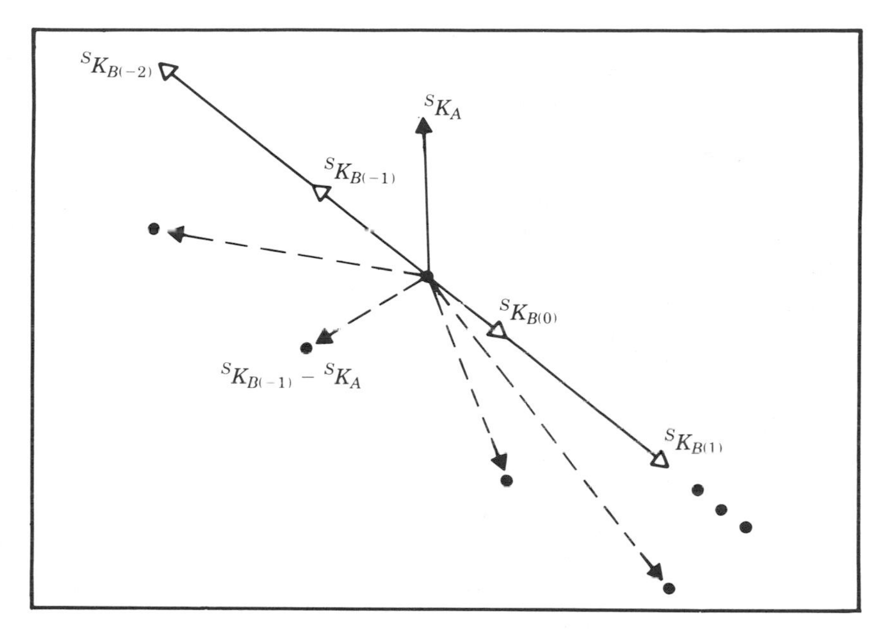

FIGURE 7.11 Choosing angle-axis representation to minimize rotation.

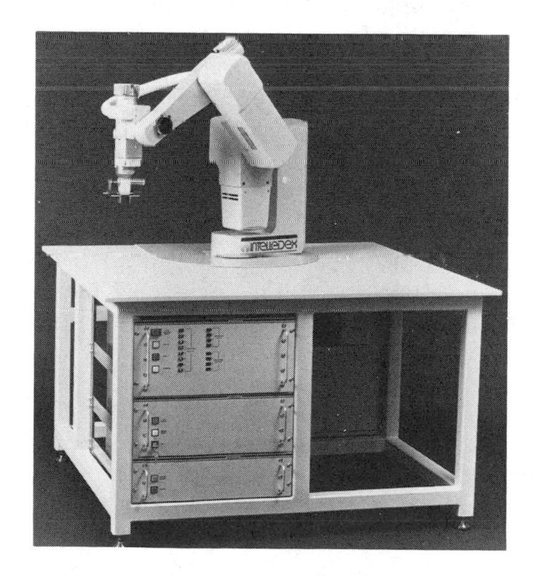

FIGURE 7.12 The Intelledex model 605T robot featuring six axes, ±0.001 inch repeatability, and ±0.002 inch accuracy.

Photo courtesy of Intelledex, Inc.

moves along Cartesian straight line paths in which interpolation of orientation is done using a represenation similar to Z-Y-Z Euler angles.

Geometric problems with Cartesian paths

Because a continuous correspondence is made between a path shape described in Cartesian space and joint positions, Cartesian paths are prone to various problems relating to workspace and singularities. Although the initial location of the manipulator and the final goal point are both within the manipulator workspace, it is quite possible that not all points lying on a straight line connecting these two points are in the workspace. This is an example of a situation in which a joint space path could easily be executed, but a Cartesian straight line path would fail.*

We saw in Chapter 5 that there are locations in the manipulator's workspace where it is impossible to choose finite joint rates that yield the desired velocity of the end-effector in Cartesian space. It should not be suprising, therefore, that there are certain paths (described in Cartesian terms) which are impossible for the manipulator to perform. If, for example, a manipulator is following a Cartesian straight line path and approaches a singular configuration of the mechanism, one or more joint velocities may increase toward infinity. Since velocities of the mechanism are upper bounded, this situation usually results in the manipulator deviating from the desired path. One approach is to scale down the overall velocity of the path to a speed where all joints stay within their velocity capabilities. In this way, although the desired temporal attributes of the path may be lost, at least the spatial aspect of the trajectory definition is adhered to.

Due to these problems with paths specified in Cartesian space, most industrial manipulator control systems support both joint space and Cartesian space path generation. The user quickly learns that because of the difficulties with Cartesian paths, joint space paths should be used as the default, with Cartesian paths used only when actually needed by the application.

7.5 Path generation at run-time

At **run-time** the path generator routine constructs the trajectory, usually in terms of θ, $\dot{\theta}$, and $\ddot{\theta}$, and feeds this information to the manipulator's

* Some robot systems would notify the user of a problem before moving the manipulator, while in some, motion would start along the path until some joint reaches its limit, at which time manipulator motion is halted.

control system. This path generator computes the trajectory at the path update rate.

Generation of joint space paths

The result of having planned a path using any of the splining methods mentioned in Section 7.3 is a set of data for each segment of the trajectory. These data are used by the path generator at run-time to calculate θ, $\dot{\theta}$, and $\ddot{\theta}$.

In the case of cubic splines, the path generator simply computes (7.3) as t is advanced. When the end of one segment is reached, a new set of cubic coefficients is recalled, t is set back to zero, and the generation continues.

In the case of linear splines with parabolic blends, the value of time, t, is checked on each update to determine whether we are currently in the linear or the blend portion of the segment. In the linear portion, the trajectory for each joint is calculated as

$$\begin{aligned}\theta &= \theta_j + \dot{\theta}_{jk} t,\\ \dot{\theta} &= \dot{\theta}_{jk},\\ \ddot{\theta} &= 0,\end{aligned} \tag{7.45}$$

where t is the time since the jth via point and $\dot{\theta}_{jk}$ was calculated at path planning time from (7.24a). In the blend region, the trajectory for each joint is calulated as

$$\begin{aligned}t_{inb} &= t - \left(\frac{1}{2}t_j + t_{jk}\right),\\ \theta &= \theta_j + \dot{\theta}_{jk}(t - t_{inb}) + \frac{1}{2}\ddot{\theta}_k t_{inb}^2,\\ \dot{\theta} &= \dot{\theta}_{jk} + \ddot{\theta}_k t_{inb},\\ \ddot{\theta} &= \ddot{\theta}_k,\end{aligned} \tag{7.46}$$

where $\dot{\theta}_{jk}$, $\ddot{\theta}_k$, t_j, and t_{jk} were calculated at path plan time by equations (7.24) through (7.28). This continues with t being reset to $\frac{1}{2}t_j$ when a new linear segment is entered, until we have worked our way through all the data sets representing the path segments.

Generation of Cartesian space paths

For the Cartesian path scheme presented in Section 7.4, we use the path generator for the linear spline with parabolic blends path. However, the values computed represent the Cartesian position and orientation rather than joint variable values, so we rewrite (7.45) and (7.46) with the symbol x representing a component of the Cartesian position and orientation vector. In the linear portion of the segment, each degree of freedom in $\mathcal{X}$ is calcuated as

$$\begin{aligned} x &= x_j + \dot{x}_{jk} t, \\ \dot{x} &= \dot{x}_{jk}, \\ \ddot{x} &= 0. \end{aligned} \tag{7.47}$$

where t is the time since the jth via point and $\dot{x}_{jk}$ was determined at path plan time using an equation analogous to (7.24a). In the blend region, the trajectory for each degree of freedom is calculated as

$$\begin{aligned} t_{inb} &= t - \left(\frac{1}{2}t_j + t_{jk}\right), \\ x &= x_j + \dot{x}_{jk}(t - t_{inb}) + \frac{1}{2}\ddot{x}_k t_{inb}^2, \\ \dot{x} &= \dot{x}_{jk} + \ddot{x}_k t_{inb}, \\ \ddot{x} &= \ddot{x}_k, \end{aligned} \tag{7.48}$$

where the quantities $\dot{x}_{jk}$, $\ddot{x}_k$, t_j, and t_{jk} were determined at plan time just as in the joint space case.

Finally, this Cartesian trajectory ($\mathcal{X}$, $\dot{\mathcal{X}}$, and $\ddot{\mathcal{X}}$) must be converted into equivalent joint space quantities. A complete analytical solution to this problem would use the inverse kinematics to calculate joint positions, the inverse Jacobian for velocities, and the inverse Jacobian plus its derivative for accelerations [5]. A simpler way often used in practice is as follows: At path update rate we convert $\mathcal{X}$ into its equivalent frame representation, ${}^S_GT$. We then use the `SOLVE` routine (see Section 4.8) to calculate the required vector of joint angles, Θ. Numerical differentiation is then used to compute $\dot{\Theta}$ and $\ddot{\Theta}$.* Thus, the algorithm is

$$
\begin{aligned}
\mathcal{X} &\rightarrow {}^S_GT, \\
\Theta(t) &= \texttt{SOLVE}({}^S_GT), \\
\dot{\Theta}(t) &= \frac{\Theta(t) - \Theta(t - \delta t)}{\delta t}, \\
\ddot{\Theta}(t) &= \frac{\dot{\Theta}(t) - \dot{\Theta}(t - \delta t)}{\delta t}.
\end{aligned}
\tag{7.49}
$$

Then, Θ, $\dot{\Theta}$, and $\ddot{\Theta}$ are supplied to the manipulator's control system.

7.6 Description of paths with a robot programming language

In Chapter 10 we will discuss **Robot programming languages** in some detail. Here, we will illustrate how various types of paths that we have discussed in this chapter might be specified in a robot language. In these examples, we use the syntax of **AL**, a robot programming language developed at Stanford University [6].

The symbols A, B, C, and D stand for variables of type "frame" in the AL language examples below. These frames specify path points which we will assume have been taught or described textually to the system. Assume the manipulator begins in position A. To move the manipulator in joint space mode along linear–parabolic-blend paths we could say

```
move ARM to C with duration = 3*seconds;
```

To move to the same position and orientation in a straight line we could say

```
move ARM to C linearly with duration = 3*seconds;
```

where the clause "linearly" denotes that Cartesian straight line motion is to be used. If duration is not important, the user can omit this specification and the system will use a default velocity, that is,

```
move ARM to C;
```

A via point can be added, and we can write

```
move ARM to C via B;
```

* This differentiation can be done noncausally for preplanned paths, resulting in better quality $\dot{\Theta}$ and $\ddot{\Theta}$. Also, many control systems do not require a $\ddot{\Theta}$ input, and so it would not be computed.

or a whole set of via points might be specified by

```
move ARM to C via B,A,D;
```

Note that in

```
move ARM to C via B with duration = 6*seconds;
```

the duration is given for the entire motion. The system decides how to split this duration between the two segments. It is possible in AL to specify the duration of a single segment; for example by

```
move ARM to C via B where duration = 3*seconds;
```

The first segment which leads to point B will have a duration of 3 seconds.

7.7 Planning paths using the dynamic model

Usually when paths are planned we use a default or a maximum acceleration at each blend point. Actually, the amount of acceleration that the manipulator is capable of at any instant is a function of the dynamics of the arm and the actuator limits. Most actuators are not characterised by a fixed maximum torque or acceleration, but rather by a torque–speed curve.

When we plan a path assuming there is a maximum acceleration at each joint or along each degree of freedom, we are making a tremendous simplification. In order to be careful not to exceed the actual capabilities of the device, this maximum acceleration must be chosen conservatively. Therefore, we are not making full use of the speed capabilities of the manipulator in paths planned by the methods introduced in this chapter.

We might ask the following question: Given a desired spatial path of the end-effector, find the timing information (which turns a description of a spatial path into a trajectory) such that the manipulator reaches the goal point in minimum time. Such problems can be solved [7]. The solution is done numerically and takes the rigid body dynamics into account as well as actuator speed-torque curves.

7.8 High-level path planning

It would be extremely convenient if we could simply tell the robot system what the desired goal point of the manipulator motion is, and let the system determine where and how many via points are required so that the goal is reached without the manipulator hitting any obstacles. In order to do this, the system must have models of the manipulator, the work area, and all potential obstacles in the area. A second manipulator may even be working in the same area and hence each arm must be considered as a moving obstacle for the other. Systems with such capabilities are not yet available commercially; however, several researchers have worked on developing such systems [8], [9].

References

[1] R. P. Paul and H. Zong, "Robot Motion Trajectory Specification and Generation," 2nd International Symposium on Robotics Research, Kyoto, Japan, August 1984.

[2] R. Taylor, "Planning and Execution of Straight Line Manipulator Trajectories," in *Robot Motion*, Brady et al., editors, MIT Press, 1983.

[3] C. DeBoor, *A Practical Guide to Splines*, Springer-Verlag, New York, 1978.

[4] D. Rogers and J. A. Adams, *Mathematical Elements for Computer Graphics*, McGraw-Hill, 1976.

[5] B. Gorla, and M. Renaud, *Robots Manipulateurs*, Cepadues - Editions, Toulouse, 1984.

[6] R. Goldman, "Design of an Interactive Manipulator Programming Environment," UMI Research Press, Ann Arbor, 1985.

[7] K. Shin and N. McKay, "Minimum-Time Control of Robotic Manipulators with Geometric Path Constraints," *IEEE Transactions on Automatic Control*, June 1985.

[8] T. Lozano-Perez, "Spatial Planning: A Configuration Space Approach," AI Memo 605, MIT Artificial Intelligence Laboratory, Cambridge, Massachusetts, 1980.

[9] R. Brooks, "Solving the Find-Path Problem by Good Representation of Free Space," *IEEE Transactions on Systems, Man, and Cybernetics*, SMC-13:190-197, 1983.

Exercises

7.1 [8] How many individual cubics are computed when a 6-jointed robot moves along a cubic spline path through two via points and stops at a goal point? How many coefficients are stored to describe these cubics?

7.2 [13] A single-link robot with a rotary joint is motionless at $\theta = -5°$. It is desired to move the joint in a smooth manner to $\theta = 80°$ in four seconds. Find the coefficients of a cubic which accomplishes this motion and brings the arm to rest at the goal. Plot the position, velocity, and acceleration of the joint as a function of time.

7.3 [14] A single-link robot with a rotary joint is motionless at $\theta = -5°$. It is desired to move the joint in a smooth manner to $\theta = 80°$ in four seconds and stop smoothly. Compute the corresponding parameters of a linear trajectory with parabolic blends. Plot the position, velocity, and acceleration of the joint as a function of time.

7.4 [30] Write a path planning routine which implements (7.25) through (7.30) in a general way for paths described by an arbitrary number of path points. For example, this routine could be used to solve Example 7.4.

7.5 [18] Sketch graphs of position, velocity, and acceleration for the 2-segment continuous acceleration spline given in Example 7.2. Sketch them for a joint for which $\theta_0 = 5.0°$, $\theta_v = 15.0°$, $\theta_g = 40.0°$, and each segment lasts 1.0 second.

7.6 [18] Sketch graphs of position, velocity, and acceleration for a 2-segment spline where each segment is a cubic, using the coefficients as given in (7.11). Sketch them for a joint where $\theta_0 = 5.0°$ for the initial point, $\theta_v = 15.0°$ is a via point, and $\theta_g = 40.0°$ is the goal point. Assume each segment has a duration of 1.0 second, and the velocity at the via point is to be 17.5 degrees/second.

7.7 [20] Calculate $\dot{\theta}_{12}$, $\dot{\theta}_{23}$, t_1, t_2, and t_3 for a 2-segment linear spline with parabolic blends (use (7.24) through (7.28)). For this joint, $\theta_1 = 5.0°$, $\theta_2 = 15.0°$, $\theta_3 = 40.0°$. Assume $t_{d12} = t_{d23} = 1.0$ second, and the default acceleration to use during blends is 80 degrees/second^2. Sketch plots of position, velocity, and acceleration of θ.

7.8 [18] Sketch graphs of position, velocity, and acceleration for the 2-segment continuous acceleration spline given in Example 7.2. Sketch them for a joint for which $\theta_0 = 5.0°$, $\theta_v = 15.0°$, $\theta_g = -10.0°$, and each segment lasts 2.0 seconds.

7.9 [18] Sketch graphs of position, velocity, and acceleration for a 2-segment spline where each segment is a cubic, using the coefficients as given in (7.11). Sketch them for a joint where $\theta_0 = 5.0°$ for the initial point, $\theta_v = 15.0°$ is a via point, and $\theta_g = -10.0°$ is the goal point. Assume each segment has a duration of 2.0 seconds, and the velocity at the via point is to be 0.0 degrees/second.

7.10 [20] Calculate $\dot{\theta}_{12}$, $\dot{\theta}_{23}$, t_1, t_2, and t_3 for a 2-segment linear spline with parabolic blends (use (7.24) through (7.28)). For this joint, $\theta_1 = 5.0°$, $\theta_2 = 15.0°$, $\theta_3 = -10.0°$. Assume $t_{d12} = t_{d23} = 2.0$ seconds, and the default acceleration to use during blends is 60 degrees/second^2. Sketch plots of position, velocity, and acceleration of θ.

7.11 [6] Give the 6×1 Cartesian position and orientation representation, ${}^{S}\mathcal{X}_{G}$, which is equivalent to ${}^{S}_{G}T$ where ${}^{S}_{G}R = ROT(\hat{Z}, 30°)$ and ${}^{S}P_{GORG} = [\,10.0 \quad 20.0 \quad 30.0\,]^T$.

7.12 [6] Give ${}^{S}T_{G}$ which is equivalent to the 6×1 Cartesian position and orientation representation ${}^{S}\mathcal{X}_{G} = [\,5.0 \quad -20.0 \quad 10.0 \quad 45.0 \quad 0.0 \quad 0.0\,]^T$.

7.13 [30] Write a program which uses the dynamic equations from Section 6.7 (the 2-link planar manipulator) to compute the time history of torques needed to move the arm along the trajectory of Exercise 7.8. What are the maximum torques required and where do they occur along the trajectory?

7.14 [32] Write a program which uses the dynamic equations from Section 6.7 (the 2-link planar manipulator) to compute the time history of torques needed to move the arm along the trajectory of Exercise 7.8. Make separate plots of the joint torques required due to inertia, velocity terms, and gravity.

7.15 [22] Do Example 7.2 when $t_{f1} \neq t_{f2}$.

Programming Exercise (Part 7)

1. Write a joint space, cubic splined path planning system. One routine which your system should include is

```
  Procedure CUBCOEF(VAR th0,thf,thdot0,thdotf: real;
VAR cc: vec4);
```

where

$$\text{th0} = \text{initial postion of } \theta \text{ at beginning of segment,}$$

$$\text{thf} = \text{final position of } \theta \text{ at segment end,}$$

$$\text{thdot0} = \text{initial velocity of segment,}$$

$$\text{thdotf} = \text{final velocity of segment.}$$

These four quantities are inputs, and "cc", an array of the four cubic coefficients, is the output.

Your program should accept up to (at least) five via point specifications in the form of tool frame, $\{T\}$, relative to station frame, $\{S\}$, in the usual user form of (x, y, ϕ). To keep life simple, all segments will have the same duration. Your system should solve for the coefficients of the cubics using some reasonable heuristic for assigning joint velocities at the via points. Hint: See option 2 in Section 7.3.

2. Write a path generator system which calculates a trajectory in joint space based on sets of cubic coefficients for each segment. It must be able to generate the multisegment path you planned in Problem 1. A duration for the segments will be specified by the user. It should produce position, velocity, and acceleration information at the path update rate, which will also be specified by the user.

3. The manipulator is the same 3-link as always. The definition of the $\{T\}$ and $\{S\}$ frames are the same as before:

$$^{W}_{T}T = [x \quad y \quad \theta] = [0.1 \quad 0.2 \quad 30.0]\,,$$

$$^{B}_{S}T = [x \quad y \quad \theta] = [0.0 \quad 0.0 \quad 0.0]\,.$$

Using a duration of 3.0 seconds per segment, plan and execute the path which starts with the manipulator at position

$$[x_1 \; y_1 \; \phi_1] = [0.758 \; 0.173 \; 0.0]\,;$$

moves through the via points

$$[x_2 \quad y_2 \quad \phi_2] = [0.6 \quad -0.3 \quad 45.0] \,,$$

$$[x_3 \quad y_3 \quad \phi_3] = [-0.4 \quad 0.3 \quad 120.0] \,;$$

and ends at the goal point (in this case same as initial point)

$$[x_4 \; y_4 \; \phi_4] = [0.758 \; 0.173 \; 0.0] \,.$$

Use a path update rate of 40 Hz, but print the position only every 0.2 seconds. Print the positions out in terms of Cartesian user form. You don't have to print out velocities or accelerations, though you might be interested to do so.

8

POSITION CONTROL OF MANIPULATORS

8.1 Introduction

Based on the material in Chapters 1 through 7 we now have the means to calculate joint-position time histories that correspond to desired end-effector motions through space. In this chapter we discuss how to cause the manipulator actually to perform these desired motions.

Very few robots use stepper motors or other actuators which can be controlled in an *open loop* fashion. Usually, manipulators are powered by actuators which output a torque or a force at each joint. In this case, we must use some kind of a control system to compute appropriate actuator commands which will realize the desired motion.

We will model a manipulator as a mechanism which is instrumented with sensors at each joint to measure the joint angle, and an actuator at each joint to apply a torque on the neighboring link. Although other physical arrangements of sensors are sometimes used, the vast majority of robots have a position sensor at each joint. Sometimes velocity sensors (tachometers) are also present at the joints. Various actuation and transmission schemes are prevalent in industrial robots, but many of these can be modelled by supposing there is a single actuator at each joint.

In order to cause the desired motion of each joint to be realized by the manipulator, we must specify a **control algorithm** which sends torque commands to the actuators. Almost always these torques are computed by continuously or intermittently using feedback from the joint sensors to compute the torque required. The use of feedback, while extremely advantageous in this and other applications, brings with it the difficult problem of control algorithm synthesis. Furthermore, relative to many control problems, the particular case of manipulator control is quite complicated.

The control problem for manipulators is inherently nonlinear. This means that much of linear control theory is not directly applicable, be it via the "classical" or "modern" approach. Nonetheless, in many proposed solutions to the nonlinear problem, one ends up using methods from linear control theory.

As in other chapters in this book, we must restrict our scope of methods considered to one or two which seem well suited to the problem at hand: mechanical manipulators. Consequently, we present one method of manipulator control, the so called **computed torque method**, in this chapter.

8.2 Control of a mass along one degree of freedom

Let us consider the position control of perhaps the simplest mechanical system, a block of mass m along a single degree of freedom. That is, a block of mass m is resting on a fricitionless track such that it is free to move in a single direction, say $\hat{X}$. Our actuator can apply to the block a force of any magnitude in the $\hat{X}$ direction. Our sensor measures the position and velocity of the block in the $\hat{X}$ direction. The open-loop equation of motion of the system is

$$f = m\ddot{x}. \tag{8.1}$$

To further simplify this already simple problem, let us assume that the mass of the block is one unit. Hence, m in equation (8.1) has value 1, and is introduced only to change units from acceleration to force. The problem is to formulate a control system which causes the block to maintain a desired position regardless of random disturbance forces.

Simple position regulation

We propose a **control law**, or algorithm, by which we will compute forces to apply to the block with the actuator. A reasonable choice is

$$f = -k_v \dot{x} + k_p(x_d - x), \tag{8.2}$$

where x_d is the desired position of the block, x is the actual position, and k_p and k_v are position and velocity **gains** of the control system. If these gains are chosen properly, the block will be controlled to maintain the desired position x_d, and disturbances will be suppressed.

Using this control law, the system responds to external disturbances as if the unit mass were attached to a spring of stiffness k_p. Oscillations of the mass and this virtual spring are damped as if the block were subject to viscous friction with coefficient k_v. In this scheme, k_p sets the **closed-loop stiffness** of the system.

Position control along a trajectory

Rather than just maintaining the block at a desired location, let us expand our controller so that the block can be made to follow a trajectory, or function of time which specifies x_d. Let us assume that a complete description of the desired position and its first two derivatives is available, namely x_d, $\dot{x}_d$, and $\ddot{x}_d$. In this case a reasonable control law is

$$f = \ddot{x}_d + k_v(\dot{x}_d - \dot{x}) + k_p(x_d - x). \tag{8.3}$$

We see that this control law is a good choice if we combine it with the equation of motion of the system given by (8.1) with $m = 1$. This leads to the system equation of motion written in **error-space**:

$$\ddot{e} + k_v \dot{e} + k_p e = 0, \tag{8.4}$$

where e is the servo error, defined as $x_d - x$. Equation (8.3) is judged to be a good control law because it yields a closed-loop system, (8.4), in which the characteristics of the suppression of errors can be set easily by choice of the values of k_p and k_v.

Equation (8.4) is a differential equation which describes the suppression of errors in our control system. By proper choice of gains k_p and k_v the solution of this differential equation can be made to be the **critical damping** case. In this situation, errors are suppressed in the fastest possible second order linear way which does not cause overshoot.

Equation (8.4) has the characteristic equation [1]

$$s^2 + k_v s + k_p = 0, \tag{8.5}$$

which has the roots

$$s_1 = -\frac{k_v}{2} + \frac{\sqrt{k_v^2 - 4k_p}}{2},$$
$$s_2 = -\frac{k_v}{2} - \frac{\sqrt{k_v^2 - 4k_p}}{2}. \tag{8.6}$$

Eq. (8.5) is critically damped when

$$k_v^2 - 4k_p = 0. \tag{8.7}$$

The roots of the closed-loop characteristic equation, s_1 and s_2, are called the closed loop **poles** of the system. The location of these poles in the real-imaginary plane determines the nature of disturbance suppression. The choice of critical damping corresponds to one particular choice for the location of the poles. Throughout the rest of this chapter we will refer to critical damping as a desirable quality. Of course, other specifications of "good" performance of a second order system could also be employed.

Figure 8.1 shows a control block diagram for the proposed trajectory following controller for a mass system with one degree of freedom.

Addition of an integral term

Many times an additional term, called an *integral* term, is added to the servo law. The control law becomes

$$f = \ddot{x}_d + k_v \dot{e} + k_p e + k_i \int e dt. \tag{8.8}$$

The term is added so that the system will have no steady-state error in the presence of biases of one kind or another. Any steady-state error

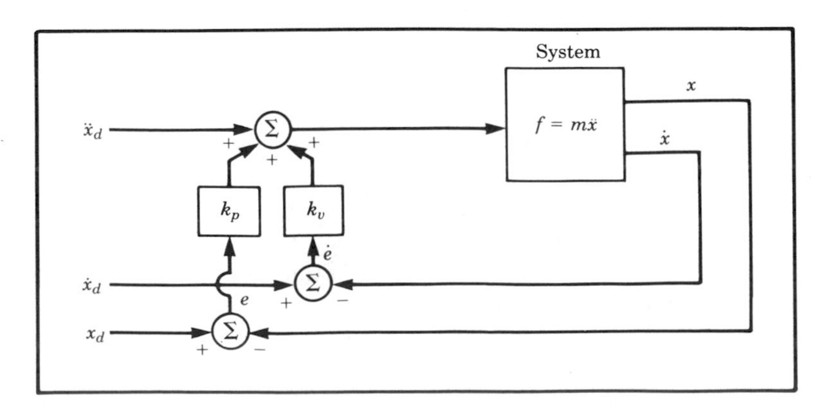

FIGURE 8.1 Block diagram of a control system for a unit mass system.

would cause the integral term to build up until it contributed a sufficient force to cause the system to move such as to reduce the steady-state error. With this control law the system becomes a third-order system, and we can solve the corresponding third-order differential equation to determine the response of the system to desired input or disturbances. Often k_i is kept quite small so that the third-order system is "close" to the second order system without this term. This form of control law is called a **PID control law**, or "proportional, integral, derivative" control law. For more details, see [2]. For simplicity, we will not show an integral term in the control laws which we will develop in this chapter.

8.3 Control law partitioning

In preparation for designing control laws for more complicated mechanical systems, let us consider the control of a spring, mass, friction system as shown in Fig. 8.2. We will "reduce" this system to the simple unit-mass problem we have already solved through a simple technique which we will call **control law partitioning**.

The open-loop equation of motion for the system is

$$m\ddot{x} + b\dot{x} + kx = f. \tag{8.9}$$

We wish to decompose the controller for this system into two parts. One part of the control law is **model based** in that it makes use of the parameters of the particular system under control. In this case, the model-based portion of the control law will make use of supposed knowledge of m, b, and k. This portion of the control law is set up such that it *reduces the system so that it appears to be a unit mass*. This will be clear when we do an example below. The second part of the control law

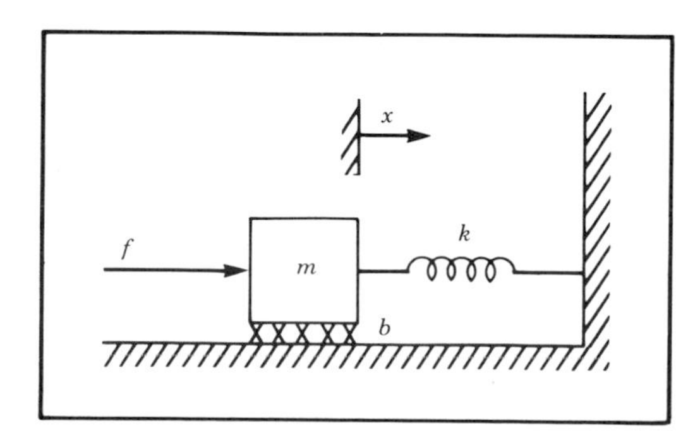

FIGURE 8.2 Mass, spring, and friction system.

is **error driven** in that it forms error signals by differencing desired and actual variables and multiplies these errors by gains. The error-driven portion of the control law is sometimes called the **servo** portion. Since the model-based portion of the control law has the effect of making the system appear to be a unit mass, the design of the servo portion is very simple—gains are chosen as if we were controlling a system composed of a single unit mass (i.e., no friction, no stiffness).

The model-based portion of the control appears in a control law of the form

$$f = \alpha f' + \beta, \tag{8.10}$$

where α and β are functions or constants and are chosen so that if f' is taken as the *new input* to the system, *the system appears to be a unit mass*. With this structure of control law, the system equation (the result of combining (8.9) and (8.10)) is

$$m\ddot{x} + b\dot{x} + kx = \alpha f' + \beta. \tag{8.11}$$

Clearly, in order to make the system appear as a unit mass from the f' input, for this particular system we should choose α and β as

$$\begin{aligned} \alpha &= m, \\ \beta &= b\dot{x} + kx. \end{aligned} \tag{8.12}$$

Making these assignments and plugging into (8.11), we have the system equation

$$\ddot{x} = f'. \tag{8.13}$$

This is the equation of motion for a unit mass. Now it is easy to design the servo portion of the control law. The servo portion computes f' as a function of errors as

$$f' = \ddot{x}_d + k_v \dot{e} + k_p e. \tag{8.14}$$

As before, k_p and k_v are computed for a system whose characteristic equation is given by (8.5), and so (8.7) gives the solution for critical damping. Figure 8.3 shows the general form of the partitioned servo law with trajectory inputs.

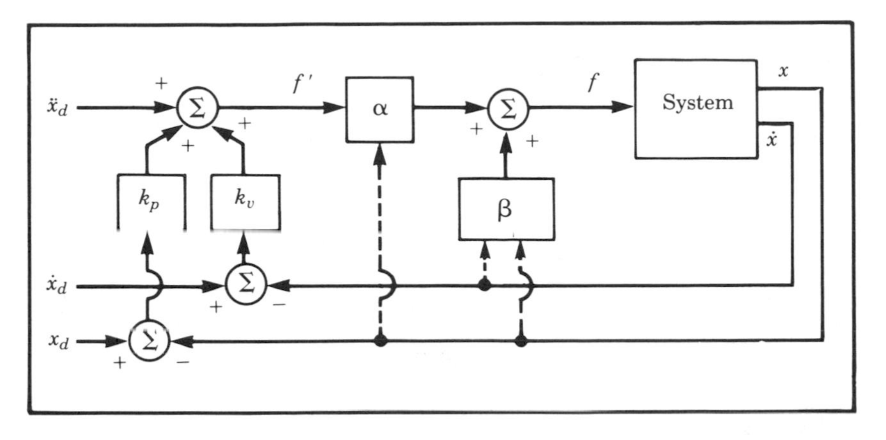

FIGURE 8.3 General form of the partitioned servo control law. Dashed lines indicate that α and β are a function of the system state, x and $\dot{x}$.

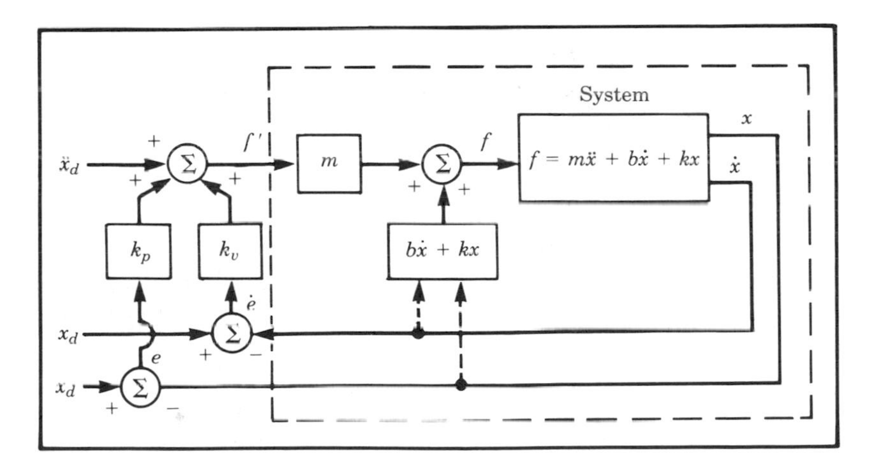

FIGURE 8.4 A control system using the partitioned servo control methodology.

EXAMPLE 8.1

Determine the proper feedback gains for the servo control law of (8.14) so that critical damping is achieved with a closed-loop stiffness of k_{CL} newtons/meter. Draw a block diagram of the complete controller which shows the partitioning of the control law.

The unit-mass system is critically damped when (8.7) is satisfied. Thus we easily compute the gains:

$$k_p = k_{CL},$$
$$k_v = 2\sqrt{k_{CL}}. \quad \textbf{(8.15)}$$

The complete control system is diagrammed in Fig. 8.4. The portion of the system which is surrounded by dashed lines can be considered a black box which has the dynamic qualities of a unit mass. Inside this black box is the real system plus the model-based portion of the control law. ■

8.4 Nonlinear and time-varying systems

In the preceding development we dealt with a linear constant coefficient differential equation. This mathematical form arose because the mass, spring, friction system of Fig. 8.2 was modelled as a linear time-invariant system. For systems whose parameters vary in time or systems which are by nature nonlinear, solutions are more difficult.

When nonlinearities are not severe, **local linearization** may be used to derive linear models which are approximations of the nonlinear equations in the neighborhood of an operating point. If the system does not operate over a small range in its variables, then as the system moves, the operating point can be moved along with it. At each new operating point, a new linearization is performed. The result of this sort of *moving linearization* is a linear but time-varying system.

Although this quasi-static linearization of the original system is useful in some analysis and design techniques, we will not make use of it in our control law synthesis procedure. Rather, we will deal with the nonlinear equations of motion directly and will not resort to linearizations in deriving a controller.

If the spring in Fig. 8.2 were not linear but instead had some nonlinear characteristic, we could consider the system quasi-statically and at each instant determine where the poles of the system are located. We would find that the poles "move" around in the real-imaginary plane as a function of the position of the block. Hence we could not select fixed gains which would keep the poles in a desirable location (for example, at critical damping). So we may be tempted to consider a more complicated control law in which the gains are time-varying (actually, varying as a function of the block's position) just such that the system is always critically damped. Essentially, this would be done by computing k_p such that the combination of the nonlinear effect of the spring would be exactly cancelled by a nonlinear term in the control law so that the overall stiffness would stay a constant at all times. Such a control scheme might be called a **linearizing** control law, since it uses a nonlinear control term to "cancel" a nonlinearity in the controlled system such that the overall closed-loop system is linear.

In our partitioned control law scheme, the servo law remains the same as always, but the model-based portion now will contain a model of the nonlinearity. Thus the model-based portion of the control performs a linearization function. This is best shown in an example.

EXAMPLE 8.2

Consider the nonlinear spring characteristic shown in Fig. 8.5. Rather than the usual linear spring relationship, $f = kx$, this spring is described by $f = qx^3$. If this spring is part of the physical system shown in Fig. 8.2, determine a control law which would keep the system critically damped with a stiffness of k_{CL}.

The open-loop equation is

$$m\ddot{x} + b\dot{x} + qx^3 = f. \tag{8.16}$$

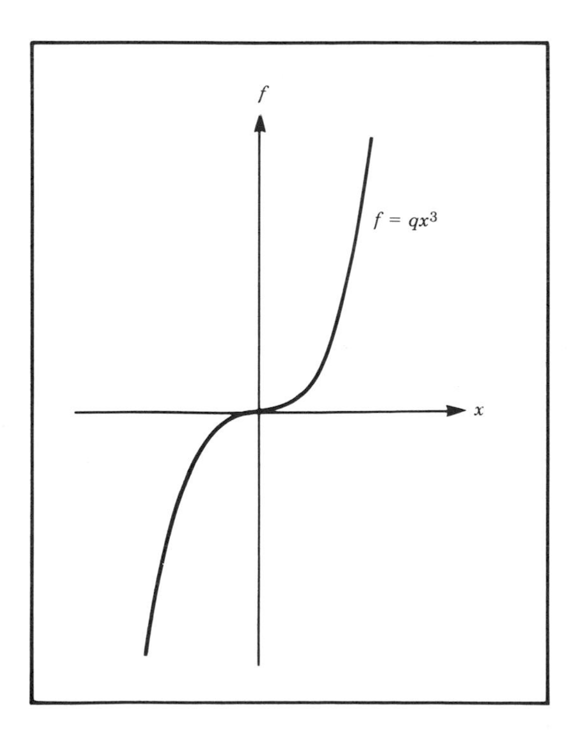

FIGURE 8.5 The force vs. distance characteristic of a nonlinear spring.

The model-based portion of the control is $f = \alpha f' + \beta$, where

$$\alpha = m,$$
$$\beta = b\dot{x} + qx^3; \tag{8.17}$$

and the servo portion is, as always

$$f' = \ddot{x}_d + k_v\dot{e} + k_p e, \tag{8.18}$$

where the values of the gains are calculated from some desired performance specification. Figure 8.6 shows a block diagram of this control system. ■

EXAMPLE 8.3

Consider the nonlinear friction characteristic shown in Fig. 8.7. Whereas linear friction is described by $f = b\dot{x}$, this **Coulomb friction** is described by $f = b_c sgn(\dot{x})$. For most of today's manipulators, the friction of the joint in its bearing (be it rotational or linear) is modelled more accurately by this nonlinear characteristic than by the simpler, linear model. If this type of friction is present in the system of Fig. 8.2, design a control system which uses a nonlinear model-based portion to critically damp the system at all times.

The open-loop equation is

$$m\ddot{x} + b_c\ sgn(\dot{x}) + kx = f. \tag{8.19}$$

The model-based portion of the control is $f = \alpha f' + \beta$, where

$$\alpha = m,$$
$$\beta = b_c\ sgn(\dot{x}) + kx; \tag{8.20}$$

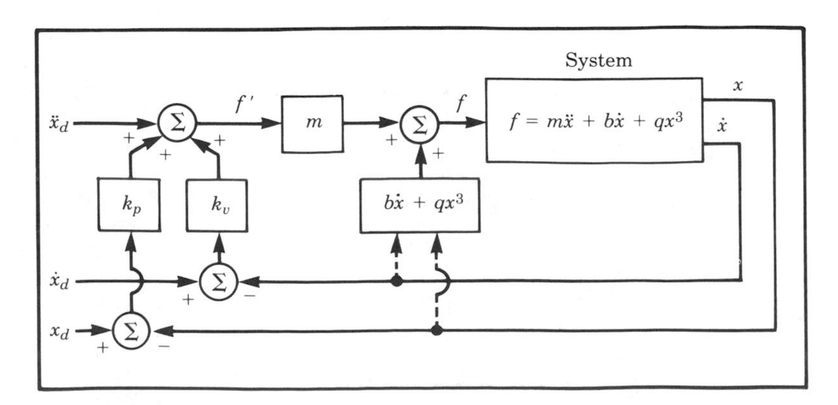

FIGURE 8.6 A nonlinear control system for a system with a nonlinear spring.

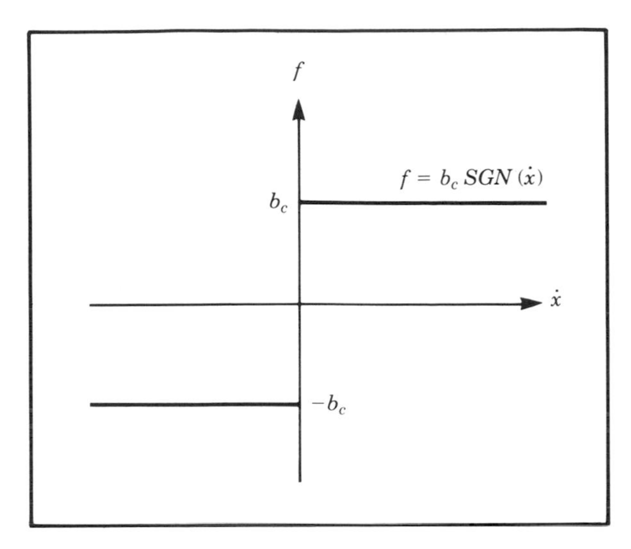

FIGURE 8.7 The force vs. velocity characteristic of Coulomb friction.

and the servo portion is

$$f' = \ddot{x}_d + k_v \dot{e} + k_p e, \tag{8.21}$$

where the values of the gains are calculated from some desired performance specification. ■

While the field of nonlinear control theory is quite difficult, we have seen that in certain simple cases it is not difficult to design a nonlinear controller. The general method used in the above simple examples is the same method we will use for the problem of manipulator control, namely:

1. Compute a nonlinear model-based control law which "cancels" the nonlinearities of the system to be controlled.
2. Reduce the system to a linear system which can be controlled using the simple linear servo law developed for the unit mass.

In some sense, the linearizing control law is implementing an *inverse model* of the system being controlled. The nonlinearities in the system cancel with those in the inverse model; this, together with the servo law, results in a linear closed loop system. Obviously, to do this cancelling, we must know the parameters and the structure of the nonlinear system. This is often a problem in practical application of this method.

EXAMPLE 8.4

Consider the single-link manipulator shown in Fig. 8.8. It has one rotational joint. The mass is considered to be located at a point at the distal end of the link, and so the moment of inertia is ml^2. There is Coulomb and viscous friction acting at the joint, and there is a load due to gravity.

The model of the manipulator is

$$\tau = ml^2\ddot{\theta} + v\dot{\theta} + c\ sgn(\dot{\theta}) + mlg\cos(\theta). \tag{8.22}$$

As always, the control system has two parts, the linearizing model-based portion and the servo law portion.

The model-based portion of the control is $f = \alpha f' + \beta$, where

$$\begin{aligned} \alpha &= ml^2, \\ \beta &= v\dot{\theta} + c\ sgn(\dot{\theta}) + mlg\cos(\theta); \end{aligned} \tag{8.23}$$

and the servo portion is, as always

$$f' = \ddot{\theta}_d + k_v\dot{e} + k_p e, \tag{8.24}$$

where the values of the gains are calculated from some desired performance specification. ■

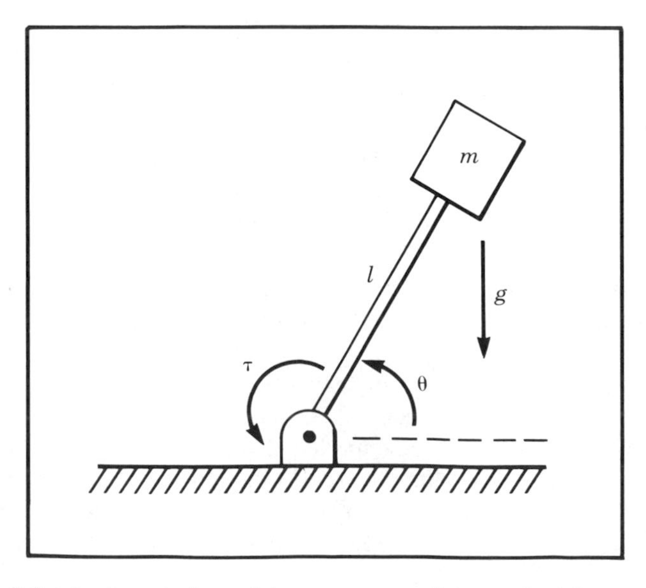

FIGURE 8.8 An inverted pendulum, or a one link manipulator.

8.5 Multi-input/multi-output control systems

Unlike the simple examples we have discussed in this chapter so far, the problem of controlling a manipulator is a multi-input/multi-output problem. We will have a vector of desired joint positions, velocities, and accelerations. The control law must compute a vector of joint actuator signals. Our basic scheme of partitioning the control into a model-based portion and a servo portion is still applicable, but will appear in a matrix–vector form. The control law is now in the form

$$F = \alpha F' + \beta, \tag{8.25}$$

where, for a system of n degrees of freedom, F, F', and β are $n \times 1$ vectors; and α is an $n \times n$ matrix. Note that the matrix α is not necessarily diagonal, but rather is chosen to **decouple** the n equations of motion. If α and β are correctly chosen, then from the F' input the system appears to be n independent unit masses. For this reason, in the multidimensional case, the model-based portion of the control law is called the **linearizing and decoupling** law. The servo law for a multidimensional system becomes

$$F' = \ddot{X}_d + K_v \dot{E} + K_p E, \tag{8.26}$$

where K_v and K_p are now $n \times n$ matrices, which are generally chosen to be diagonal with constant gains on the diagonal. E and $\dot{E}$ are $n \times 1$ vectors of errors in position and velocity, respectively.

8.6 The control problem for manipulators

In the case of manipulator control, we developed a model and the corresponding equations of motion in Chapter 6. As we saw, these equations are quite complicated. The rigid body dynamics have the form:

$$\tau = M(\Theta)\ddot{\Theta} + V(\Theta, \dot{\Theta}) + G(\Theta), \tag{8.27}$$

where $M(\Theta)$ is the $n \times n$ inertia matrix of the manipulator, $V(\Theta, \dot{\Theta})$ is an $n \times 1$ vector of centrifugal and Coriolis terms, and $G(\Theta)$ is an $n \times 1$ vector of gravity terms. Each element of $M(\Theta)$ and $G(\Theta)$ is a complicated function which depends on Θ, the position of all the joints of the manipulator. Each element of $V(\Theta, \dot{\Theta})$ is a complicated function of both Θ and $\dot{\Theta}$.

Additionally, we may incorporate a model of friction (or other nonrigid-body effects). Assuming that our model of friction is a function

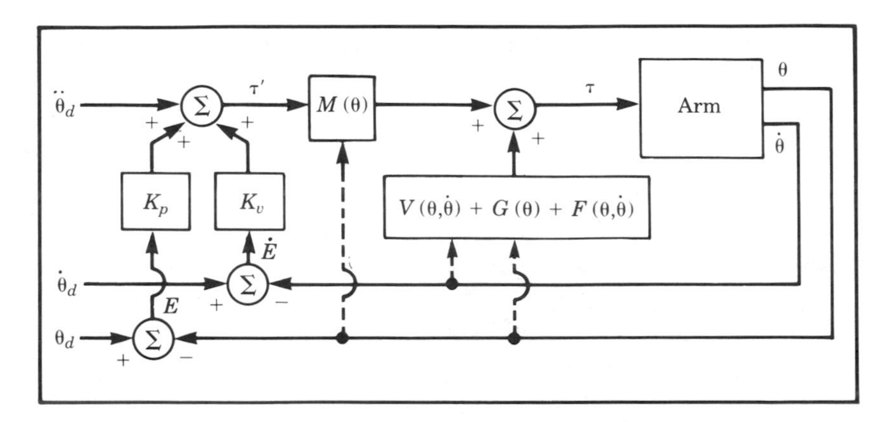

FIGURE 8.9 A model-based manipulator control system.

of joint positions and velocities, we add a term, $F(\Theta, \dot{\Theta})$, to (8.27) to yield the model

$$\tau = M(\Theta)\ddot{\Theta} + V(\Theta, \dot{\Theta}) + G(\Theta) + F(\Theta, \dot{\Theta}). \tag{8.28}$$

The problem of controlling a complicated system like (8.28) can be handled by the partitioned controller scheme we have introduced in this chapter. In this case, we have

$$\tau = \alpha\tau' + \beta, \tag{8.29}$$

where τ is the $n \times 1$ vector of joint torques. We choose

$$\begin{aligned} \alpha &= M(\Theta), \\ \beta &= V(\Theta, \dot{\Theta}) + G(\Theta) + F(\Theta, \dot{\Theta}), \end{aligned} \tag{8.30}$$

with the servo law

$$\tau' = \ddot{\Theta}_d + K_v\dot{E} + K_pE, \tag{8.31}$$

where

$$E = \Theta_d - \Theta. \tag{8.32}$$

The resulting control system is shown in Fig. 8.9.

Figure 8.10 shows the block diagram of a possible practical implementation of the decoupling and linearizing position control system. The dynamic model is expressed in its *configuration space* form so that the dynamic parameters of the manipulator appear as functions of manipulator position only. These functions might then be computed by a *background* process or by a second control computer. In this architecture,

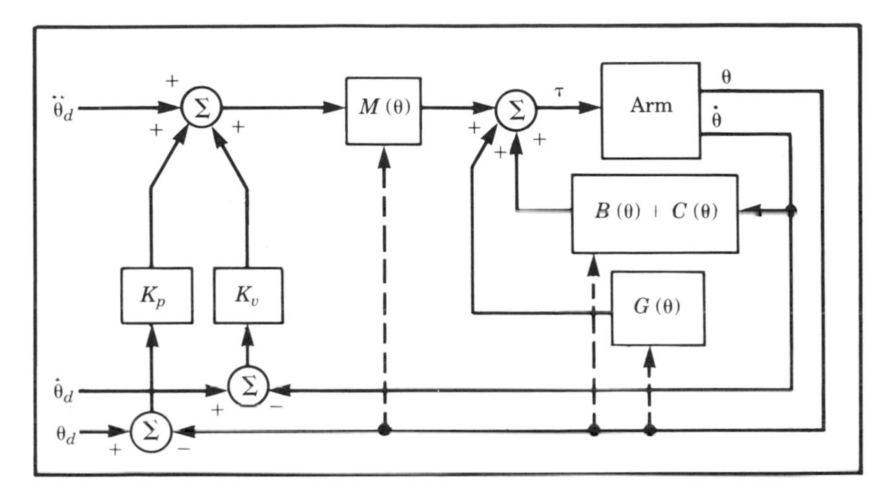

FIGURE 8.10 An implementation of the model-based manipulator control system.

the dynamic parameters can be updated at a rate slower than the rate of the closed loop servo. For example, the background computation might proceed at 60 Hz; whereas the closed loop servo is computed at 200 Hz.

If we have good knowledge of all the parameters of the manipulator (i.e., link lengths, inertias, friction coefficients) then the control scheme of Fig. 8.10 will yield a control system which keeps each joint of the manipulator critically damped over all configurations of the arm. Since our model will never be exact, this will never be perfectly true, but for all practical purposes it may be. One way to analyze the effect of error in the model is to indicate a vector of disturbance torques acting at the joints. In Fig. 8.10 we have indicated these disturbances as an input. The disturbance torques arise from any unmodelled or incorrectly modelled terms in the dynamic equations. If a friction term is not included in the model-based control law, the disturbance torques would largely be caused by friction effects. These unknown torques also include the effects of resonances and all "noise."

Writing the system error equation with inclusion of the unknown disturbances, we arrive at

$$\ddot{E} + K_v \dot{E} + K_p E = M^{-1}(\Theta)\, \tau_d, \tag{8.33}$$

where τ_d is the vector of disturbance torques at the joints. The left-hand side of (8.33) is uncoupled, but from the right hand side we see that a disturbance on any particular joint will introduce errors at all the other joints, since $M(\Theta)$ is not, in general, diagonal.

8.7 Practical considerations

In developing the decoupling and linearizing control in the last few sections, we have implicitly made a few assumptions which are rarely true in practice. In this section we discuss the practical problems faced by the control engineer in designing manipulator control systems, and the present day industrial solutions.

Lack of knowledge of parameters

The major problem is that the values of the parameters in the model are often not known accurately. This is particularly true of friction effects. In fact, it is usually extremely difficult to know the structure of the friction model, let alone the parameter values. Finally, if the manipulator has some portion of its dynamics which are not repeatable, because, for example, they change as the robot ages, it is difficult to have good parameter values in the model at all times.

By nature, most robots will be picking up various parts and tools. When a robot is holding a tool, the inertia and the weight of the tool change the dynamics of the manipulator. In an industrial situation, the mass properties of the tools may be known—in this case they can be accounted for in the modelled portion of the control law. When a tool is grasped, the inertia matrix and center of mass of the last link of the manipulator can be updated to a new value which represents the combined effect of the last link plus tool. However, in many applications the mass properties of objects that the manipulator picks up are not generally known, so that maintenance of an accurate dynamic model is impossible.

Since in general our model of the manipulator will never be perfect, we define the following notation: $\hat{M}(\Theta)$ is our model of the manipulator inertia matrix, $M(\Theta)$. Likewise, $\hat{V}(\Theta,\dot{\Theta})$, $\hat{G}(\Theta)$, and $\hat{F}(\Theta,\dot{\Theta})$ are our models of the velocity terms, gravity terms, and friction terms of the actual mechanism. Perfect knowledge of the model would mean:

$$\begin{aligned} \hat{M}(\Theta) &= M(\Theta), \\ \hat{V}(\Theta,\dot{\Theta}) &= V(\Theta,\dot{\Theta}), \\ \hat{G}(\Theta) &= G(\Theta), \\ \hat{F}(\Theta,\dot{\Theta}) &= F(\Theta,\dot{\Theta}). \end{aligned} \qquad \textbf{(8.34)}$$

Therefore, although the manipulator dynamics are given by

$$\tau = M(\Theta)\ddot{\Theta} + V(\Theta,\dot{\Theta}) + G(\Theta) + F(\Theta,\dot{\Theta}), \qquad \textbf{(8.35)}$$

our control law computes

$$\begin{aligned} \tau &= \alpha\tau' + \beta, \\ \alpha &= \hat{M}(\Theta), \\ \beta &= \hat{V}(\Theta,\dot{\Theta}) + \hat{G}(\Theta) + \hat{F}(\Theta,\dot{\Theta}). \end{aligned} \tag{8.36}$$

Decoupling and linearizing will not therefore be perfectly accomplished when parameters are not known exactly. Writing the closed-loop equation for the system, we have:

$$\begin{aligned} &\ddot{E} + K_v\dot{E} + K_pE \\ &\quad = \hat{M}^{-1}\left[\left(M - \hat{M}\right)\ddot{\Theta} + \left(V - \hat{V}\right) + \left(G - \hat{G}\right) + \left(F - \hat{F}\right)\right], \end{aligned} \tag{8.37}$$

where the arguments of the dynamic functions are not shown for brevity. Note that if (8.34) were true (i.e., the model were exact), then the right-hand side of (8.37) would be zero and the errors would disappear. When the parameters are not known exactly, the mismatch between actual and modelled parameters will cause servo errors to be excited in the rather complicated way given by (8.37).

Time required to compute the model

In all our considerations of the partitioned control law strategy, we have implicitly assumed that the entire system was running in continuous time, and that the computations in the control law require zero time for their computation. Given any amount of computation, with a large enough computer we can do the computations sufficiently fast that this is a reasonable approximation; however, the expense of the computer may make the scheme economically unfeasible. In the manipulator control case, the entire dynamic equation of the manipulator, (8.36), must be computed in the control law. These computations are quite involved and consequently, as discussed in Chapter 6, there has been a great deal of interest in developing fast computational schemes to compute them in an efficient way. As computer power becomes more and more affordable, control laws which require a great deal of computation will become more practical.

Almost all manipulator control systems are now performed in digital circuitry and are run at a certain **sampling rate**. This means that the position (and possibly other) sensors are read at discrete points in time. Based on the value read, an actuator command is computed and sent to the actuator. Thus reading sensors and sending actuator commands are not done continuously, but rather at a finite sampling rate.

To analyze the effect of delay due to computation and finite sample rate, we must use tools from the field of **discrete time control**. In discrete time, differential equations turn into difference equations, and a complete set of tools has been developed to answer questions about stability and pole placement for these systems. Discrete time control theory is beyond the scope of this book, although to do serious work in the area of manipulator control, many of the concepts from discrete time systems are essential. See [3] and other references.

Although important, ideas and methods from discrete time control theory are often difficult to apply to the case of manipulator control. Whereas we have managed to write a complicated differential equation of motion for the manipulator dynamic equation, a discrete time equivalent is impossible to obtain in general. This is because, for a general manipulator, the only way to solve for the evolution of the manipulator for a given set of initial conditions, an input, and a finite interval is by numerical integration (as we saw in Chapter 6). Discrete time models are possible if we are willing to use series solutions to the differential equations, or if we make approximations. However, if we need to make approximations to develop a discrete model, then it is not clear whether we have a better model than we have when just using the continuous model and making the approximation of infinite computing power. Suffice it to say that analysis of the discrete time manipulator control problem is difficult, and usually simulation is resorted to in order to judge the effect that a certain sample rate will have on performance.

We will generally assume that the computations can be performed quickly enough and often enough that the continuous time approximation is valid. This raises the question of how quick is quick enough? There are several points which need to be considered in choosing a sufficiently fast servo rate:

Tracking reference inputs: The frequency content of the desired or reference input places an absolute lower bound on the sample rate. The sample rate must be at least twice the bandwidth of reference inputs. This is usually not the limiting factor.

Disturbance rejection: In disturbance rejection, an upper bound on performance is given by a continuous time system. If the sample period is longer than the correlation time of the disturbance effects (assuming a statistical model for random disturbances), then these disturbances will not be suppressed. In one example [3] it appears that a good rule of thumb is that the sample period be 10 times shorter than the correlation time of the noise.

Anti-aliasing: Anytime an analog sensor is used in a digital control scheme, there will be a problem with aliasing unless the sensor's output is strictly band limited. In the usual case, sensors do not have a band limited output, and so sample rate should be chosen such that the amount of energy which appears in the aliased signal is small.

Structural resonances: We have not included bending modes in our characterization of a manipulator's dynamics. All real mechanisms have finite stiffness and so will be subject to various kinds of vibrations. If it is important to suppress these vibrations (and it often is) we must choose a sample rate which is at least twice the natural frequency of these resonances. A rate 10 times higher is recommended.

In Fig. 8.9, the model-based control portion of the control law is "in the servo loop" in that signals "flow" through that black box with each tick of the servo clock. If we wish to select a sample rate of 200 Hz, then the dynamic model of the manipulator must be computed at this rate. Another possible control system is shown in Fig. 8.11. Here, the model-based control is "outside" the servo loop. It would be possible to have a fast inner servo loop which just consists of multiplying errors by gains, and the model based torques could be added at a slower rate.

Unfortunately, the scheme of Fig. 8.11 does not provide complete decoupling. If we write the system equations we will find that the error equation of this system is

$$\ddot{E} + M^{-1}(\Theta)K_v\dot{E} + M^{-1}(\Theta)K_pE = 0. \quad \mathbf{(8.38)}$$

Clearly, as configuration of the arm changes, the effective closed-loop gain changes, and the quasi-static poles move around in the real-imaginary plane as the arm changes configuration. However, equation (8.38) could be used as a starting point to consider designing a **robust**

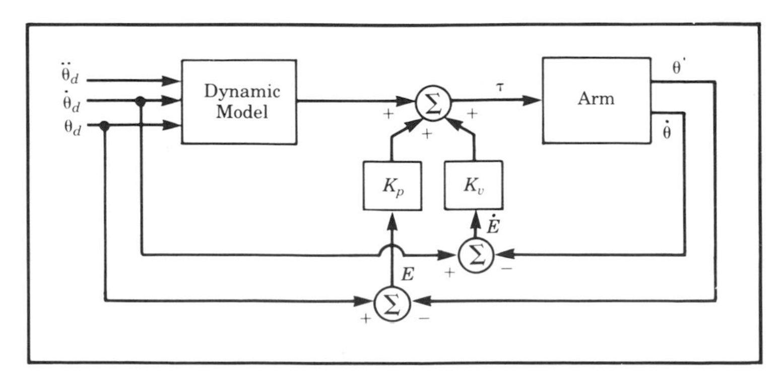

FIGURE 8.11 Control scheme with the model-based portion "outside" the servo loop.

controller. That is, to find a good set of constant gains such that despite the "motion" of the poles, they are guaranteed to remain in reasonably favorable locations. Alternatively, it may be possible to compute variable gains which change with configuration of the arm so that the system's quasi-static poles remain in fixed positions.

Note that in the system of Fig. 8.11 the dynamic model is computed as a function of the desired path only, and so when the desired path is known in advance, values could be computed "off-line" before motion begins. At run time, the precomputed torque histories would then be read out of memory. Likewise, if time-varying gains are computed, they too could be computed beforehand and stored. Hence such a scheme could be quite inexpensive computationally at run time and thus achieve a high servo rate.

8.8 Present industrial robot control systems

Because of the problems with having good knowledge of parameters, it is not clear whether it makes sense to go to the trouble of computing a complicated model-based control law for manipulator control. The expense of the computer power needed to compute the model of the manipulator at a sufficient rate may not be worthwhile, especially when lack of knowledge of parameters may nullify the benefits of such an approach. Manufacturers of industrial robots have decided, probably for economic reasons, that attempting to use a complete manipulator model in the controller is not worthwhile. Instead, present day manipulators are controlled with very simple control laws which are generally completely error driven. It is still instructive to consider these simpler control schemes within the context of our partitioned controller structure. An industrial robot with a high performance servo system is shown in Fig. 8.12.

Individual joint PID control

Most present industrial robots have a control scheme that in our notation would be described by

$$\begin{aligned} \alpha &= I, \\ \beta &= 0, \end{aligned} \tag{8.39}$$

where I is the $n \times n$ identity matrix. The servo portion is

$$\tau' = \ddot{\Theta}_d + K_v \dot{E} + K_p E + K_i \int E dt, \tag{8.40}$$

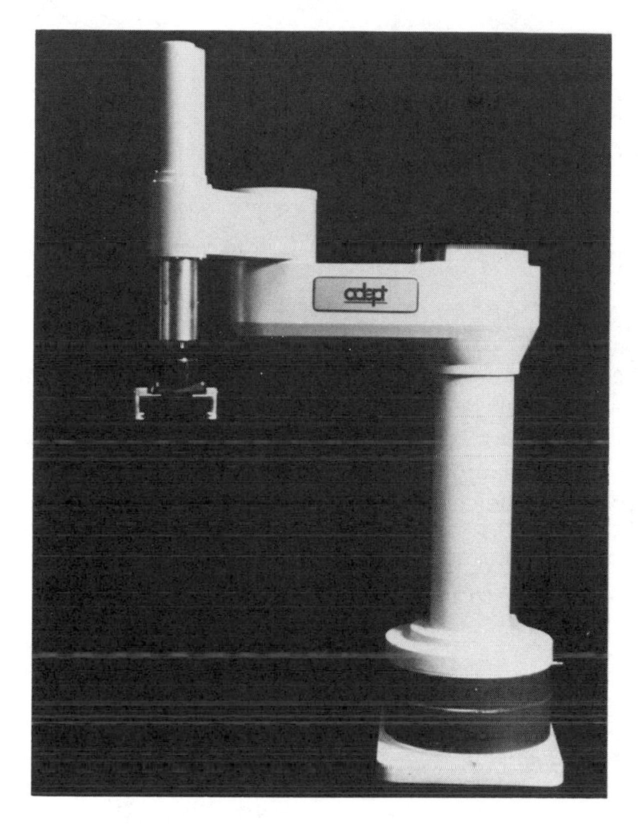

FIGURE 8.12 The Adept One, a direct drive robot by Adept Technology, Inc.

where K_v, K_p, and K_i are constant diagonal matrices. In many cases, $\ddot{\Theta}_d$ is not available, and this term is simply set to zero. That is, most simple robot controllers do not use a model-based component *at all* in their control law. This type of PID control scheme is simple because each joint is controlled as a separate control system. Often, one microprocessor per joint is used to implement (8.40).

The performance of a manipulator controlled in this way is not simple to describe. Since no decoupling is being done, the motion of each joint affects the other joints. These interactions cause errors which are suppressed by the error driven control law. It is impossible to select fixed gains which will critically damp the response to disturbances for all configurations. Therefore, "average" gains are chosen which approximate critical damping in the center of the robot's workspace. In various extreme configurations of the arm, the system becomes either underdamped or overdamped. Depending on the details of the mechanical design of the robot, these effects may be fairly small, and control is good. In such systems, it is important to keep the gains as high as possible so that these inevitable disturbances will be quickly suppressed. The upper limit on gain values is dictated by many things such as servo rate, noise in the sensors, and structural resonances.

Exactly how errors are suppressed by this controller, while not easy to intuit, can be stated by writing (8.37) with $\hat{V} = \hat{G} = \hat{F} = 0$ and $\hat{M} = I$. Here, we also assume $K_i = 0$, and we write

$$\ddot{E} + K_v \dot{E} + K_p E = (M - I)\ddot{\Theta} + V + G + F, \qquad \textbf{(8.41)}$$

where the arguments of the dynamic quantities have been dropped for brevity. A manipulator controlled in this way will damp out disturbances in different ways depending on the configuration of the arm. Additionally, the manipulator's own motion will continually excite servo errors because of the coupling between joints. While (8.41) doesn't look theoretically pleasing, it is representative of how nearly all industrial robots are controlled today. The use of sufficiently high K_v and K_p allows these robots to be controlled reasonably well.

"Effective joint inertia" individual joint PID control

Another possible control scheme would be described by

$$\begin{aligned} \alpha &= M'(\Theta), \\ \beta &= 0, \end{aligned} \qquad \textbf{(8.42)}$$

where M' is an $n \times n$ diagonal matrix with functions of configuration on the diagonal. The servo portion is

$$\tau' = \ddot{\Theta}_d + K_v \dot{E} + K_p E + K_i \int E dt, \qquad \textbf{(8.43)}$$

Where K_v, K_p, and K_i are constant diagonal matrices. Here, the mass matrix of the manipulator is implemented in a simplified manner by not computing the off-diagonal elements. In this model the joints are seen as independent systems whose inertia changes with configuration. This changing inertia is modelled to try to (at least partially) cancel the effect of changing inertia.

Such a system will be easier to keep *near* critical damping, but will still have to continually suppress disturbances which result from coupling between the joints of the arm.

Inertial decoupling

Some robots have a control scheme that in our notation would be described by

$$\alpha = \hat{M}(\Theta),$$
$$\beta = 0, \tag{8.44}$$

where $\hat{M}$ is an $n \times n$ model of the inertia matrix of the manipulator. The servo portion is

$$\tau' = \ddot{\Theta}_d + K_v\dot{E} + K_pE + K_i\int Edt, \tag{8.45}$$

where K_v, K_p, and K_i are constant diagonal matrices. Assuming that the model of the mass matrix is exact, the closed-loop error space equation for this kind of manipulator controller is given by

$$\ddot{E} + K_v\dot{E} + K_pE = \hat{M}^{-1}(V + G + F), \tag{8.46}$$

where the arguments of the dynamic quantities have been dropped for brevity. Since the gravity terms will tend to cause static positioning errors, some robot manufacturers include a gravity model, $G(\theta)$, in the control law (that is, $\beta = \hat{G}(\Theta)$ in our notation).

Various approximations of decoupling control

There are various ways to simplify the dynamic equations of a particular manipulator. After the simplification, a simpler decoupling and linearizing law could be derived. A usual simplification might be to disregard components of torque due to the velocity terms—that is, to model only the inertial and gravity terms. Often, friction models are not included in the controller since friction is so hard to model correctly. Sometimes the inertia matrix is simplified so that it accounts for the major coupling between axes but not for minor cross-coupling effects.

8.9 Cartesian based control systems

In this section we introduce the notion of **Cartesian based control**. Although such approaches are not currently used in industrial robots, there is activity at several research institutions on such schemes.

Comparison with joint based schemes

In all the control schemes for manipulators which we have discussed so far, we assumed that the desired trajectory was available in terms of time histories of joint position, velocity, and acceleration. Given that these desired inputs were available, we designed **joint based control** schemes, that is, schemes in which we develop trajectory errors by differencing desired and actual quantities expressed in joint space. Very often we wish the manipulator end-effector to follow straight lines or other path shapes described in Cartesian coordinates. As we saw in Chapter 7, it is possible to compute the time histories of the joint space trajectory which correspond to Cartesian straight line paths. Figure 8.13 shows this approach to manipulator trajectory control. A basic feature of the approach is the **trajectory conversion** process which is used to compute the joint trajectories. This is then followed by some kind of joint based servo scheme as we have been studying.

The trajectory conversion process is quite difficult (in terms of computational expense) if it is to be done analytically. The computations which would be required are

$$\begin{aligned} \Theta_d &= INVKIN(\mathcal{X}_d), \\ \dot{\Theta}_d &= J^{-1}(\Theta)\dot{\mathcal{X}}_d, \\ \ddot{\Theta}_d &= \dot{J}^{-1}(\Theta)\dot{\mathcal{X}}_d + J^{-1}\ddot{\mathcal{X}}_d. \end{aligned} \tag{8.47}$$

To the extent that such a computation is done at all in present-day systems, usually just the solution for Θ_d is performed using the inverse kinematics, and then the joint velocities and accelerations are computed numerically by first and second differences, as in (7.49). However, such numerical differentiation introduces noise and introduces a lag unless it can be done with a non-causal filter.* Therefore, we are interested in

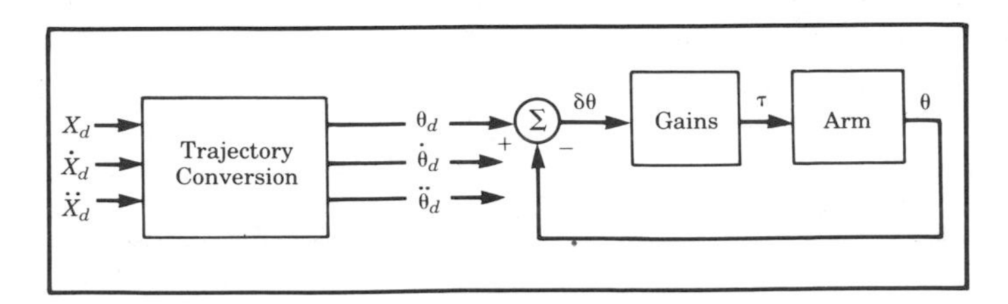

FIGURE 8.13 A joint based control scheme with Cartesian path input.

* Numerical differentiation introduces a lag unless it can be based on past, present, and future values. When the entire path is preplanned, this kind of noncausal numerical differentiation can be done.

either finding a less computationally expensive way of computing (8.47), or suggesting a different control scheme in which this information is not needed.

An alternative approach is shown in Fig. 8.14. Here, the sensed position of the manipulator is immediately transformed by means of the kinematic equations into a Cartesian description of position. This Cartesian description is then compared to the desired Cartesian position in order to form errors in Cartesian space. Control schemes which are based on forming errors in Cartesian space are called **Cartesian based control** schemes.

The trajectory conversion process is replaced by some kind of coordinate conversion inside the servo loop. Note that Cartesian-based controllers must perform many computations in the loop because of the kinematics and other transformations which are now "inside the loop." This may be a drawback of the Cartesian-based methods, since the resulting system may run at a lower sampling frequency compared to joint based systems (given the same size computer). This would, in general, degrade the stability and disturbance-rejection capabilities of the system.

Intuitive schemes of Cartesian control

One possible control scheme which comes to mind rather intuitively is shown in Fig. 8.15. Here Cartesian position is compared to the desired position to form an error, δX, in Cartesian space. This error, which may be presumed small if the control system is doing its job, may be mapped into a small displacement in joint space by means of the inverse Jacobian. The resulting errors in joint space, $\delta\theta$ are then multiplied by gains to compute torques which will tend to reduce these errors. Note that Fig. 8.15 shows a simplified controller in that, for clarity, the velocity

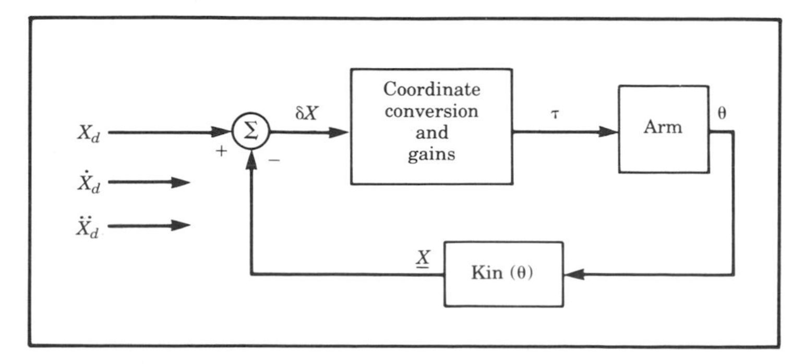

FIGURE 8.14 The concept of a Cartesian-based control scheme.

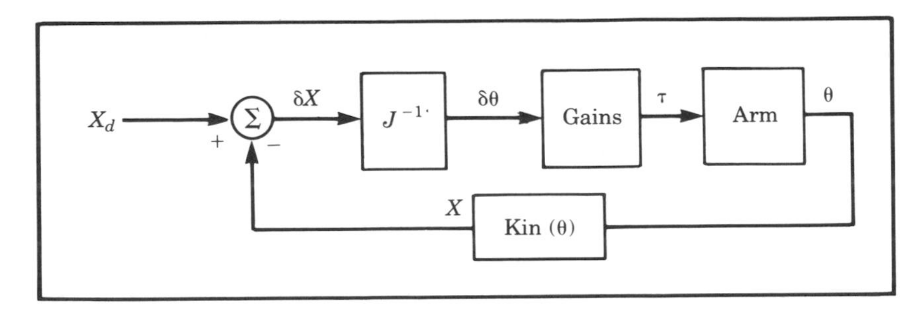

FIGURE 8.15 The inverse Jacobian Cartesian control scheme.

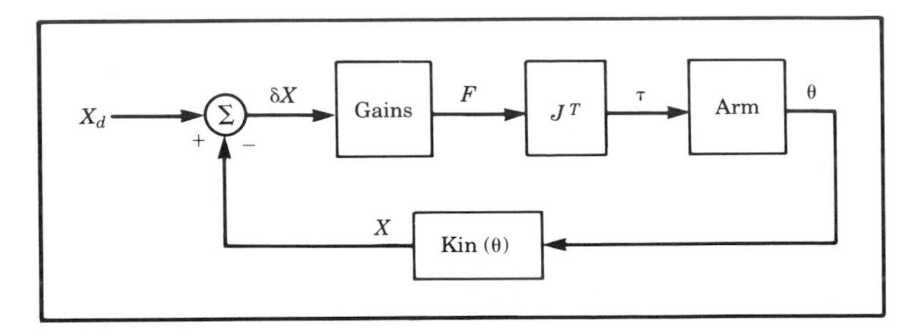

FIGURE 8.16 The transpose Jacobian Cartesian control scheme.

feedback has not been shown. It could be added in a straightforward manner. We will call this scheme the **inverse Jacobian controller**.

Another scheme which may come to mind is shown in Fig. 8.16. Here the Cartesian error vector is multiplied by a gain to compute a Cartesian force vector. This can be thought of as a Cartesian force which, if applied to the end-effector of the robot, would push the end-effector in a direction which would tend to reduce the Cartesian error. This Cartesian force vector (actually a force-torque vector) is then mapped through the Jacobian transpose in order to compute the equivalent joint torques which would tend to reduce the observed errors. We will call this scheme the **transpose Jacobian controller**.

The *inverse Jacobian controller* and the *transpose Jacobian controller* have both been arrived at intuitively. We cannot be sure that such arrangements would be stable, let alone perform well. It is also curious that the schemes are extremely similar except the one contains the Jacobian's inverse, and the other contains its transpose. Remember, the inverse is not equal to the transpose in general (only in the case of a strictly Cartesian manipulator does $J^T = J^{-1}$). The exact dynamic performance of such systems (if expressed in a second-order error space equation for example) is very complicated. It turns out that both schemes will work, but not well. Both can be made stable by appropriate gain selection, including some form of velocity feedback which was not shown in Figs. 8.15 and 8.16. While both will work, neither is *correct* in the sense that we cannot choose fixed gains which will result in fixed closed-loop poles. The dynamic response of such controllers will vary with arm configuration.

Cartesian decoupling scheme

For Cartesian-based controllers, like joint based controllers, good performance would be characterised by constant error dynamics over all configurations of the manipulator. Since errors are expressed in Cartesian space in Cartesian-based schemes, this means that we would like to design a system which, over all possible configurations, would suppress Cartesian errors in a critically damped fashion.

Just as we achieved good control with a joint based controller which was based on a linearizing and decoupling model of the arm, we can do the same for the Cartesian case. However, we must now write the dynamic equations of motion of the manipulator in terms of Cartesian variables. This can be done, as discussed in Chapter 6. The resulting form of the equations of motion is quite analogous to the joint space version. The rigid body dynamics can be written:

$$\mathcal{F} = M_x(\Theta)\ddot{\mathcal{X}} + V_x(\Theta, \dot{\Theta}) + G_x(\Theta), \tag{8.48}$$

where $\mathcal{F}$ is a force-torque vector acting on the end-effector of the robot, and $\mathcal{X}$ is an appropriate Cartesian vector representing position and orientation of the end-effector [4]. Analogous to the joint space quantities, $M_x(\Theta)$ is the mass matrix in Cartesian space, $V_x(\Theta, \dot{\Theta})$ is a vector of velocity terms in Cartesian space, and $G_x(\Theta)$ is a vector of gravity terms in Cartesian space.

Just as we did in the joint based case, we can use the dynamic equations in a decoupling and linearizing controller. Since (8.48) computes $\mathcal{F}$, a Cartesian force vector which should be applied to the hand, we will also need to use the transpose of the Jacobian in order to implement the control. That is, after $\mathcal{F}$ is calculated by (8.48), since we cannot actually cause a Cartesian force to be applied to the end-effector, we instead compute the joint torques needed to effectively apply this force:

$$\tau = J^T(\Theta)\ \mathcal{F}. \tag{8.49}$$

Figure 8.17 shows a Cartesian arm control system using complete dynamic decoupling. Note that the arm is preceeded by the Jacobian transpose. Notice that the controller of Fig. 8.17 allows Cartesian paths to be described directly with no need for trajectory conversion.

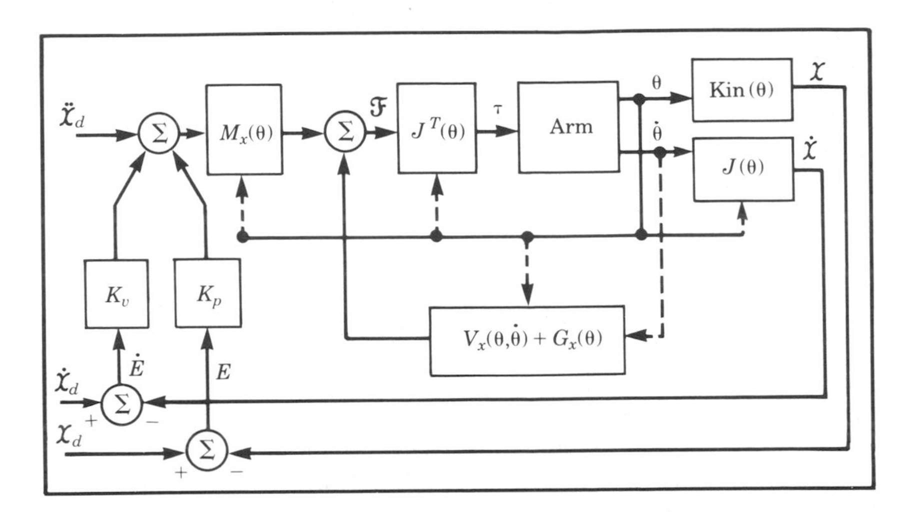

FIGURE 8.17 The Cartesian model based control scheme.

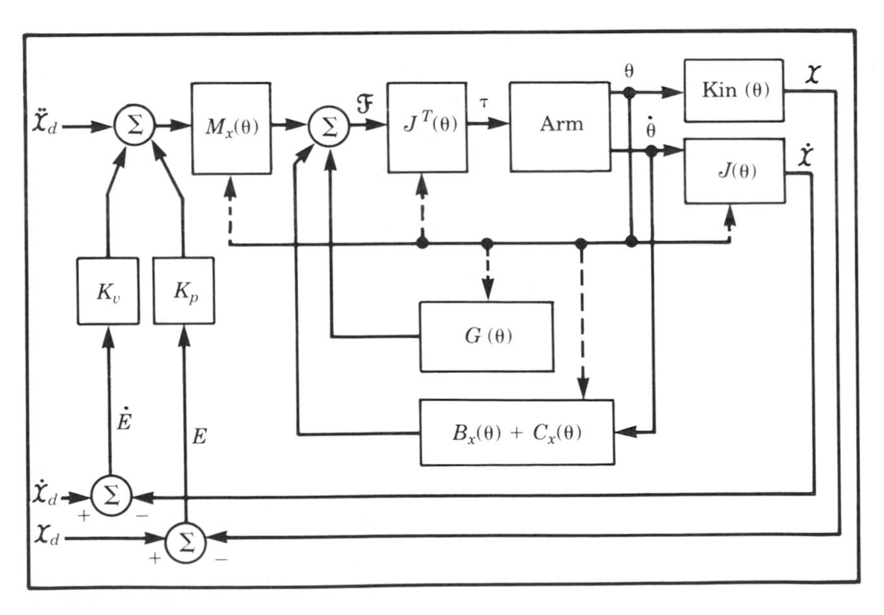

FIGURE 8.18 An implementation of the Cartesian model-based control scheme.

As in the joint space case, a practical implementation may best be achieved through use of a dual-rate control system. Figure 8.18 shows a block diagram of a Cartesian-based decoupling and linearizing controller in which the dynamic parameters are written as functions of manipulator position only. These dynamic parameters are updated at a rate slower than the servo rate by a background process or a second control computer. This is appropriate because we desire a fast servo to maximize disturbance rejection and stability. Since the dynamic parameters are functions of manipulator position only, they need be updated at a rate related only to how fast the manipulator is changing configuration. This rate probably need be no higher than 100 Hz, while servo rates up to 500 Hz may be desirable [4].

8.10 Adaptive control

In our discussion of model-based control, we noted that often parameters of the manipulator are not known exactly. When the parameters in the model do not match the parameters of the real device, servo errors will result, as made explicit in (8.37). These servo errors could be used to drive some adaptation scheme which attempts to update the values of the model parameters until the errors disappear. Several such adaptive schemes have been proposed.

An ideal adaptive scheme might be one such as the one in Fig. 8.19. Here, we are using a model-based control law as developed in this chapter. There is an adaptation process which, based on observation of manipulator state and servo errors, readjusts the parameters in the

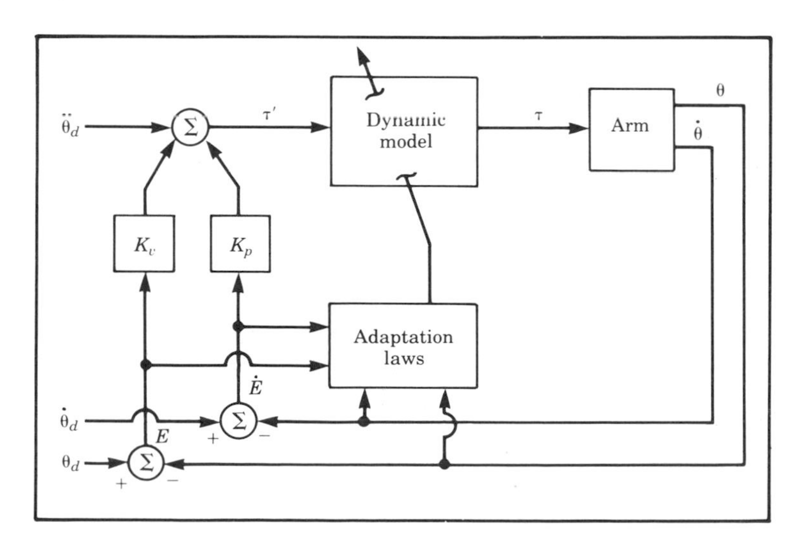

FIGURE 8.19 The concept of an adaptive manipulator controller.

nonlinear model until the errors disappear. Such a system would *learn* its own dynamic properties.

The design of such systems is thought to be a difficult problem. This is a current area of research [5].

References

[1] W. Boyce and R. DiPrima, *Elementary Differential Equations,* Third Edition, Wiley, 1977.

[2] G. Franklin, J. Powell, and A. Emami-Naeini, *Feedback Control of Dynamic Systems,* Addison-Wesley, 1985.

[3] G. Franklin and J. Powell, *Digital Control of Dynamic Systems,* Addison-Wesley, 1980.

[4] O. Khatib, "The Operational Space Formulation in Robot Manipulator Control," 15th ISIR, Tokyo, Japan, September 11–13, 1985.

[5] J. Craig, P. Hsu, and S. Sastry, "Adaptive Control of Mechanical Manipulators," *IEEE Conference on Robotics and Automation*, San Francisco, Calif., April 1986.

Exercises

8.1 [15] Give the nonlinear control equations for an α, β-partitioned controller for the system:

$$\tau = \left(2\sqrt{\theta} + 1\right) \ddot{\theta} + 3\dot{\theta}^2 - \sin(\theta).$$

Choose gains so that this system is always critically damped with $k_{CL} = 10$.

8.2 [15] Give the nonlinear control equations for an α, β-partitioned controller for the system:

$$\tau = 5\theta\dot{\theta} + 2\ddot{\theta} - 13\dot{\theta}^3 + 5.$$

Choose gains so that this system is always critically damped with $k_{CL} = 10$.

8.3 [19] Draw a block diagram showing a joint space controller for the 2-link arm from Section 6.7 such that the arm is critically damped over its entire workspace. Show the equations inside the blocks of a block diagram.

8.4 [20] Draw a block diagram showing a Cartesian space controller for the 2-link arm from Section 6.7 such that the arm is critically damped over its entire workspace. See Example 6.5. Show the equations inside the blocks of a block diagram.

8.5 [18] Design a trajectory-following control system for the systems whose dynamics are given by

$$\tau_1 = m_1 l_1^2 \ddot{\theta}_1 + m_1 l_1 l_2 \dot{\theta}_1 \dot{\theta}_2,$$

$$\tau_2 = m_2 l_2^2 (\ddot{\theta}_1 + \ddot{\theta}_2) + v_2 \dot{\theta}_2.$$

Do you think these equations could represent a real system?

8.6 [17] For the control system designed for the one-link manipulator in Example 8.4, give an expression for the steady-state position error as a function of error in the mass parameter. Let $\psi_m = m - \hat{m}$. The result should be a function of l, g, θ, ψ_m, and k_p. For what position of the manipulator is this maximum?

8.7 [26] For the two degree of freedom mechanical system of Fig. 8.20, design a controller which can cause x_1 and x_2 to follow trajectories and suppress disturbances in a critically damped fashion.

8.8 [30] Consider the dynamic equations of the 2-link manipulator from Section 6.7 in configuration space form. Derive expressions for the sensitivity of the computed torque value versus small deviations in Θ. Can you say something about how often the dynamics should be recomputed in a controller like that of Fig. 8.10 as a function of average joint velocities expected during normal operations?

8.9 [32] Consider the dynamic equations of the 2-link manipulator from Example 6.5 in Cartesian configuration space form. Derive expressions for the sensitivity of the computed torque value versus small deviations in Θ. Can you say something about how often the dynamics should be recomputed in a controller like that of Fig. 8.18 as a function of average joint velocities expected during normal operations?

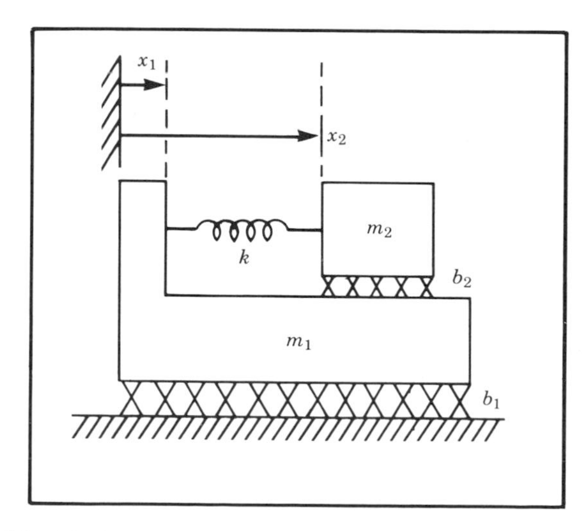

FIGURE 8.20 Mechanical system with two degrees of freedom.

8.10 [15] Design a control system for the system

$$f = 5x\dot{x} + 2\ddot{x} - 12.$$

Choose gains so that this system is always critically damped with a closed-loop stiffness of 20.

Programming Exercise (Part 8)

1. We wish to simulate a simple control system for the 3-link planar arm. This control system will be like present-day industrial robots in that we will not use dynamic decoupling. In our notation we have

$$\alpha = I,$$

$$\beta = 0,$$

$$\tau' = K_v \dot{E} + K_p E.$$

We will use

$$K_p = \begin{bmatrix} 175.0 & 0.0 & 0.0 \\ 0.0 & 110.0 & 0.0 \\ 0.0 & 0.0 & 20.0 \end{bmatrix}.$$

Choose a diagonal K_v so that when the arm is in the center of its workspace (say, $\Theta = (60.0, -80.0, 20.0)$ degrees) it is approximately critically damped. To select K_v values you might want to consider the values in the mass matrix at this configuration and use (8.12). Of course, there is no perfect selection.

Use the simulation routine **UPDATE** to simulate a discrete-time servo running at 100 Hz. That is, calculate the error-driven control law at 100 Hz, not at the frequency of the numerical integration process. Test the control scheme on the following tests

a) Start the arm at $\Theta = (60, -110, 20)$ and command it to stay there until *time* = 3.0, when the setpoints should instantly change to $\Theta = (60, -50, 20)$. That is, give a step input of 60 degrees to joint 2. Record the error-time history for each joint.

b) Control the arm to follow the cubic spline trajectory from Programming Exercise, Part 6. Record the error-time history for each joint.

2. Repeat the same tests with a new controller which uses a complete dynamic model of the 3-link to decouple and linearize the system. For this case, use

$$K_p = \begin{bmatrix} 100.0 & 0.0 & 0.0 \\ 0.0 & 100.0 & 0.0 \\ 0.0 & 0.0 & 100.0 \end{bmatrix}.$$

Choose a diagonal K_v which guarantees critical damping over all configurations of the arm. Compare the results with those obtained with the simpler controller.

9

FORCE CONTROL OF MANIPULATORS

9.1 Introduction

While position control is appropriate when a manipulator is following a trajectory through space, when any contact is made between the end-effector and the manipulator's environment, position control may not suffice. Consider a manipulator washing a window with a sponge. Due to the compliance of the sponge, it may be possible to regulate the force applied to the window by controlling the position of the end-effector relative to the glass. If the sponge is very compliant, and/or the position of the glass is known very accurately, this technique would work quite well.

However, if the stiffness of the end-effector, tool, or environment is high, it becomes increasingly difficult to perform operations in which the manipulator contacts a surface. Instead of a sponge, imagine that the manipulator is scraping paint off a glass surface using a rigid scraping tool. If there is any uncertainty in the position of the glass surface, or errors in the position servo of the manipulator, this task would become impossible. Either the glass would be broken, or the manipulator would wave the scraping tool over the glass with no contact taking place.

In both the washing and scraping tasks, it would be more reasonable not to specify the position of the plane of the glass, but rather *to specify a force which is to be maintained normal to the surface.*

More so than in previous chapters, in this chapter we present methods which are not yet employed by industrial robots, except in an extremely simplified way. The major thrust of the chapter is to introduce the **hybrid position/force controller** which is one formalism through which industrial robots may someday be controlled in order to perform tasks requiring force control. However, regardless of which method(s) emerge as practical for industrial application, many of the concepts introduced in this chapter will certainly remain valid.

9.2 Application of industrial robots to assembly tasks

The bulk of the industrial robot population work in relatively simple applications such as spot welding, spray painting, and pick and place operations. Force control has already appeared in a few applications; for example, some robots are already capable of simple force control which allows them to do tasks such as grinding and deburring. Apparently the next big area of application will be to assembly line tasks in which one or more parts are mated. In such *parts-mating* tasks, monitoring and control of the forces of contact is extremely important.

Precise control of manipulators in the face of uncertainties and variations in their work environments is a prerequisite to application of robot manipulators to assembly operations in industry. It seems that by providing manipulator hands with sensors that can give information about the state of manipulation tasks, important progress can be made toward using robots for assembly tasks. Currently, the dexterity of manipulators remains quite low and continues to limit their application in the automated assembly area.

The use of manipulators for assembly tasks requires that the precision with which parts are positioned with respect to one another be quite high. Current industrial robots are often not accurate enough for these tasks, and building robots that are may not make sense. Manipulators of greater precision can be achieved only at the expense of size, weight, and cost. The ability to measure and control contact forces generated at the hand, however, offers a possible alternative for extending the effective precision of a manipulator. Since relative measurements are used, absolute errors in the position of the manipulator and the manipulated objects are not as important as they would be in a purely position controlled system. Since small variations in relative position generate large contact forces when parts of moderate stiffness interact, knowledge and control of these forces can lead to a tremendous increase in effective positional accuracy.

9.3 Force sensors

A variety of devices have been designed to measure forces of contact between a manipulator's end-effector and the environment which it is contacting. Most such sensors make use of sensing elements called **strain gauges**, of either the semiconductor or the metal foil variety. These strain gauges are bonded to a metal structure and produce an output proportional to the strain in the metal. In this type of force sensor design the issues addressed by the designer are

1. How many sensors are needed to resolve the desired information?
2. How are the sensors mounted relative to each other on the structure?
3. What structure allows good sensitivity while maintaining stiffness?
4. How can protection against mechanical overload be built into the device?

There are three places where such sensors are usually placed on a manipulator:

1. At the joint actuators. These sensors measure the torque or force output of the actuator itself. These are useful for some control schemes, but usually do not provide good sensing of contact between the end-effector and the environment.
2. Between the end-effector and last joint of the manipulator. These sensors are usually refered to as **wrist sensors**. They are a mechanical structure instrumented with strain gauges which can measure the forces and torques acting on the end-effector. Typically, these sensors are capable of measuring from three to six components of the force/torque vector acting on the end-effector.
3. At the "fingertips" of the end-effector. Usually, these **force sensing fingers** have built-in strain gauges to measure from one to four components of force acting at each fingertip.

As an example, Fig. 9.1 shows a drawing of the internal structure of a popular style of wrist force sensor designed by Scheinman [1]. Bonded

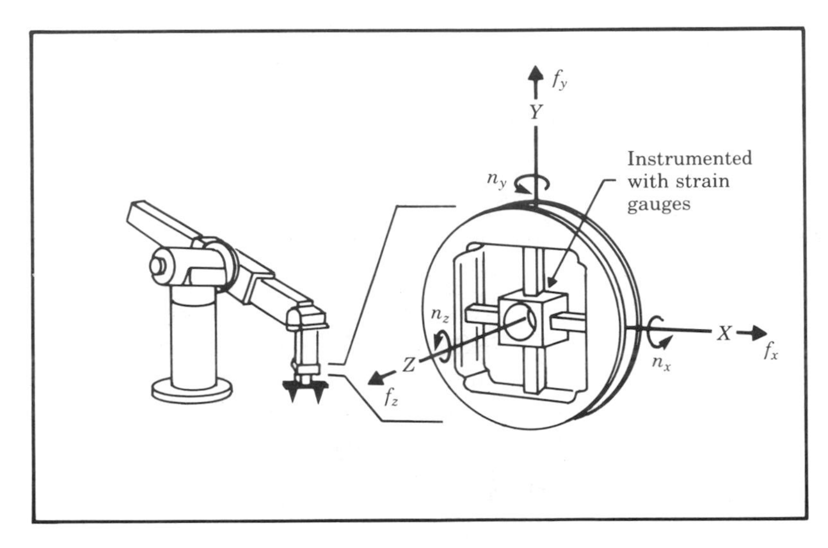

FIGURE 9.1 The internal structure of a typical force-sensing wrist.

to the cross-bar structure of the device are eight pairs of semiconductor strain gauges. Each pair is wired in a voltage divider arrangment. Each time the wrist is queried, eight analog voltages are digitized and read into the computer. Calibration schemes have been designed with which to arrive at a constant 6×8 *calibration matrix* which maps these eight strain measurements into the force-torque vector, $\mathcal{F}$, acting on the end-effector. The sensed force-torque vector can be transformed to a reference frame of interest as we saw in Example 5.7.

Additional detail on design of sensors is beyond the intended scope of this book. In this chapter we will assume that a suitable force transducer is present, and from now on, we will concern ourselves with how such force information might be used in a control algorithm.

9.4 A framework for control in partially-constrained tasks

The approach presented in this chapter is based on a framework for control in situations in which motion of the manipulator is partially constrained due to contact with one or more surfaces [2], [3].

Every manipulation task can be broken down into subtasks that are defined by a particular contact situation occuring between the manipulator end-effector (or tool) and the work environment. With each such subtask we may associate a set of constraints, called the **natural constraints**, that result from the particular mechanical and geometric characteristics of the task configuration. For instance, a hand in contact with a stationary, rigid surface is not free to move through that surface, and hence a *natural* position constraint exists. If the surface is frictionless, the hand is not free to apply arbitrary forces tangent to the surface, and hence a *natural* force constraint exists.

In general, for each subtask configuration a **generalized surface** can be defined with position constraints along the normals to this surface and force constraints along the tangents. These two types of constraint, force and position, partition the degrees of freedom of possible end-effector motions into two orthogonal sets that must be controlled according to different criteria.

Figure 9.2 shows two representative tasks along with their associated natural constraints. Notice that in each case, the task is described in terms of a frame $\{C\}$, the so-called **constraint frame**, which is located in a task relevant location. According to the task, $\{C\}$ may be fixed in the environment or may move with the end-effector of the manipulator. In Fig. 9.2(a), the constraint frame is attached to the crank as shown and

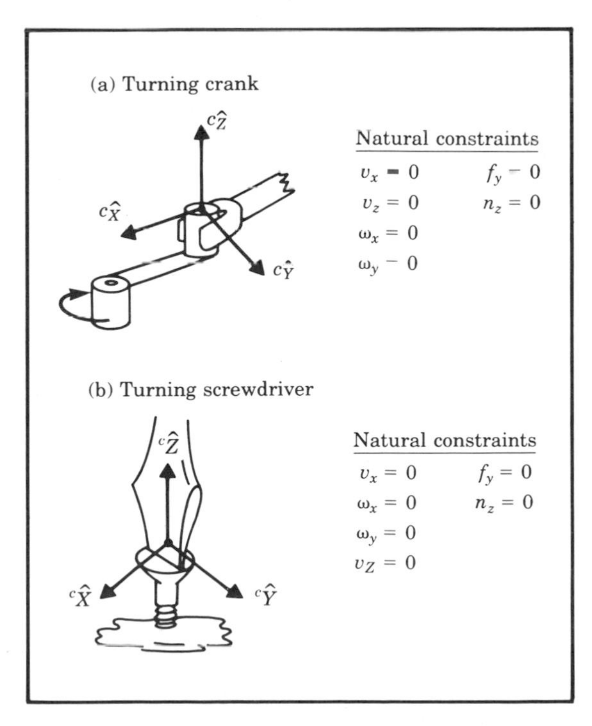

FIGURE 9.2 The natural constraints for two different tasks.

moves with the crank with the $\hat{X}$ direction always directed toward the pivot point of the crank. Friction acting at the fingertips insures a secure grip on the handle, which is on a spindle so that it may rotate relative to the crank arm. In Fig. 9.2(b), the constraint frame is attached to the tip of the screwdriver and moves with it as the task proceeds. Notice that in the $\hat{Y}$ direction the force is constrained to be zero since the slot of the screw would allow the screwdriver to slip out in that direction. In these examples, a given set of constraints remains true throughout the task. In more complex situations, the task is broken into subtasks for which a constant set of natural constraints can be identified.

In Fig. 9.2, position constraints have been indicated by giving values for components of velocity of the end-effector, $\mathcal{V}$, described in frame $\{C\}$. We could just as well have indicated position constraints by giving expressions for position, rather than velocities. However, in many cases it is simpler to specify a position contraint as a "velocity equals zero" constraint. Likewise, force constraints have been specified by giving values to components of the force-torque vector, $\mathcal{F}$, acting on the end-effector described in frame $\{C\}$. Note that when we say *position constraints* we mean position and/or orientation constraints, and when we say *force constrants* we mean force and/or torque constraints. The term *natural constraints* is used to indicate that these constraints arise naturally from the particular contacting situation. They have nothing to do with the desired or intended motion of the manipulator.

Additional constraints, called **artificial constraints**, are introduced in accordance with the natural constraints to specify desired motions or force application. That is, each time the user specifies a desired trajectory in either position or force, an artificial constraint is defined. These constraints also occur along the tangents and normals of the generalized constraint surface; but unlike natural constraints, artificial force constraints are specified along surface normals, and artificial position constraints along tangents—hence consistency with the natural constraints is preserved.

Figure 9.3 shows the natural and artificial constraints for two tasks. Note that when a natural position constraint is given for a particular degree of freedom in $\{C\}$, an artificial force constraint should be specified, and vice versa. Any given degree of freedom in the constraint frame is at any instant controlled to meet either a position or a force constraint.

Assembly strategy is a term which refers to a sequence of planned artificial constraints that will cause the task to proceed in a desirable manner. Such strategies must include methods by which the system can detect a change in the contacting situation so that transitions in the natural constraints can be tracked. With each such change in natural constraints, a new set of artificial constraints is recalled from the set of assembly strategies and enforced by the control system. Methods for automatically choosing the constraints for a given assembly task await

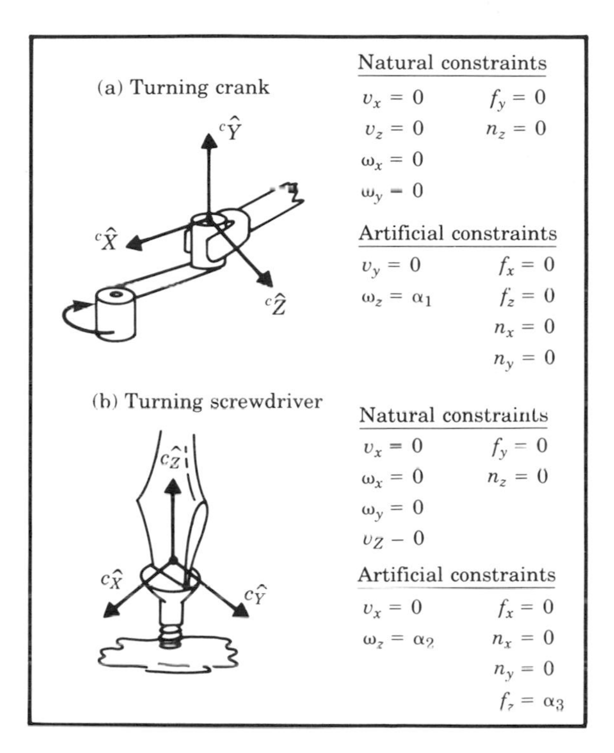

FIGURE 9.3 The natural and artificial constraints for two tasks.

further research. In this chapter we will assume that a task has been analyzed in order to determine the natural constraints, and that a human planner has determined an **assembly strategy** with which to control the manipulator.

Note that we will usually ignore friction forces between contacting surfaces in our analysis of tasks. This will suffice for our introduction to the problem, and in fact will yield strategies which work in many cases. Usually friction forces of sliding are acting in directions chosen to be position controlled, and so these forces appear as disturbances to the position servo and are overcome by the control system.

EXAMPLE 9.1

Figure 9.4(a–d) shows an assembly sequence used to put a round peg into a round hole. The peg is brought down onto the surface to the left of the hole and then slid along the surface until it drops into the hole. It is then inserted until the peg reaches the bottom of the hole, at which time the assembly is complete. Each of the four indicated contacting situations defines a subtask. For each of the subtasks shown, give the natural and artificial constraints. Also indicate how the system senses the change in the natural constraints as the operation proceeds.

First, we will attach the constraint frame to the peg as shown in Fig. 9.4(a). In Fig. 9.4(a), the peg is in free space, and so the natural constraints are

$$^{C}\mathcal{F} = 0. \tag{9.1}$$

Therefore the artificial constraints in this case consitute an entire position trajectory which moves the peg in the ${}^{C}\hat{Z}$ direction toward the surface. For example

$$^{C}\mathcal{V} = \begin{bmatrix} 0 \\ 0 \\ v_{approach} \\ 0 \\ 0 \\ 0 \end{bmatrix}, \tag{9.2}$$

where $v_{approach}$ is the speed with which to approach the surface.

In Fig. 9.4(b), the peg has reached the surface. To detect that this has happened, we observe the force in the ${}^{C}\hat{Z}$ direction. When this sensed force exceeds a threshold, we sense contact, which implies a new contacting situation with a new set of natural constraints. Assuming that the contacting situation is as shown in Fig. 9.4(b), the peg is not free to move in ${}^{C}\hat{Z}$, or to rotate about ${}^{C}\hat{X}$ or ${}^{C}\hat{Y}$. In the other three degrees of freedom it is not free to apply forces, and hence the natural

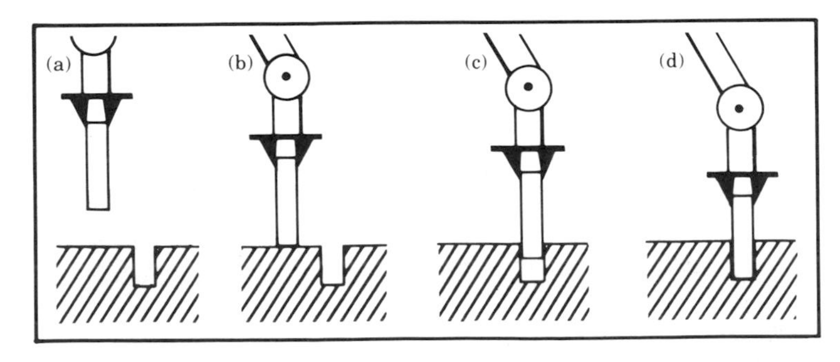

FIGURE 9.4 The sequence of four contacting situations for peg insertion.

constraints are

$$\begin{aligned} {}^{C}v_z &= 0, \\ {}^{C}\omega_x &= 0, \\ {}^{C}\omega_y &= 0, \\ {}^{C}f_x &= 0, \\ {}^{C}f_y &= 0, \\ {}^{C}n_z &= 0. \end{aligned} \tag{9.3}$$

The artificial constraints describe the strategy of sliding along the surface in the ${}^{C}\hat{X}$ direction while applying small forces to insure that contact is maintained. Thus we have

$$\begin{aligned} {}^{C}v_x &= v_{slide}, \\ {}^{C}v_y &= 0, \\ {}^{C}w_z &= 0, \\ {}^{C}f_z &= f_{contact}, \\ {}^{C}n_x &= 0, \\ {}^{C}n_y &= 0. \end{aligned} \tag{9.4}$$

where $f_{contact}$ is the force applied normal to the surface as the peg is slid, and v_{slide} is the velocity with which to slide across the surface.

In Fig. 9.4(c), the peg has fallen slightly into the hole. This situation is sensed by observing the velocity in the ${}^{C}\hat{Z}$ direction and waiting for it to cross a threshold (to become nonzero in the ideal case). When this is observed, it signals that once again the natural constraints have changed, and thus our strategy (as embodied in the artificial constraints) must change. The new natural constraints are

$$\begin{aligned} {}^{C}v_x &= 0, \\ {}^{C}v_y &= 0, \\ {}^{C}\omega_x &= 0, \\ {}^{C}\omega_y &= 0, \\ {}^{C}f_z &= 0, \\ {}^{C}n_z &= 0. \end{aligned} \tag{9.5}$$

We choose the artificial constraints to be

$$
\begin{aligned}
{}^{C}v_z &= v_{insert}, \\
{}^{C}w_z &= 0, \\
{}^{C}f_x &= 0, \\
{}^{C}f_y &= 0, \\
{}^{C}n_x &= 0, \\
{}^{C}n_y &= 0,
\end{aligned}
\qquad (9.6)
$$

where v_{insert} is the velocity at which the peg is inserted into the hole. Finally, the situation shown in Fig. 9.4(d) is detected when the force in the ${}^{C}\hat{Z}$ direction increases above a threshold. ■

It is interesting to note that changes in the natural constraints are always detected by observing the position or force variable that is *not* being controlled. For example, to detect the transition from Fig. 9.4(b) to Fig. 9.4(c) we monitor the velocity in ${}^{C}\hat{Z}$ while we are controlling force in ${}^{C}\hat{Z}$. To determine when the peg has hit the bottom of the hole, we monitor ${}^{C}f_z$ although we are controlling ${}^{C}v_z$.

Determining assembly strategies for fitting more complicated parts together is quite complex. We have also neglected the effects of friction in our simple analysis of this task. The development of automatic planning systems which include the effects of friction and may be applied to practical situations is still a research topic [4], [5].

9.5 The hybrid position/force control problem

Figure 9.5 shows two extreme examples of contacting situations. In Fig. 9.5(a), the manipulator is moving through free space. In this case the natural constraints are all force constraints—namely, since there is nothing to react against, all forces are constrained to be zero.* With an arm having six degrees of freedom, we are free to move in six degrees of freedom in position, but we are unable to exert forces in any direction. Figure 9.5(b) shows the extreme situation of a manipulator with its end-effector glued to a wall. In this case, the manipulator is subject to six natural position constraints since it is not free to be repositioned. However, the manipulator is free to exert forces and torques to the object with six degrees of freedom.

* It is important to remember that we are concerned here with *forces of contact* between end-effector and environment, not inertial forces

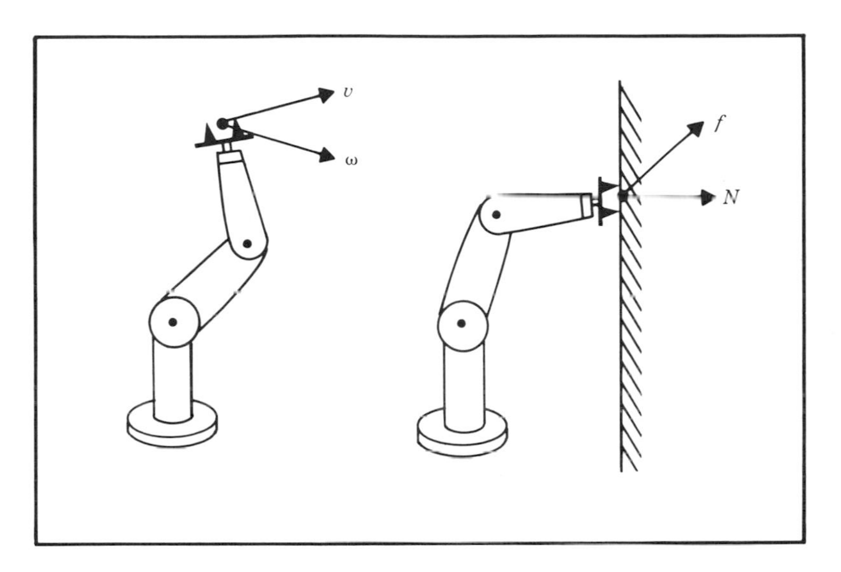

FIGURE 9.5 The two extremes of contacting situations. The manipulator on the left is moving in free space where no reaction surface exits. The manipulator on the right is glued to the wall so that no free motion is possible.

In Chapter 8 we studied the position control problem which applies to the situation of Fig. 9.5(a). Since the situation of Fig. 9.5(b) does not occur very often in practice, we usually must consider force control in the context of partially constrained tasks in which degrees of freedom of the system are subject to position control, and others are subject to force control. Thus, in this chapter we are interested in considering **hybrid position/force control** schemes.

The hybrid position/force controller must solve three problems:

1. Position control of a manipulator along directions in which a natural force constraint exists.
2. Force control of a manipulator along direction in which a natural position constraint exists.
3. A scheme to implement the arbitrary mixing of these modes along orthogonal degrees of freedom of an arbitrary frame, $\{C\}$.

9.6 Force control of a mass-spring

In Chapter 8 we began our study of the complete position control problem with the study of the very simple problem of controlling a single block of mass. We were then able to use a model of the manipulator in such a way that the problem of controlling the entire manipulator became equivalent to controlling n independent masses (for a manipulator with n joints). In a similar way, we begin our look at force control by controlling the force applied by a simple single degree of freedom system.

In considering forces of contact we must make some model of the environment upon which we are acting. For our purposes of conceptual development we will use a very simple model of interaction between a controlled body and the environment. We model contact with an environment as a spring—that is, we assume our system is rigid and the environment has some stiffness, k_e.

Let us consider the control of a mass attached to a spring as in Fig. 9.6. We will also include an unknown disturbance force, f_{dist}, which might be thought of as modeling unknown friction or cogging in the manipulator's gearing. The variable we wish to control is the force acting on the environment, f_e, which is the force acting in the spring,

$$f_e = k_e x. \tag{9.7}$$

The equation describing this physical system is

$$f = m\ddot{x} + k_e x + f_{dist}, \tag{9.8}$$

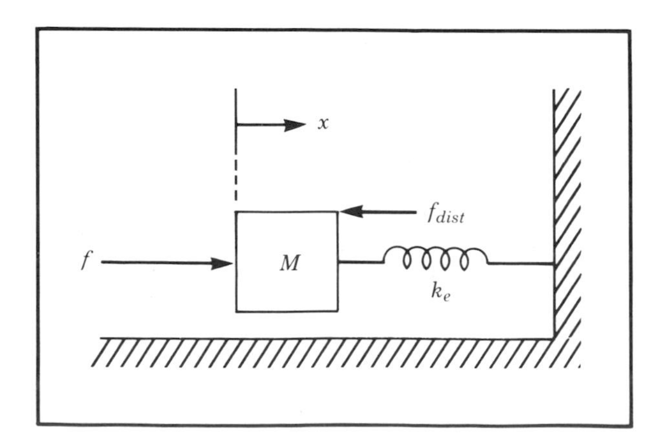

FIGURE 9.6 A spring-mass mystem.

or, written in terms of the variable we wish to control, f_e, we have

$$f = mk_e^{-1}\ddot{f}_e + f_e + f_{dist}. \quad \mathbf{(9.9)}$$

Using the partitioned controller concept, we use

$$\alpha = mk_e^{-1},$$
$$\beta = f_e + f_{dist}$$

to arrive at the control law,

$$f = mk_e^{-1}\left[\ddot{f}_d + k_{vf}\dot{e}_f + k_{pf}e_f\right] + f_e + f_{dist}, \quad \mathbf{(9.10)}$$

where $e_f = f_d - f_e$ is the force error between the desired force, f_d, and the sensed force on the environment, f_e. If we could compute (9.10), we would have the closed loop system

$$\ddot{e}_f + k_{vf}\dot{e}_f + k_{pf}e_f = 0. \quad \mathbf{(9.11)}$$

However, we cannot use knowledge of f_{dist} in our control law, and so (9.10) is not feasible. We might leave that term out of the control law, but a steady state analysis shows that there is a better choice, especially when the stiffness of the environment, k_e, is high (which is the usual situation).

If we choose to leave the f_{dist} term out of our control law, equate (9.9) and (9.10), and do a steady state analysis by setting all time derivatives to zero, we find

$$e_f = \frac{-f_{dist}}{\alpha}, \quad \mathbf{(9.12)}$$

where $\alpha = mk_e^{-1}k_{pf}$, the effective force feedback gain. However, if we choose to use f_d in the control law (9.10) in place of the term $f_e + f_{dist}$, we find the steady state error to be

$$e_f = -\frac{f_{dist}}{1+\alpha}. \quad \mathbf{(9.13)}$$

When the environment is stiff, as is often the case, α may be small, and so the steady state error calculated in (9.13) is quite an improvement over that of (9.12). Therefore we suggest the control law

$$f = mk_e^{-1}\left[\ddot{f}_d + k_{vf}\dot{e}_f + k_{pf}e_f\right] + f_d. \quad \mathbf{(9.14)}$$

Figure 9.7 shows a block diagram of the closed-loop system using the control law (9.14).

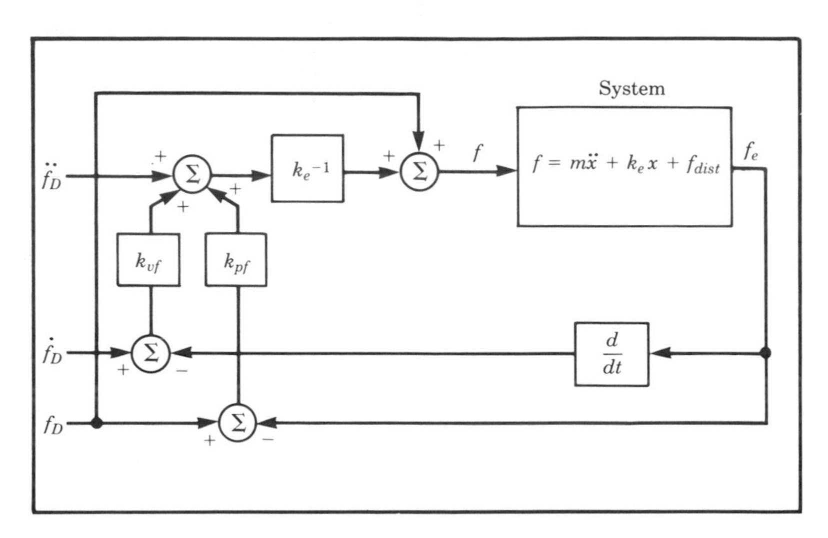

FIGURE 9.7 A force control system for the spring-mass system.

Generally, practical considerations change the implementation of a force control servo quite a bit from the ideal shown in Fig. 9.7. First, usually force trajectories are constants; that is, we are usually interested in controlling the contact force to be at some constant level. Applications in which contact forces should follow some arbitrary function of time are rare. Therefore, the $\dot{f}_d$ and $\ddot{f}_d$ inputs of the control system are very often permanently set to zero. Another reality is that sensed forces are quite "noisy," and numerical differentiation to compute $\dot{f}_e$ is ill-advised. However, since $f_e = k_e x$, we can obtain the derivative of the force on the environment as $\dot{f}_e = k_e \dot{x}$. This is much more realistic in that most manipulators have means of obtaining good measures of velocity. Having made these two pragmatic choices, we write the control law as

$$f = m\left[k_{pf} k_e^{-1} e_f - k_{vf}\dot{x}\right] + f_d, \tag{9.15}$$

with the corresponding block diagram shown in Fig. 9.8.

Note that an interpretation of the system of Fig. 9.8 is that force errors generate a setpoint for an inner velocity control loop with gain k_{vf}. Some force control laws also include an integral term to improve steady state performance.

An important remaining problem is that the stiffness of the environment, k_e, appears in our control law but is often unknown, and perhaps changes from time to time. However, often an assembly robot is dealing with rigid parts, and k_e could be guessed to be quite high. Generally

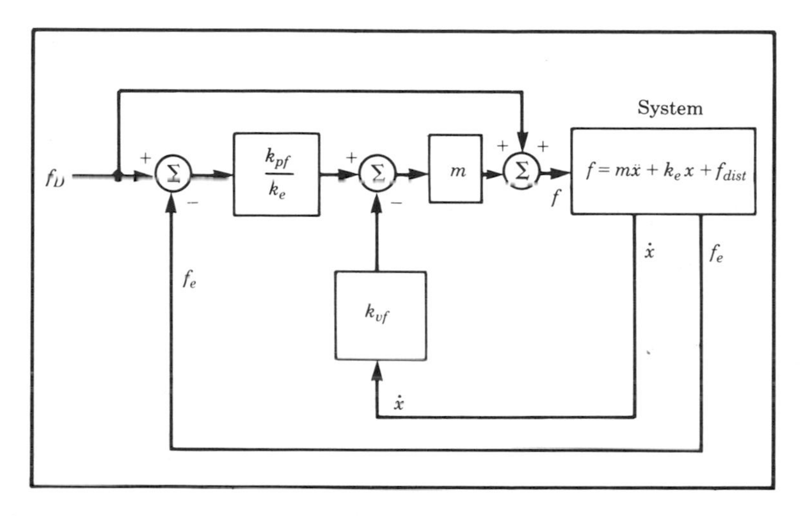

FIGURE 9.8 A practical force control system for the spring-mass.

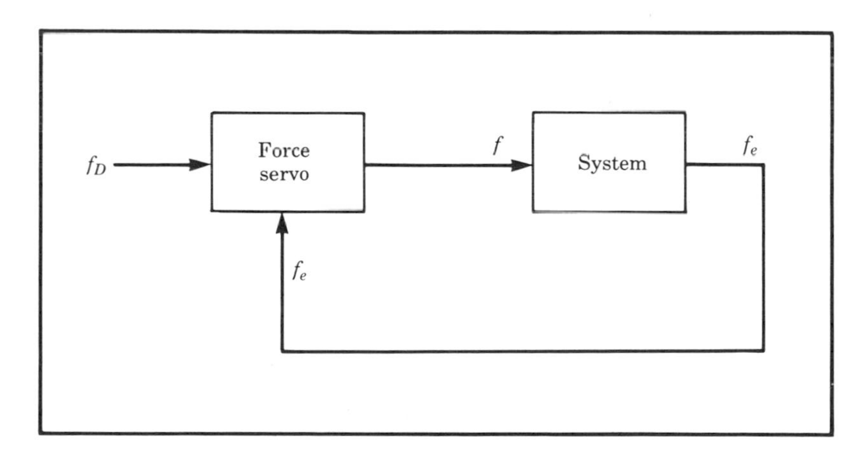

FIGURE 9.9 The force control servo as a black box.

this assumption is made, and gains are chosen such that the system is somewhat robust with respect to variations in k_e.

Our purpose in constructing a control law to control the force of contact has been to show one suggested structure and to expose a few issues. For the remainder of this chapter, we will simply assume that such a force controlling servo could be built, and abstract it away into a black box, as shown in Fig. 9.9.

9.7 The hybrid position-force control scheme

In this section we introduce an architecture for a control system which implements the hybrid position/force controller.

A Cartesian manipulator aligned with $\{C\}$

We will first consider the simple case of a manipulator having three degrees of freedom with prismatic joints acting in the $\hat{Z}$, $\hat{Y}$, and $\hat{X}$ directions. For simplicity, we will assume that each link has mass m and slides in frictionless bearings. Let us also assume that the joint motions are lined up exactly with the constraint frame, $\{C\}$. The end-effector is in contact with a surface with stiffness k_e which is oriented with its normal in the $-{}^C\hat{Y}$ direction. Hence force control is required in that direction and position control in the ${}^C\hat{X}$ and ${}^C\hat{Z}$ directions. See Fig. 9.10.

In this case, the solution to the hybrid position/force control problem is clear. We should control joints 1 and 3 with the position controller developed for a unit mass in Chapter 8. Joint 2 (operating in the $\hat{Y}$ direction) should be controlled with the force controller developed in Section 9.5. We could then supply a position trajectory in the ${}^C\hat{X}$ and ${}^C\hat{Z}$ directions while independently supplying a force trajectory (perhaps just a constant) in the ${}^C\hat{Y}$ direction.

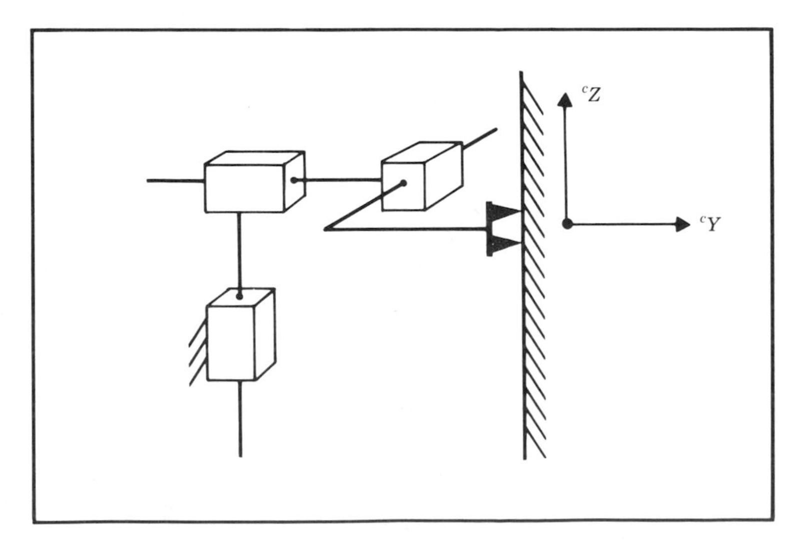

FIGURE 9.10 A Cartesian manipulator with three degrees of freedom contacting a surface.

If we wish to be able to switch the nature of the constraint surface such that its normal might also be $\hat{X}$ or $\hat{Z}$, we can slightly generalize our Cartesian arm control system as follows. We build the structure of the controller such that we may specify a complete position trajectory in all three degrees of freedom and also a force trajectory in all three degrees of freedom. Of course, we can't control so as to meet these six constraints at any one time, but rather, we will set modes to indicate which components of which trajectory will be followed at any given time.

Consider the controller shown in Fig. 9.11. Here we indicate the control of all three joints of our simple Cartesian arm in a single diagram by showing both the position controller and the force controller. The matrices S and S' have been introduced to control which mode—position or force—is used to control each joint of the Cartesian arm. The S matrix is diagonal with ones and zeros on the diagonal. Where a one is present in S, a zero is present in S' and position control is in effect. Where a zero is present in S, a one is present in S' and force control is in effect. Hence the matrices S and S' are simply switches which set the mode of control to be used with each degree of freedom in $\{C\}$. In accordance with the setting of S, there are always three components of the trajectory being controlled, though the relative mix between position control and force control is arbitrary. The other three components of desired trajectory and associated servo errors are being ignored. Hence when a certain degree of freedom is under force control, position errors on that degree of freedom are ignored.

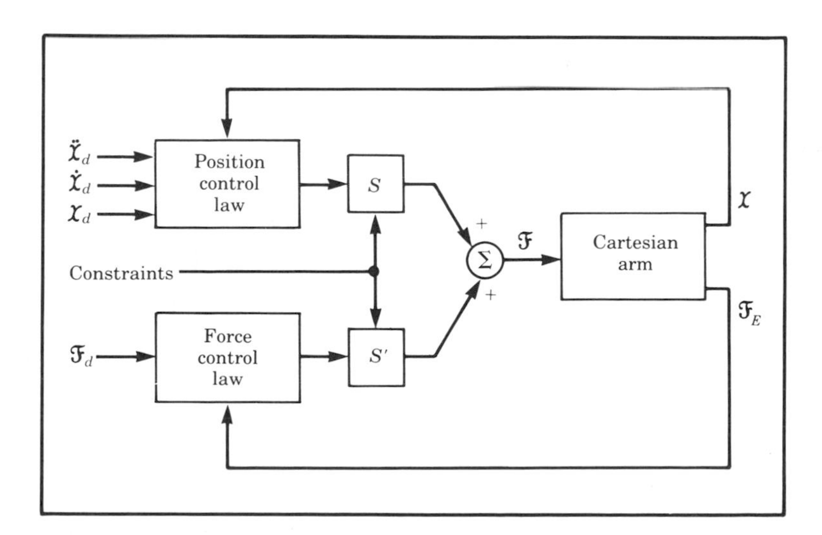

FIGURE 9.11 The hybrid controller for a 3DOF Cartesian arm.

EXAMPLE 9.2

For the situation shown in Fig. 9.10 with motions in the ${}^{C}\hat{Y}$ direction constrained by the reaction surface, give the matrices S and S'.

Because the $\hat{X}$ and $\hat{Z}$ components are to be position controlled, we enter ones on the diagonal of S corresponding to these two components. This will cause the position servo to be active in these two directions, and the input trajectory will be followed. Any position trajectory input for the $\hat{Y}$ component will be ignored. The S' matrix has the ones and zeros on the diagonal inverted, and hence we have

$$S = \begin{bmatrix} 1 & 0 & 0 \\ 0 & 0 & 0 \\ 0 & 0 & 1 \end{bmatrix},$$

$$S' = \begin{bmatrix} 0 & 0 & 0 \\ 0 & 1 & 0 \\ 0 & 0 & 0 \end{bmatrix}. \quad \blacksquare \tag{9.16}$$

Figure 9.11 shows the hybrid controller for the special case that the joints line up exactly with the constraint frame, $\{C\}$. In the following subsection we use techniques studied in previous chapters to generalize the controller to work with general manipulators and for an arbitrary $\{C\}$. However, in the ideal case, the system performs as if the manipulator had an actuator "lined up" with each of the degrees of freedom in $\{C\}$.

A general manipulator

Generalizing the hybrid controller shown in Fig. 9.11 so that a general manipulator may be used is straightforward using the concept of Cartesian based control. In Chapter 6 we discussed how the equations of motion of a manipulator could be written in terms of Cartesian motion of the end-effector, and in Chapter 8 we saw how such a formulation might be used to achieve decoupled Cartesian position control of a manipulator. The major idea is that through use of a dynamic model written in Cartesian space, it is possible to control so that the combined system of the actual manipulator and computed model appear as a set of independent, uncoupled unit masses. Once this decoupling and linearizing is done, we can apply the simple servo which we have already developed in Section 9.5.

Figure 9.12 shows the compensation based on the formulation of the manipulator dynamics in Cartesian space such that the manipulator appears as a set of uncoupled unit masses. For use in the hybrid control scheme, the Cartesian dynamics and the Jacobian are written in the

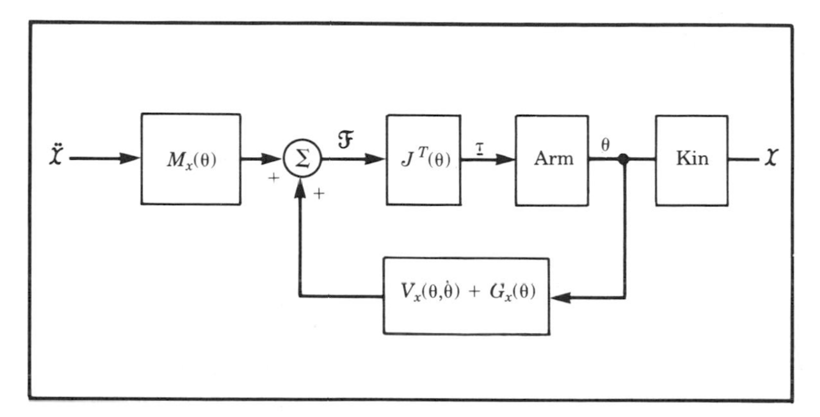

FIGURE 9.12 The Cartesian decoupling scheme introduced in Chapter 8.

constraint frame, $\{C\}$. Likewise, the kinematics are computed with respect to the constraint frame.

Since we have designed the hybrid controller for a Cartesian manipulator which is aligned with the constraint frame, and because the Cartesian decoupling scheme provides us with a system with the same input–output properties, we only need to combine the two to generate the generalized hybrid position/force controller.

Figure 9.13 shows the block diagram of the hybrid controller for a general manipulator. Note that the dynamics are written in the constraint frame, as is the Jacobian. The kinematics are written to include the transformation of coordinates into the constraint frame, and the sensed forces are likewise transformed into $\{C\}$. Servo errors are calculated in $\{C\}$, and control modes within $\{C\}$ are set through proper choice of S. Figure 9.14 shows a manipulator being controlled by such a system.

Adding variable stiffness

Controlling a degree of freedom in strict position or force control represents control at two ends of the spectrum of servo stiffness. An ideal position servo is infinitely stiff and rejects all force disturbances acting on the system. Likewise, an ideal force servo exhibits zero stiffness and maintains the desired force application regardless of position disturbances. It may be useful to be able to control the end-effector to exhibit stiffnesses other than zero or infinite. In general, we may wish to control the **mechanical impedance** of the end-effector [7], [8].

In our analysis of contact, we have imagined that the environment is very stiff. When we contact a stiff environment, we use zero-stiffness

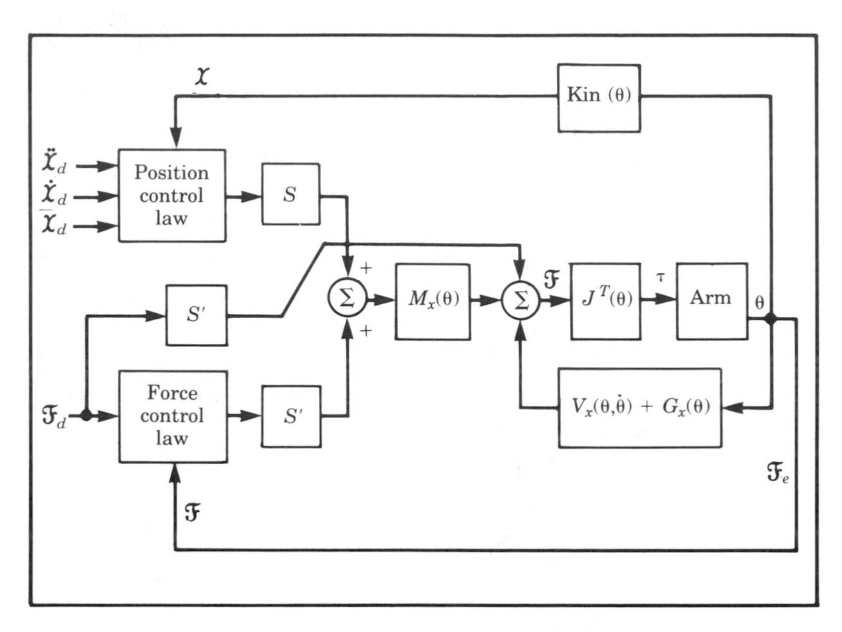

FIGURE 9.13 The hybrid position/force controller for a general manipulator. For simplicity, the velocity feedback loop has not been shown.

force control. When we contact zero-stiffness (moving in free space) we use high-stiffness position control. Hence it appears that controlling the end-effector to exhibit a stiffness which is approximately the inverse of the local environment is perhaps a good strategy. Therefore, in dealing with parts such as plastic or springs, we may wish to set servo stiffness to other than zero or infinite.

Within the framework of the hybrid controller, this is done simply by using position control and lowering the position gain corresponding to the appropriate degree of freedom in $\{C\}$. Generally, if this is done, the corresponding velocity gain is lowered so that that degree of freedom remains critically damped. The ability to change both position and velocity gains of the position servo along the degees of freedom of $\{C\}$ allows the hybrid position/force controller to implement a generalized impedance of the end-effector [9]. However, in many practical situations we are dealing with the interaction of stiff parts, so that pure position control or pure force control is desired.

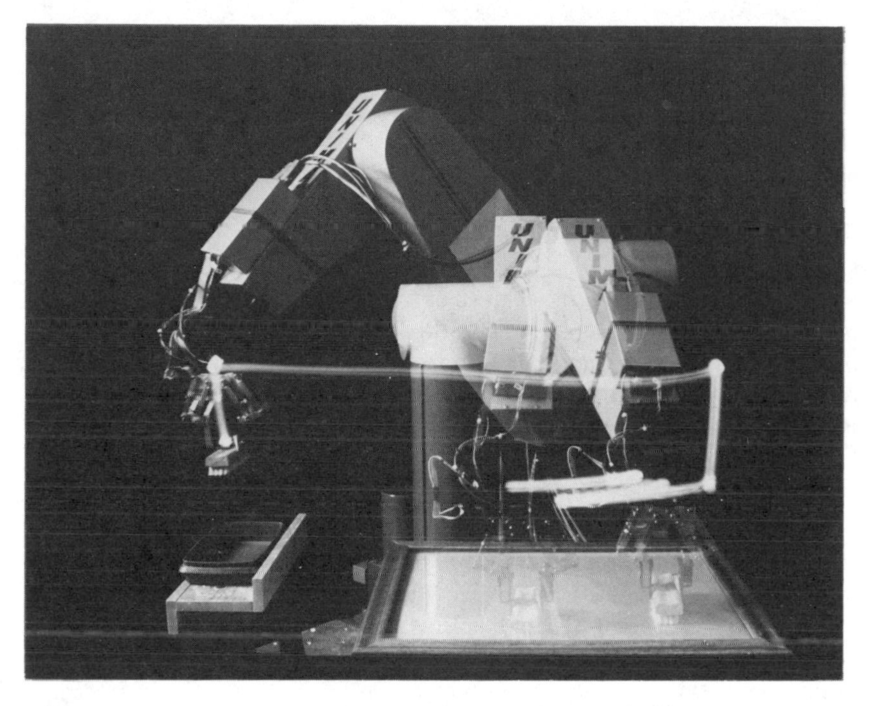

FIGURE 9.14 A PUMA 560 manipulator washes a window under control of the COSMOS system developed by O. Khatib at Stanford University. These experiments use force sensing fingers and a control structure similar to that of Fig. 9.13 [6].

9.8 Present industrial robot control schemes

True force control, such as the hybrid position/force controller introduced in this chapter, does not exist today in industrial robots. Among the problems of practical implementation are the rather large amount of computation required, lack of accurate parameters for the dynamic model, lack of rugged force sensors, and the burden of difficulty placed on the user of specifying a position/force strategy.

Passive compliance

Extremely rigid manipulators with very stiff position servos are ill-suited to tasks in which parts come into contact and contact forces are generated. In such situations parts are often jammed or damaged. Ever since early experiments with manipulators attempting to do assembly, it was realized that to the extent that the robots could perform such tasks, it was only thanks to the compliance of the parts, of the fixtures, or of the arm itself. This ability of one or more parts of the system to "give" a little was often enough to allow the successful mating of parts.

Once this was realized, devices were specially designed which introduced compliance into the system on purpose. The most successful such device is the RCC or *remote center compliance* device developed at Draper Labs [10]. The RCC was cleverly designed such that it introduced the "right" kind of compliance which allowed certain tasks to proceed smoothly and rapidly with little or no chance of jamming. The RCC is essentially a spring with six degrees of freedom which is inserted between the manipulator's wrist and the end-effector. By setting the stiffnesses of the six springs, various amounts of compliance can be introduced. Such schemes are called **passive compliance** schemes, and are used in industrial applications of manipulators in some tasks.

Compliance through softening position gains

Rather than achieving compliance in a passive, and therefore fixed, way it is possible to devise schemes in which the apparent stiffness of the manipulator is altered through adjustment of the gains of a position control system. A few industrial robots do something of this type for applications such as grinding in which contact with a surface needs to be maintained, but delicate force control is not required.

A particularly interesting approach has been suggested by Salisbury [8]. In this scheme, the position gains in a joint based servo system are modified in such a way that the end-effector appears to have a certain stiffness along Cartesian degrees of freedom. That is, consider a general spring with six degrees of freedom. Its action could be described by

$$\mathcal{F} = K_{px} \delta \mathcal{X}, \tag{9.17}$$

where K_{px} is a diagonal 6×6 matrix with three linear stiffnesses followed by three torsional stiffnesses on the diagonal. How could we make the end-effector of a manipulator exhibit this stiffness characteristic?

Recalling the definition of the manipulator Jacobian, we have

$$\delta \mathcal{X} = J(\Theta) \delta \Theta. \tag{9.18}$$

We have, combining with (9.17),

$$\mathcal{F} = K_{px} J(\Theta) \delta \Theta. \tag{9.19}$$

From static force considerations we have

$$\tau = J^T(\Theta) \mathcal{F}, \tag{9.20}$$

which, combined with (9.19), yields

$$\tau = J^T(\Theta)K_{px}J(\Theta)\delta\Theta. \tag{9.21}$$

Here, the Jacobian is usually written in the tool frame. Equation (9.21) is an expression for how joint torques should be generated as a function of small changes in joint angles, $\delta\Theta$, in order that the manipulator end-efector behave as a Cartesian spring with six degrees of freedom.

Whereas a simple joint based position controller might use the control law

$$\tau = K_p E + K_v \dot{E}, \tag{9.22}$$

where K_p and K_v are constant diagonal gain matrices, and E is servo error defined as $\Theta_d - \Theta$, Salisbury suggests using

$$\tau = J^T(\Theta)\ K_{px} J(\Theta)\ E + K_v \dot{E}, \tag{9.23}$$

where K_{px} is the desired stiffness of the end-effector in Cartesian space. For a manipulator with six degrees of freedom, K_{px} is diagonal with the six values on the diagonal representing the three translational and three rotational stiffnesses that the end-effector is to exhibit. Essentially, through use of the Jacobian, a Cartesian stiffness has been transformed to a joint-space stiffness.

Force sensing

Force sensing allows a manipulator to detect contact with a surface and, based on this sensation, to take some action. For example, the term **guarded move** is sometimes used to mean the strategy of moving under position control until a force is felt, then halting motion. Additionally, force sensing can be used to weigh objects which the manipulator lifts. This can be used as a simple check during a parts handling operation—to insure that a part was acquired, or that the appropriate part was acquired.

Some commercially available robots come equipped with force sensors in the end-effector. These robots can be programmed to stop motion or to take other action when a force threshold is exceeded, and some can be programmed to weigh objects which are grasped in the end-effector.

References

[1] V. Scheinman, "Design of a Computer Controlled Manipulator," M.S. Thesis, Mechanical Engineering Dept., Stanford University, 1969.

[2] M. Mason, "Compliance and Force Control for Computer Controlled Manipulators," M.S. Thesis, MIT AI Laboratory, May 1978.

[3] M. Raibert and J. Craig, "Hybrid Position/Force Control of Manipulators," *ASME Journal of Dynamic Systems, Measurement, and Control*, June 1981.

[4] M. Mason, "Automatic Planning of Fine Motions: Correctness and Completeness," IEEE International Conference on Robotics, Atlanta, March 1984.

[5] T. Lozano-Perez, M. Mason, and R. Taylor, "Automatic Synthesis of Fine-Motion Strategies for Robots," 1st International Symposium of Robotics Research, Bretton Woods, New Hampshire, August 1983.

[6] O. Khatib, "The Operational Space Formulation in Robot Manipulator Control," 15th ISIR, Tokyo, Japan, September 11–13, 1985.

[7] D. Whitney, "Force Feedback Control of Manipulator Fine Motions," Proceedings Joint Automatic Control Conference, San Francisco, Calif., 1976.

[8] J.K. Salisbury, "Active Stiffness Control of a Manipulator in Cartesian Coordinates," 19th IEEE Conference on Decision and Control, December 1980.

[9] J.K. Salisbury and J. Craig, "Articulated Hands: Force Control and Kinematic Issues," *International Journal of Robotics Research*, Vol. 1, Number 1.

[10] S. Drake, "Using Compliance in Lieu of Sensory Feedback for Automatic Assembly," Ph.D. Thesis, Mechanical Engineering Dept., MIT, September 1977.

Exercises

9.1 [12] Give the natural constraints present for a peg of square cross section sliding into a hole of square cross section. Show your definition of $\{C\}$ in a sketch.

9.2 [10] Give the artificial constraints (i.e., the trajectory) you would suggest in order to cause the peg in Exercise 9.1 to slide further into the hole without jamming.

9.3 [20] Show that using the control law (9.14) with a system given by (9.9) results in the error-space equation

$$\ddot{e}_f + k_{vf}\dot{e}_f + (k_{pf} + m^{-1}k_e)e_f = m^{-1}k_e f_{dist},$$

and hence that choosing gains to provide critical damping is only possible if the stiffness of the environment, k_e, is known.

9.4 [17] Given

$$^{A}_{B}T = \begin{bmatrix} 0.866 & -0.500 & 0.000 & 10.0 \\ 0.500 & 0.866 & 0.000 & 0.0 \\ 0.000 & 0.000 & 1.000 & 5.0 \\ 0 & 0 & 0 & 1 \end{bmatrix}.$$

If the force-torque vector at the origin of $\{A\}$ is

$$
{}^{A}\mathcal{F} = \begin{bmatrix} 0.0 \\ 2.0 \\ -3.0 \\ 0.0 \\ 0.0 \\ 4.0 \end{bmatrix},
$$

find the 6×1 force-torque vector with reference point the origin of $\{B\}$.

9.5 [17] Given

$$
{}^{A}_{B}T = \begin{bmatrix} 0.866 & 0.500 & 0.000 & 10.0 \\ -0.500 & 0.866 & 0.000 & 0.0 \\ 0.000 & 0.000 & 1.000 & 5.0 \\ 0 & 0 & 0 & 1 \end{bmatrix}.
$$

If the force-torque vector at the origin of $\{A\}$ is

$$
{}^{A}\mathcal{F} = \begin{bmatrix} 6.0 \\ 6.0 \\ 0.0 \\ 5.0 \\ 0.0 \\ 0.0 \end{bmatrix},
$$

find the 6×1 force-torque vector with reference point the origin of $\{B\}$.

9.6 [18] Describe in English how you accomplish the insertion of a book into a narrow crack between books on your crowded bookshelf.

9.7 [20] Give the natural and artificial constraints for the task of closing a hinged door with a manipulator. Make any reasonable assumptions needed. Show your definition of $\{C\}$ in a sketch.

9.8 [20] Give the natural and artificial constraints for the task of uncorking a bottle of champagne with a manipulator. Make any reasonable assumptions needed. Show your definition of $\{C\}$ in a sketch.

9.9 [41] For the stiffness servo system of Section 9.8, we have made no claim that the system is stable. Assume that (9.23) is used as the servo portion of a decoupled and linearized manipulator (so the n joints appear as unit masses). Prove that the controller is stable for any K_v which is positive definite.

9.10 [48] For the stiffness servo system of Section 9.8, we have made no claim that the system is or can be critically damped. Assume that (9.23) is used as the servo portion of a decoupled and linearized manipulator (so the n joints appear as unit masses). Is it possible to design a K_p which is a function of Θ which causes the system to be critically damped over all configuratons?

Programming Exercise (Part 9)

1. Implement a Cartesian stiffness control system for the 3-link planar manipulator by using the control law (9.23) to control the simulated arm. Use the Jacobian written in frame {3}.

For the manipulator in position $\Theta = [60.0 \ -90.0 \ 30.0]$ and K_{px} of the form

$$K_{px} = \begin{bmatrix} k_{small} & 0.0 & 0.0 \\ 0.0 & k_{big} & 0.0 \\ 0.0 & 0.0 & k_{big} \end{bmatrix},$$

simulate the application of the following static forces:

a) a one Newton force acting at the origin of {3} in the $\hat{X}_3$ direction, and

b) a one Newton force acting at the origin of {3} in the $\hat{Y}_3$ direction.

The values of k_{small} and k_{big} should be found experimentally. Use a large value of k_{big} for high stiffness in the $\hat{Y}_3$ direction, and a low value of k_{small} for low stiffness in the $\hat{X}_3$ direction. What are the steady state deflections in the two cases?

10

ROBOT PROGRAMMING LANGUAGES AND SYSTEMS

10.1 Introduction

In this chapter we consider the interface between the human user and an industrial robot. It is by means of this interface that a user takes advantage of all the underlying mechanics and control algorithms which we have studied in previous chapters.

The sophistication of the user interface is becoming extremely important as manipulators and other programmable automation are applied to more and more demanding industrial applications. It turns out that the nature of the user interface is a very important concern. In fact, much of the challenge of the design and use of industrial robots focuses on this aspect of the problem.

Robot manipulators differentiate themselves from fixed automation by being "flexible," which means programmable. Not only are the movements of manipulators programmable, but through the use of sensors and communications with other factory automation, manipulators can *adapt* to variations as the task proceeds.

In considering the programming of manipulators, it is important to remember that they are typically only a minor part of an automated process. The term **workcell** is used to describe a local collection of equipment which may include one or more manipulators, conveyor systems, parts feeders, and fixtures. At the next higher level, workcells might be interconnected in factory-wide networks so that a central control computer can control the overall factory flow. Hence, the programming of manipulators is often considered within the broader problem of programming a variety of interconnected machines in an automated factory workcell.

10.2 The three levels of robot programming

There have been many styles of user interface developed for programming robots. Before the rapid proliferation of microcomputers in industry, robot controllers resembled the simple sequencers often used to control fixed automation. Modern approaches focus on computer programming, and issues in programming robots include all the issues faced in general computer programming, and more.

Teach by showing

Early robots were all programmed by a method that we will call **teach by showing**, which involved moving the robot to a desired goal point and recording its position in a memory which the sequencer would read during playback. During the teach phase, the user would guide the robot by hand, or through interaction with a **teach pendant**. Teach pendants are hand-held button boxes which allow control of each manipulator joint or of each Cartesian degree of freedom. Some such controllers allow testing and branching so that simple programs involving logic can be entered. Some teach pendants have alphanumeric displays and are approaching hand-held terminals in complexity. Figure 10.1 shows an operator using a teach pendant to program a large industrial robot.

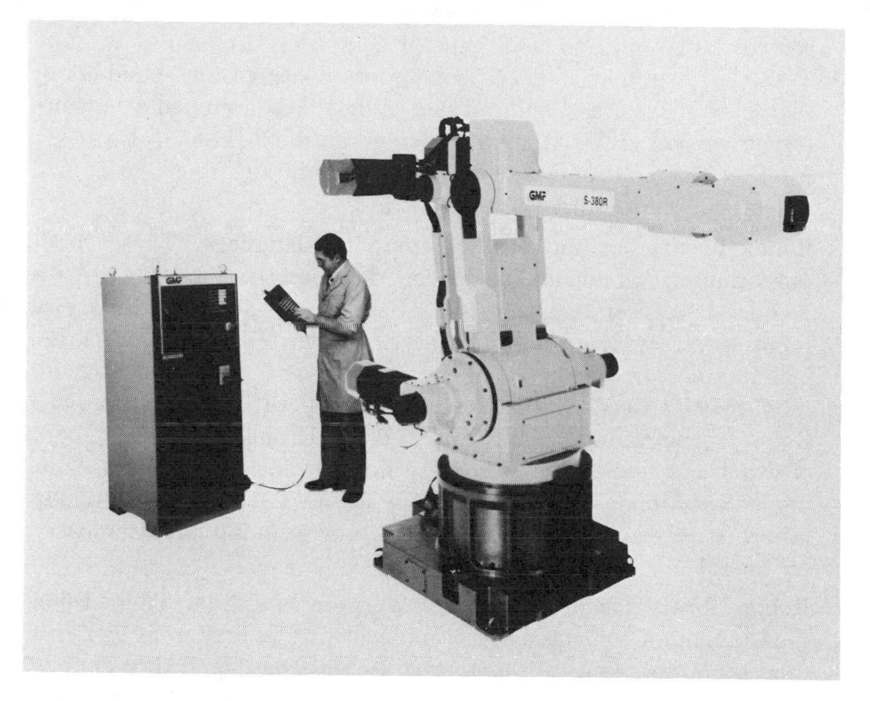

FIGURE 10.1 The GMF S380 is often used in automobile-body spot-welding applications. Here an operator uses a teach pendant interface to program the manipulator.

Photo courtesy of GMFanuc Corp.

Explicit robot programming languages

With the arrival of inexpensive and powerful computers, the trend has has been increasingly toward programming robots via programs written in computer programming languages. Usually these computer programming languages have special features which apply to the problems of programming manipulators, and so are called **robot programming languages**. Most of the systems which come equipped with a robot programming language have also retained a teach-pendant style interface as well.

Robot programming languages have taken on many forms as well. We will split them into three categories as follows:

1. **Specialized manipulation languages**. These robot programming languages have been built by developing a completely new language which, while addressing robot-specific areas, may or may not be considered a general computer programming language. An example is the VAL language developed to control the industrial robots of Unimation, Inc [1]. VAL was developed especially as a manipulator

control language, and as a general computer language it is quite weak. For example, it does not support floating-point numbers or character strings, and subroutines cannot pass arguments. A more recent version, VAL II, now provides most of these features [2]. Another example of a specialized manipulation language is AL, developed at Stanford University [3].

2. **Robot library for an existing computer language**. These robot programming languages have been developed by starting with a popular computer language (e.g., Pascal) and adding a library of robot-specific subroutines. The user then writes a Pascal program making use of frequent calls to the predefined subroutine package for robot-specific needs. Examples include AR-BASIC from American Robot Corporation [4] and ROBOT-BASIC from Intelledex [5], both of which are essentially subroutine libraries for a standard BASIC implementation. JARS, developed by NASA's Jet Propulsion Laboratory, is an example of such a robot programming language based on Pascal [6].
3. **Robot library for a new general-purpose language**. These robot programming languages have been developed by first creating a new general purpose language as a programming base, and then supplying a library of predefined robot-specific subroutines. Examples of such robot programming languages are AML developed by IBM [7], and RISE developed at Silma, Inc. [8].

Studies of actual application programs for robotic workcells have shown that a large percentage of the language statements are not robot-specific [7]. Instead, a great deal of robot programming has to do with initialization, logic testing and branching, communication, etc. For this reason, a trend may develop to move away from developing special languages for robot programming, and toward developing extensions to general languages, as in categories 2 and 3 above.

Task-level programming languages

The third level of robot programming methodology is embodied in **task-level programming languages**. These are languages which allow the user to command desired subgoals of the task directly, rather than to specify the details of every action the robot is to take. In such a system, the user is able to include instructions in the application program at a significantly higher level than in an explicit robot programming language. A task-level robot programming system must have the ability to perform many planning tasks automatically. For example, if an instruction to "grasp the bolt" is issued, the system must plan a path of the manipulator which avoids collision with any surrounding obstacles, automatically choose a good grasp location on the bolt, and grasp it. In contrast, in an explicit robot programming language, all these choices must be made by the programmer.

The border between explicit robot programming languages and task-level programming languages is quite distinct. Incremental advances are being made to explicit robot programming languages which help to ease programming, but these enhancements cannot be counted as components of a task-level programming system. True task-level programming of manipulators does not exist yet, but is an active topic of research [9], [10].

10.3 A sample application

Figure 10.2 shows an automated workcell which completes a small subassembly in a hypothetical manufacturing process. The workcell consists of a conveyor under computer control which delivers a workpiece. A camera connected to a vision system is used to locate the workpiece on the conveyor. There is an industrial robot (a PUMA 560 is pictured) equipped with a force sensing wrist. A small feeder located on the work surface supplies another part to the manipulator. A computer controlled press may be loaded and unloaded by the robot, and finished assemblies are placed in a pallet.

The entire process is controlled by the manipulator's controller in a sequence as follows:

1. The conveyor is signalled to start, and is stopped when the vision system reports that a bracket has been detected on the conveyor.
2. The vision system determines the bracket's position and orientation on the conveyor and inspects the bracket for defects such as the wrong number of holes.
3. Using the output of the vision system, the manipulator grasps the bracket with a specified force. The distance between the fingertips is checked to insure that the bracket has been properly grasped. If it has not, the robot moves out of the way and the vision task is repeated.
4. The bracket is placed in the fixture on the work surface. At this point, the conveyor can be signalled to start again for the next bracket. That is, steps 1 and 2 can begin in parallel with the following steps.
5. A pin is picked from the feeder and inserted partway into a tapered hole in the bracket. Force control is used to perform this insertion and to perform simple checks on its completion. If the pin feeder

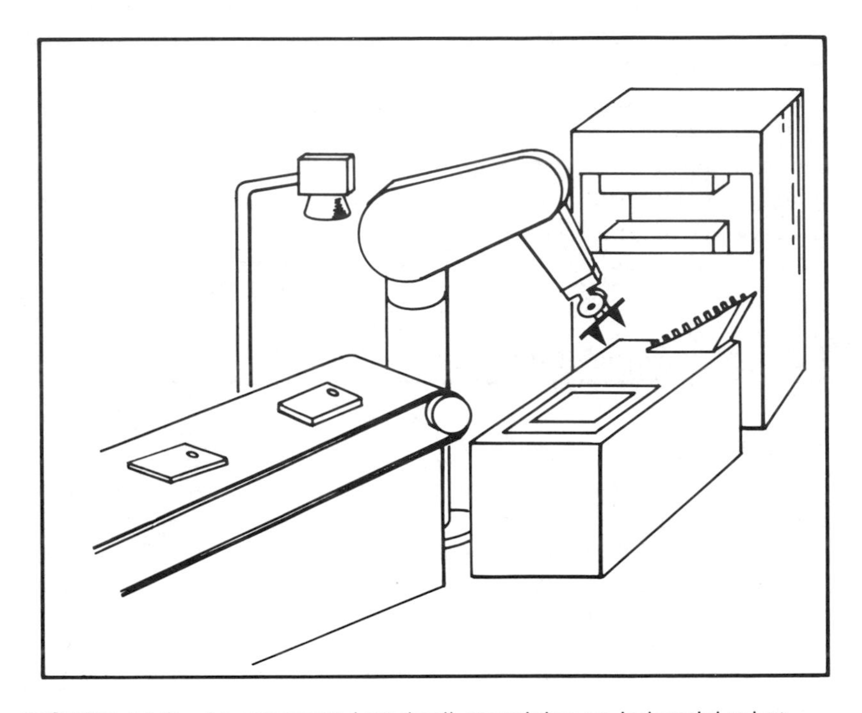

FIGURE 10.2 An automated workcell containing an industrial robot.

was empty, an operator is notified and the manipulator waits until commanded to resume by the operator.

6. The bracket–pin assembly is grasped by the robot and placed in the press.

7. The press is commanded to actuate, and presses the pin the rest of the way into the bracket. The press signals that it has completed, and the bracket is placed back into the fixture for a final inspection.

8. Using force sensing the assembly is checked for proper insertion of the pin. The manipulator senses the reaction force when it presses sideways on the pin, and can do several checks to determine how far the pin protrudes from the bracket.

9. If the assembly is judged to be good, the robot places the finished part into the next available pallet location. If the pallet is full, the operator is signalled. If the assembly is bad, it is dropped into the trash bin.

10. Wait for step 2 (started earlier in parallel) to complete, then go to step 3.

This is an example of a task that is possible (though slightly challenging) for today's industrial robots. It should be clear that the definition of such a process through "teach by showing" techniques is probably not feasible. For example, in dealing with pallets, it is laborious to have to teach all the pallet compartment locations; it is much preferable to teach only the corner location and then compute the others making use of the dimensions of the pallet. Further, specifying interprocess signalling and setting up parallelism using a typical teach pendant or a menu-style interface is usually not possible at all. This kind of application necessitates a robot programming language approach to process description. On the other hand, this application is too complex for any existing task-level languages to deal with directly. It is typical of the great many applications which must be addressed with an explicit robot programming approach. We will keep this sample application in mind as we discuss features of robot programming languages.

10.4 Requirements of a robot programming language

World modeling

Since manipulation programs must by definition involve moving objects in three dimensional space, it is clear that any robot programming language needs a means of describing such actions. The most common element of robot programming languages is the existence of special **geometric types**. For example, *types* are introduced which are used to represent joint angle sets, as well as Cartesian positions, orientations, and frames. Predefined operators which can manipulate these types often are available. The "standard frames" introduced in Chapter 3 might serve as a possible model of the world: All motions are described as tool frame relative to station frame, with goal frames being constructed from arbitrary expressions involving geometric types.

Given a robot programming environment which supports geometric types, the robot and other machines, parts, and fixtures can be modeled by defining named variables associated with each object of interest. Figure 10.3 shows part of our example workcell with frames shown attached in task-relevant locations. Each of these frames would be represented with a variable of type "frame" in the robot program.

In many robot programming languages, this ability to define named variables of various geometric types and refer to them in the program forms the basis of the world model. Note that the physical shapes

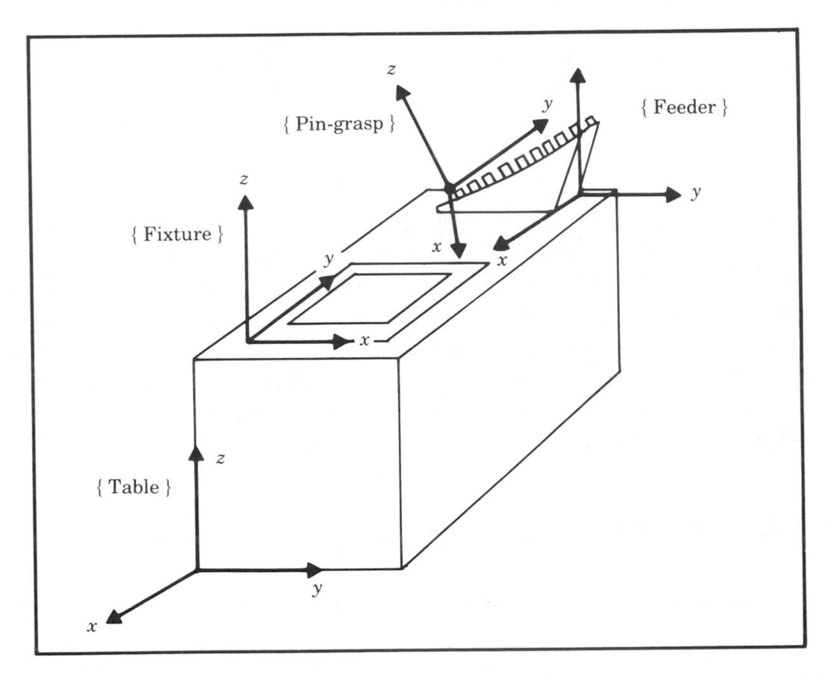

FIGURE 10.3 Often a workcell is modeled only by a set of frames which are attached to relevant objects.

of the objects are not part of such a world model, and neither are surfaces, volumes, masses, or other properties. The extent to which objects in the world are modeled is one of the basic design decisions made when designing a robot programming system. Most present-day systems support only the style just described.

Some world-modeling systems allow the notion of **affixments** between named objects [3], [8]. That is, the system can be notified that two or more named objects have become "affixed" and from then on, if one object is explicitly moved with a language statement, any objects affixed to it are moved as well. Thus, in our application, once the pin has been inserted into the hole in the bracket, the system would be notified (via a language statement) that these two objects have become affixed. Subsequent motions of the bracket (that is, changes to the value of the frame variable "bracket") would cause the value stored for variable "pin" to be updated as well.

Ideally, a world-modeling system would include much more information about the objects with which the manipulator has to deal, and about the manipulator itself. For example, consider a system in which objects are described with CAD-style models which represent the spatial shape of an object by giving definitions of its edges, surfaces, or volume. With such data available to the system, it begins to become possible to implement many of the features of a task-level programming system.

One such feature is **automatic collision detection**, in which the system, using models of the manipulator and objects surrounding it, can warn the user about collisions that would result if a requested motion were carried out. One step beyond automatic collision detection is the **automatic path planning** problem of computing a collision-free path for a manipulator in a cluttered workcell when only the final goal destination is specified. Unless care is taken, such computations are so complex as to be prohibitive. Most of the recent research has focused on how to reduce the computational burden through clever representations of empty and occupied space [10].

Motion specification

A very basic function of a robot programming language is to allow the description of desired motions of the robot. Through the use of motion statements in the language, the user interfaces to path planners and generators of the style discussed in Chapter 7. Motion statements allow the user to specify via points and the goal point, and whether to use joint-interpolated motion or Cartesian straight-line motion. Additionally, the user may have control over the speed or duration of a motion.

To illustrate various syntaxes for motion primitives, we will consider the following example manipulator motions: 1) move to position "goal1," then 2) move in a straight line to position "goal2," then 3) move without stopping through "via1" and come to rest at "goal3." Assuming all of these path points had already been taught or described textually, this program segment would be written as follows.

In VAL II:

```
move goal1
moves goal2
move via1
move goal3
```

In AL (here controlling the manipulator "garm"):

```
move garm to goal1;
move garm to goal2 linearly;
move garm to goal3 via via1;
```

In Intelledex Robot Basic:

```
10 move goal1
20 move straight goal2
30 cpon
40 move via1
50 move goal3
60 cpoff
```

Most languages have similar syntax for simple motion statements like these. Differences in the basic motion primitives from one robot programming language to another become more apparent if we consider features such as:

1. The ability to do math on structured types like frames, vectors, and rotation matrices.
2. The ability to describe geometric entities like frames in several different convenient representations—with the ability to convert between representations.
3. The ability to give constraints on the duration or velocity of a particular move. For example, many systems allow the user to set the speed to a fraction of maximum. Fewer allow the user to specify a desired duration, or a desired maximum joint velocity directly.
4. The ability to specify goals relative to various frames, including frames defined by the user and frames in motion (on a conveyor for example).

Flow of execution

As in more conventional computer programming languages, a robot programming system allows the user to specify the flow of execution. That is, concepts such as testing and branching, looping, calls to subroutines, and even interrupts are generally found in robot programming languages.

More so than in many computer applications, parallel processing is generally important in automated workcell applications. First of all, very often two or more robots are used in a single workcell and work simultaneously to reduce the cycle time of the process. But even in single-robot applications such as the one shown in Fig. 10.2, there is other workcell equipment which must be controlled by the robot controller in a parallel fashion. Hence signal and wait primitives are often found in robot programming languages, and occasionally more sophisticated parallel execution constructs are provided [3].

Another frequent occurrence is the need to monitor various processes with some kind of sensor. Then, either by interrupt or through polling, the robot system must be able to respond to certain events which are detected by the sensors. The ability easily to specify such **event monitors** is afforded by some robot programming languages [2], [3].

Programming environment

As with any computer languages, a good programming environment helps to increase programmer productivity. Manipulator programming is difficult and tends to be very interactive, with a lot of trial and error. If the user were forced to continually repeat the "edit-compile-run" cycle of compiled languages, productivity would be low. Therefore, most robot programming languages are now *interpreted* so that individual language statements can be run one at a time during program development and debugging. Typical programming support such as text editors, debuggers, and a file system are also required.

Sensor integration

An extremely important part of robot programming has to do with interaction with sensors. The system should have the minimum capability to query touch and force sensors and use the response in if-then-else constructs. The ability to specify event monitors to watch for transitions on such sensors in a *background* mode is also very useful.

Integration with a vision system allows the vision system to send the manipulator system the coordinates of an object of interest. For example, in our sample application, a vision system locates the brackets on the conveyor belt and returns to the manipulator controller their position and orientation relative to the camera. Since the camera's frame is known relative to the station frame, a desired goal frame for the manipulator can be computed from this information.

Some sensors may be part of other equipment in the workcell. For example, some robot controllers can use input from a sensor attached to a conveyor belt so that the manipulator can track the belt's motion and acquire objects from the belt as it moves [2].

The interface to force control capabilities as discussed in Chapter 9 comes through special language statements which allow the user to specify force strategies [3]. Such force control strategies are by necessity an integrated part of the manipulator control system—the robot programming language simply serves as an interface to those capabilities. Programming robots which make use of active force control may require other special features, such as the ability to display force data collected during a constrained motion [3].

In systems which support active force control, the description of the desired force application may become part of the motion specification. The AL language describes active force control in the motion primitives by specifying six components of stiffness (three translational and three rotational) and a bias force. In this way, the manipulator's apparent stiffness is programmable. To apply a force, usually the stiffness is set to zero in that direction, and a bias force is specified. For example:

```
move garm to goal
  with stiffness=(80,80,0,100,100,100)
  with force=20*ounces along zhat;
```

10.5 Problems peculiar to robot programming languages

While advances in recent years have helped, programming robots is still difficult. Robot programming shares all the problems of conventional computer programming, plus some additional difficulties caused by effects of the physical world [11].

Internal world model versus external reality

A central feature of a robot programming system is the world model that is maintained internally in the computer. Even when this model is quite simple, there are ample difficulties in assuring that it matches the physical reality that it attempts to model. Discrepancies between internal model and external reality result in poor or failed grasping of objects, collisions, and a host of more subtle problems.

This correspondence between internal model and the external world must be established for the program's initial state and must be maintained throughout its execution. During initial programming or debugging it is generally up to the user to suffer the burden of insuring that the state represented in the program corresponds to the physical state of the workcell. Unlike more conventional programming, where only internal variables need to be saved and restored to re-establish a former situation, in robot programming, physical objects must usually be repositioned.

Besides the uncertainty inherent in each object's position, the manipulator itself is limited to a certain degree of accuracy. Very often steps in an assembly will require the manipulator to make motions requiring greater precision than it is capable of. A common example of this is inserting a pin into a hole where the clearance is an order of magnitude less than the positional accuracy of the manipulator. To further complicate matters, the manipulator's accuracy usually varies over its workspace.

In dealing with those objects whose locations are not known exactly, it is essential to somehow refine the positional information. This can sometimes be done with sensors, e.g., vision, touch, or by using appropriate force strategies for constrained motions.

During debugging of manipulator programs, it is very useful to be able to modify the program and then back up and try a procedure again. Backing up entails restoring the manipulator and objects being manipulated to a former state. However, in working with physical objects, it is not always easy, or even possible, to undo an action. Some examples are the operations of painting, riveting, drilling, or welding which cause a physical modification of the objects being manipulated. It may therefore be necessary for the user to get a new copy of the object to replace the old, modified one. Further, it is likely that some of the operations just prior to the one being retried will also need to be repeated to establish the proper state required before the desired operation can be successfully retried.

Context sensitivity

Bottom-up programming is a standard approach to writing a large computer program in which one develops small, low level pieces of a program and then puts them together into larger pieces, eventually resulting in a completed program. For this method to work it is essential that the small pieces be relatively insensitive to the language statements that precede them and that there are no assumptions concerning the context with which these program pieces execute. For manipulator progamming this is often not the case; code which worked reliably when tested in isolation frequently fails when placed in the context of the larger program. These problems generally arise from dependencies on manipulator configuration and speed of motions.

Manipulator programs may be highly sensitive to initial conditions, for example, the initial manipulator position. In motion trajectories, the starting position will influence the trajectory that will be used for the motion. The initial manipulator position may also influence the velocity with which the arm will be moving during some critical part of the motion. For example, these statements are true for manipulators which follow cubic spline joint space paths studied in Chapter 7. While these effects might be dealt with by proper programming care, such problems may not arise until after the initial language statements have been debugged in isolation and are then joined with statements preceding them.

Because of insufficient manipulator accuracy, a program segment written to perform an operation at one location is likely to need to be tuned (i.e., positions retaught and the like) to make it work at a different location. Changes in location within the workcell result in changes in the manipulator's configuration in reaching goal locations. Such attempts at relocating manipulator motions within the workcell test the accuracy of the manipulator kinematics and servo system, and frequently problems arise. Such relocation may cause a change in the manipulator's kinematic configuration, for example, from left shoulder to right shoulder, or from elbow up to elbow down. Moreover, these changes in configuration may cause large arm motions during what had previously been a short, simple motion.

The nature of the spatial shape of trajectories is likely to change as paths are located in different portions of the manipulator's workspace. This is particularly true of joint space trajectory methods, but use of Cartesian path schemes can also lead to problems if singularities are nearby.

When testing a manipulator motion for the first time it is often wise to have the manipulator move slowly. This allows the user a chance to stop the motion if it appears to be about to cause a collision. It also allows the user to inspect the motion closely. After the motion has undergone some initial debugging at a slower speed it is then desirable to increase motion speeds. In doing so many of the aspects of the motion may change. Due to limitations in most manipulator control systems, greater servo errors are to be expected in following the quicker trajectory. Also, in force control situations involving contact with the environment, speed changes can completely change force strategies required for success.

The manipulator's configuration also affects the delicacy and accuracy of forces that may be applied with it. This is a function of how well conditioned the Jacobian of the manipulator is at a certain configuration, which is generally difficult to consider when developing robot programs [12].

Error recovery

Another direct consequence of working with the physical world is that objects may not be exactly where they should be and hence motions that deal with them may fail. Part of manipulator programming involves attempting to take this into account and making assembly operations as robust as possible, but, even so, errors are likely; and an important aspect of manipulator programming is how to recover from these errors.

Almost any motion statement in the user's program can fail, sometimes for a variety of reasons. Some of the more common causes are:

objects shifting or dropping out of the hand, an object missing from where it should be, jamming during an insertion, not being able to locate a hole, and so on.

The first problem that arises for error recovery is identifying that an error has indeed occurred. Because robots generally have quite limited sensing and reasoning capabilities, *error detection* is often difficult. In order to detect an error, a robot program must contain some type of explicit test. This test might involve checking the manipulator's position to see that it lies in the proper range; for example, when doing an insertion, no change in position might indicate jamming, while too much change might indicate that the hole was missed entirely or the object has slipped out of the hand. If the manipulator system has some type of visual capabilities then it might take a picture and check for the presence or absence of an object, and, if the object is present, determine its location. Other checks might involve force, such as weighing the load being carried to check that the object is still there and has not been dropped, or checking that a contact force remains within certain bounds during a motion.

Since every motion statement in the program may potentially fail, these explicit checks can be quite cumbersome and can take up more space than the rest of the program. Rather than attempt to deal with all possible errors, which is extremely difficult, usually just the few statements which seem most likely to fail are checked. The process of determining which portions of a robot application program are likely to fail is one which requires a certain amount of interaction and partial testing with the robot during the program development stage.

Once an error has been detected, an attempt can be made to recover from it. This can be done totally by the manipulator under program control, or it may involve manual intervention by the user, or some combination of the two. In any event, the recovery attempt may in turn result in new errors. It is easy to see how code to recover from errors can become the major part of the manipulator program.

The use of parallelism in manipulator programs can further complicate recovery from errors. When several processes are running concurrently and one causes an error to occur, it may or may not affect other processes. In many cases it will be possible to back up the offending process, while allowing the others to continue. Other times it will be necessary to reset several or all of the running processes.

Simulation and off-line programming

The term **off-line programming** is used to mean the programming of physical equipment such as manipulators and other workcell devices without access to the equipment when the programming is performed.

There are very important economic reasons why this capability is quite important. Consider a plant which is using hundreds of robots to manufacture an item which must undergo a large design change (perhaps a new model is brought out each year). If robot programming can only take place using the manipulator itself, then the entire assembly line process must be halted in order to reprogram the manipulators and other workcell devices. In contrast, if the process could be programmed off line, then actual manufacturing would only have to be interrupted briefly while the new programs are down-loaded into the workcell control computers.

In order to accomplish the off-line programming of manipulators, a robot programming system must be extended sufficiently that an application can be developed without the use of actual workcell equipment for other than very final testing. Such systems have only recently been developed. They are usually built around a graphic simulation capability, which allows manipulator motions to be displayed through computer graphical animation. A sufficient graphic capability allows for programs to be developed as if the actual workcell devices are available. Figure 10.4 shows the screen of one such off-line robot programming system.

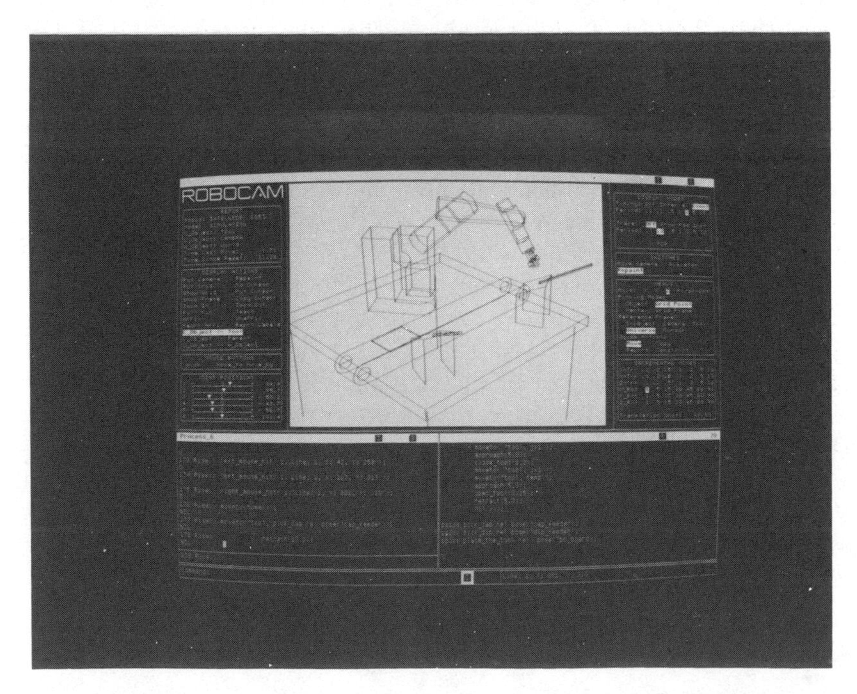

FIGURE 10.4 Silma's RoboCam system extends a robot programming language with graphics and simulation capability so that programming can be accomplished off line [8].

One difficulty with off-line programming systems is the proliferation of different robot programming languages [13]. If the off-line programming system is to be used with a variety of robot models, it must support several different language conventions. Either the off-line system's user must program in different languages (always using the one corresponding to the *native* language of the robot being simulated), or the off-line programming system must provide language translation software. If the off-line system provides language translation capability, the user can program various robots in a common language [14].

A central feature of off-line programming systems for industrial manipulators is the connection to CAD data bases. Manufacturing data which was generated by the designer of a part or subsystem may be of direct use by the robot program. Hence a robot program which accomplishes drilling and/or deburring, for example, may be almost directly generated from a CAD description of the part. In these application areas, the distinction between industrial manipulators and numerically controlled machines may become fuzzy.

Off-line programming of robots offers other potential benefits which are just beginning to be appreciated by industrial robot users. We have discussed many of the problems associated with robot programming, and most have to do with the fact that an external, physical workcell is being manipulated by the robot program. This makes backing up to try different strategies tedious. Programming of robots in simulation offers a way of keeping the bulk of the programming work strictly internal to a computer—until the application is nearly complete. Thus, many of the problems peculiar to robot programming tend to diminish.

Problems involving the internal model matching the external reality exist in any off-line programming systems. After downloading a completed application program, positions of parts, feeders, and fixtures will usually need to be retaught using traditional teach by showing techniques. In some cases, the off-line programming system can be temporarily connected to the robot in an *on-line* mode, so that these final program corrections can be made. In this way, although the complete task was not programmed off-line, a significant portion was. In some industries, if only a small portion of the work can be performed off-line, it still makes economic sense to do so.

References

[1] B. Shimano, "VAL: A Versatile Robot Programming and Control System," Proceedings of COMPSAC 1979, Chicago, November 1979.

[2] B. Shimano, C. Geschke, and C. Spalding, "VAL - II: A Robot Programming Language and Control System," SME Robots VIII Conference, Detroit, June 1984.

[3] S. Mujtaba and R. Goldman, "AL Users' Manual," Third Edition, Stanford Dept. of Computer Science, Report No. STAN-CS-81-889, December 1981.

[4] A. Gilbert et al., *AR-BASIC: An Advanced and User Friendly Programming System for Robots*, American Robot Corp., June 1984.

[5] Intelledex, Inc., "Model 605 Operator's Manual," Section 3, Intelledex, Inc., 1983.

[6] J. Craig, "JARS – JPL Autonomous Robot System: Documentation and Users Guide," JPL Interoffice memo, September 1980.

[7] R. Taylor, P. Summers, and J. Meyer, "AML: A Manufacturing Language," *International Journal of Robotics Research*, Vol. 1, No. 3, Fall 1982.

[8] M. Burke and J. Craig, "The RoboCam User's Manual - Version 1.2," Silma, Inc., Los Altos, Calif., January 1985.

[9] R. Taylor, "A Synthesis of Manipulator Control Programs from Task-Level Specifications," Stanford University AI Memo 282, July 1976.

[10] T. Lozano-Perez, "Automatic Planning of Manipulator Transfer Movements," *IEEE Transactions on Man, Systems, and Cybernetics*, Vol. SMC-11, No. 10, October, 1981.

[11] R. Goldman, *Design of an Interactive Manipulator Programming Environment*, UMI Research Press, Ann Arbor, 1985.

[12] Salisbury, J.K. and Craig, J.J., "Articulated Hands: Force Control and Kinematic Issues," *International Journal of Robotics Research*, Vol.1, No. 1, Spring 1982.

[13] W. Gruver, B. Soroka, J. Craig, and T. Turner, "Commercially Available Robot Programming Languages," 13th ISIR, Chicago, April 1983.

[14] J. Craig, "Anatomy of an Off-line Programming System," *SME Robotics Today*, February 1985.

Exercises

10.1 [15] Write a robot program (use any of [1] through [8]) to pick a block up from location A and place it in location B.

10.2 [20] Describe tieing your shoelace in simple English commands that might form the basis of a robot program.

10.3 [32] Design the syntax of a new robot programming language. Include ways to give duration or speeds to motion trajectories, I/O statements to peripherals, commands to control the gripper, and force sensing (i.e., guarded move) commands. You can skip force control and parallelism (see Exercise 10.4).

10.4 [28] Extend the specification of a new robot programming language that you started in Exercise 10.3 by adding force control syntax and syntax for parallelism.

10.5 [38] Write a program in a commercially available robot programming language to perform the application outlined in Section 10.3. Make any reasonable assumptions concerning I/O connections and other details. Use any of the references [1] through [8] or other for details of syntax.

10.6 [28] Use any robot language (for example, references [1] through [8]) and write a general routine for unloading an arbitrarily sized pallet. The routine should keep track of indexing through the pallet, and signal a human operator when the pallet is empty. Assume the parts are unloaded onto a conveyor belt.

10.7 [35] Use any robot language (for example, references [1] through [8]) and write a general routine for unloading an arbitrarily sized source pallet and loading an arbitrarily sized destination pallet. The routine should keep track of indexing through the pallets, and signal a human operator when the source pallet is empty or when the destination pallet is full.

10.8 [35] Use the AL language [3] and write a program which uses force control to fill a cigarette box with 20 cigarettes. Assume that the manipulator has an accuracy of about 0.25 inch, so force control should be used for many operations. The cigarettes are presented on a conveyor belt with a vision system returning their coordinates.

10.9 [35] Use any robot language (for example, references [1] through [8]) and write a program to assemble the hand-held portion of a standard telephone. The six components (handle, microphone, speaker, two caps, and cord) arrive in a *kit*, that is, a special pallet holding one of each kind of part. Assume there is a fixture into which the handle can be placed which holds it. Make any other reasonable assumptions needed.

10.10 [33] Write an AL program [3] which uses two manipulators. One, called GARM has a special end-effector designed to hold a wine bottle. The other arm, BARM, will hold a wine glass, and is equipped with a force sensing wrist which can be used to signal GARM to stop pouring when it senses the glass is full.

Programming Exercise (Part 10)

1. Create a user interface to the other programs you have developed by writing a few subroutines in Pascal. Once these routines are defined, a "user" could write a Pascal program which contains calls to these routines to perform a 2-D robot application in simulation.

Define primitives which allow the user to set station and tool frames:

```
setstation(SrelB:vec3);

settool(TrelW:vec3);
```

where "`SrelB`" gives the station frame relative to the base frame of the robot and "`TrelW`" defines the tool frame relative to the wrist frame of the manipulator. Define motion primitives:

```
moveto(goal:vec3);

moveby(increment:vec3);
```

where "goal" is a specification of the goal frame relative to the station frame and "increment" is a specification of a goal frame relative to the current tool frame. Allow multisegment paths to be described when the user first calls the "pathmode" function, then specifies motions to via points, and finally says "runpath." For example:

```
pathmode;    (* enter path mode *)
moveto(goal1);
moveto(goal2);
runpath;    (* execute the path without stopping at goal1 *)
```

Write a simple "application" program and have your system print the location of the arm every n seconds.

INDEX